A Naturalist's Sojourn in Jamaica

PHILIP HENRY GOSSE

CAMBRIDGE
UNIVERSITY PRESS

CAMBRIDGE
UNIVERSITY PRESS

University Printing House, Cambridge, CB2 8BS, United Kingdom

Published in the United States of America by Cambridge University Press, New York

Cambridge University Press is part of the University of Cambridge.
It furthers the University's mission by disseminating knowledge in the pursuit of education, learning and research at the highest international levels of excellence.

www.cambridge.org
Information on this title: www.cambridge.org/9781108063739

This edition first published 1851
This digitally printed version 2013

ISBN 978-1-108-06373-9 Paperback

The original edition of this book contains a number of colour plates, which have been reproduced in black and white. Colour versions of these images can be found online at www.cambridge.org/9781108063739

CAMBRIDGE LIBRARY COLLECTION

Books of enduring scholarly value

Life Sciences

Until the nineteenth century, the various subjects now known as the life sciences were regarded either as arcane studies which had little impact on ordinary daily life, or as a genteel hobby for the leisured classes. The increasing academic rigour and systematisation brought to the study of botany, zoology and other disciplines, and their adoption in university curricula, are reflected in the books reissued in this series.

A Naturalist's Sojourn in Jamaica

The English naturalist Philip Henry Gosse (1810–88) travelled to Jamaica in 1844 and stayed for eighteen months to observe the diverse wildlife there. Upon his return he described his findings in a trilogy of books. The first two examined the island's birds – he has been hailed as the 'father of Jamaican ornithology' – but he used the present work, first published in 1851, to describe all the other forms of life on the island, from beetles to fruit trees. Lamenting that natural history was too often presented as a 'science of dead things', Gosse made his investigations come alive in this work by writing it in a diary form, discussing what he encountered as his journey progressed, and providing a number of illustrations. His lively and engaging style won him a wide audience, and this work remains an important early example of popular natural history.

Cambridge University Press has long been a pioneer in the reissuing of out-of-print titles from its own backlist, producing digital reprints of books that are still sought after by scholars and students but could not be reprinted economically using traditional technology. The Cambridge Library Collection extends this activity to a wider range of books which are still of importance to researchers and professionals, either for the source material they contain, or as landmarks in the history of their academic discipline.

Drawing from the world-renowned collections in the Cambridge University Library and other partner libraries, and guided by the advice of experts in each subject area, Cambridge University Press is using state-of-the-art scanning machines in its own Printing House to capture the content of each book selected for inclusion. The files are processed to give a consistently clear, crisp image, and the books finished to the high quality standard for which the Press is recognised around the world. The latest print-on-demand technology ensures that the books will remain available indefinitely, and that orders for single or multiple copies can quickly be supplied.

The Cambridge Library Collection brings back to life books of enduring scholarly value (including out-of-copyright works originally issued by other publishers) across a wide range of disciplines in the humanities and social sciences and in science and technology.

P.H. GOSSE, DEL.[t] M. & N. HANHART IMP.[t]

BLUEFIELDS

A NATURALIST'S SOJOURN IN JAMAICA.

BY

PHILIP HENRY GOSSE, A.L.S., &c.

ASSISTED BY RICHARD HILL, ESQ., COR. M.Z.S. LOND.,
MEM. COUNC. ROY. SOC. AGRIC. OF JAMAICA.

. . . . Thy desire, which tends to know
The works of God, thereby to glorify
The great Work-Master, leads to no excess
That reaches blame, but rather merits praise'
The more it seems excess. MILTON.

LONDON:
LONGMAN, BROWN, GREEN, AND LONGMANS.
1851.

PREFACE.

NATURAL History is far too much a science of dead things; a *necrology*. It is mainly conversant with dry skins furred or feathered, blackened, shrivelled, and hay-stuffed; with objects, some admirably beautiful, some hideously ugly, impaled on pins, and arranged in rows in cork drawers; with uncouth forms, disgusting to sight and smell, bleached and shrunken, suspended by threads and immersed in spirit (in defiance of the aphorism, that "he who is born to be hanged will never be drowned") in glass bottles. These distorted things are described; their scales, plates, feathers counted; their forms copied, all shrivelled and stiffened as they are; their colours, changed and modified by death or partial decay, carefully set down; their limbs, members, and organs measured, and the results recorded in thousandths of an inch; two names are given to every one; the whole is enveloped in a mystic cloud of Græco-Latino-English phraseology (often barbaric enough); — and this is Natural History!

Of the hundred thousand animals which are considered as "known to naturalists," it is probably much within the mark to assert that ninety thousand are "known" only in such sort as is described above.

What should we think if the world were to collect from Egypt the tens of thousands of mummies that are said to be entombed in the mighty catacombs of that country, and having placed them in museums should appoint learned men minutely to measure their differing features and limbs, to describe their appearance with exactitude, and to depict their portraits in all the leathery blackness of their physiognomy; then to give each a name, and record the whole in a book;—what should we think if the world would call this Egyptian *History?*

It is manifest that there is not an iota of *History* in either the one or the other. For History is the record of the actions of men, their relations to other men, the circumstances in which they acted, their characters, the influence of their lives upon society, their connexion with the times preceding and following their own, and other points of interest, not one of which could be gathered from a description of their dead and preserved bodies, though ever so exact and minute. So, that alone is worthy to be called *Natural History*, which investigates and records the condition of living things, of things in a state of nature; if animals, of *living* animals:—which tells of their "sayings and doings," their varied notes and utterances, songs and cries; their actions, in ease and under the pressure of circumstances; their affections and passions, towards their young, towards each other, towards other animals, towards man: their various arts and devices, to protect their progeny, to procure food, to escape from their enemies, to defend themselves from attacks; their ingenious resources for concealment; their stratagems to overcome their victims; their modes of bringing forth, of feeding, and of training, their offspring; the relations of their structure to their wants and habits; the countries in which they dwell; their connexion with the inanimate world around them, mountain or plain, forest or field, barren heath or bushy dell, open savanna or wild hidden glen, river, lake, or sea: — this would be indeed *zoology, i. e.* the science of *living* creatures. And if we have their portraits, let us have them drawn from the life, while the bright eyes are glancing, and the flexible features express the emotions of the mind within, and the hues, so often fleeting and evanescent, exist in their unchanged reality, and the attitudes are full of the elegance and grace that free, wild nature assumes.

The author would not be misunderstood. He is far from despising the labours of those who describe and catalogue the specimens that travellers send to the cabinets of Europe. Careful and minute descriptions, accurate admeasurements, and distinctive names are absolutely indispensable to science; *but they must not be confounded with science itself.* Valuable as these details are, they form only the stepping-stone by which we arrive at the knowledge of animals; or rather they are the cumbrous machinery by which that knowledge is preserved and communicated to the world. The knowledge of the name, position, and size of every rope in a ship is absolutely necessary to a seaman, for without it there could be no precision in command, no co-operation in obedience;—but surely it is not *seamanship;* and he who should suppose himself a skilful navigator, *with only such knowledge*, though acquired with the most minute pains, by actual study of a ship lying in port, would find himself egregiously mistaken, when he came to battle with sky and sea, tempests and billows and currents, quicksands, and bristling rocks, and the breakers of a lee shore.

Let closet-science take its true place as the handmaid of Natural History; arranging and appropriating the observations of the true naturalist, and enabling him to record them with precision. The former may be compared to the shelves, drawers, and pigeon-holes of a cabinet, carefully arranged, affording a place for every thing; the latter to a room-full of valuable objects and curiosities, thrown promiscuously in a heap. The objects themselves are almost unavailable until they be arranged in the shelves and drawers appropriated to them; yet it is obvious that the shelves are for the sake of the objects, not the objects for the sake of the shelves. The means, however, are too often treated as if they were the end.

The efforts of many naturalists, naturalists in the proper sense of the word, have been and are directed zealously towards the right object, a real knowledge of animals. Many have followed in the steps of the venerable Gilbert White, and have discovered thousands of facts of the highest interest, which they have communicated in graphic language. We consequently possess the living portraitures of many animals, drawn from life, and depicted with a master's hand. Most of the animals of Europe, at least the *Vertebrata*, have been more or less studied *at home*, and the Birds of America have found worthy biographers. But if we look at the expanded world beside, how little is really known of its living treasures; how little even of the zoology of England's vast colonial possessions!

The writer of the present volume has endeavoured to add a trifle to the amount of zoological knowledge. He lately paid a visit to Jamaica, one of the loveliest islands of the tropics, where the eighteen months of his sojourn were almost exclusively devoted to Natural History. The memoirs presented in the following pages may claim at least one excellence, they were drawn up verbatim on the spot, in the midst of the animated beings which they describe; they are, generally, not the results of brief and transient observations of their subjects, but of a protracted acquaintance with them, in which feature after feature was delineated, and line after line was added, from time to time. The Author has aimed to do more than merely give a record of the habits and instincts of animals; he has essayed to describe (as well as feeble words may attain to do it) somewhat of the glory and loveliness of the scenes in which they dwell; and he has endeavoured to do this with a kind of panoramic effect, so that the reader might have before his mind a succession of pictures, as it were, of a beauteous tropic island.

In the arrangement of the Work the form of a Journal has been maintained to such an extent as to give a slight thread of continuity to the whole. It is not, however, a Diary; chronological sequence having been always made to yield to the superior advantage of unity and completeness in the exhibition of the various subjects. The Author has grouped together all the information that he had collected on each subject, though obtained at different times; and thus the memoirs generally take the form of *monographs* more or less complete.

The following pages are enriched with many papers from another pen. The Author considers it one of the happiest reminiscences of a visit unusually pleasant, that it gave him the acquaintance of a gentleman whose talents and acquirements would have done honour to any country, but whose excellencies as a man of science, as a gentleman, and as a Christian, shine with peculiar lustre in the comparative seclusion of his native island. The Author has long had the privilege of his correspondence;— he enjoys the still higher privilege of calling him friend. It is with no small gratification that on the title-page of this volume, he can again associate with his own, the honoured name of Richard Hill, of Spanish Town.

In the progress of the work through the press the memoirs communicated by this gentleman have greatly accumulated. The Author (or rather as far as these are concerned, the compiler) has felt loath to withhold the valuable information collected by the zeal and industry of his friend, and so kindly placed at his disposal; though, being contained in letters, a considerable number of which are dated subsequently to his own departure from the island, they still further attenuate that "thread of continuity" alluded to above, which before was sufficiently slender. The Author ventures to hope that the greater degree of completeness thus attained in the survey of the Jamaican *Fauna*, will atone for this lack of constructive unity.

Twenty-four new species of animals are described

in the following pages, distributed nearly equally through the classes of Mammalia, Reptiles, and Fishes. The notices of Birds will be found but slight, a volume having been already devoted to the Author's researches on this beautiful and interesting class of animals.

The Author begs to express his thanks to the officers in the zoological and botanical departments of the British Museum, for much valuable assistance, kindly and courteously rendered; and to all the friends both in England and Jamaica, to whom he has been indebted for suggestions or information.

De Beauvoir Square, London,
September, 1851.

CONTENTS.

PLATES.

THE

NATURALIST IN JAMAICA.

INCIDENTS OF THE VOYAGE.

On the 20th of October, 1844, I sailed from the Thames for Jamaica. The early portion of the voyage presented nothing so worthy of notice as the rapid change of climate. Three weeks' run exchanged the chilly fogs and frosts of a London autumn for an atmosphere so hot, that even a sheet on one's bed at night was too oppressive to be endured. The dark grey waves of the north, swelling and surging with hoarse roar and crests of driving foam, were left behind; and were replaced by the delightful stillness of the tropic sea, where the sun's rays piercing into the depth, imparted a beautiful azure tint, approaching to pale Antwerp blue, to the transparent water; and the little crests of the rippling surface, as they rose and fell, and broke into drops beneath the unclouded light, glittered and sparkled gloriously.

The sight of a large Whale, doubtless a Rorqual (*Balænoptera*), as I distinctly saw his dorsal, wallowing in his huge unwieldiness, and making "the deep to be hoary," was an incident which, though far from new to me, was not without interest;—I looked on the vastest of known animals. On another day, when about 160 miles S.W. of Madeira, the sailors, with a surface line, "caught the Bonito," a beautiful pearly fish of the mackarel family, with rainbow stripes on the sides. The bait, which proved too enticing for him, was the same as that with which his more familiar, but not less beautiful, cousin is often taken, a piece of red rag. The stomach was found to be distended with a multitude of small Snipe-fishes (*Centriscus*), all of the same size, about two inches and a half long. A living specimen of the same Snipe-fish was drawn up on the same day in a bucket of water. The *Centriscus* is described by Risso as rarely wandering far from the shore, and as delighting in the mud at the bottom of the shoal sea. But the facts just mentioned suggest very different habits. The Bonito is well known to be a surface swimmer; and his morning's meal having been exclusively made of the *Centrisci*, combines, with the living specimen lifted in the bucket, to prove that the latter is also a surface-species, while the locality shows it to be pelagic.

In mid-ocean, eleven hundred miles from the nearest point of land, a large Turtle, probably of the Loggerhead species (*Chelone caretta*), swimming on the surface, was disturbed in his recreation by the approach of the ship, and dived with a splash into the security of the clear depths below. And as we approached the lovely Archipelago, toward which my eyes and thoughts were continually turning, we saw, more and more frequently, that little mimic ship, that sailors delight to call the Portuguese man-of-war (*Physalisa pelagica*), floating and tossing upon the waves. Some of these appeared to me unusually beautiful (though I have on former occasions seen them in great numbers, once, in the Gulf, being nearly a whole day traversing a fleet of them), the "sail" being of a rich rosy pink hue, and the bladder, or "hull," glittering in the sun with a glassy brilliancy.

An occurrence of much more zoological interest, however, the sight of a very rare, if not quite new, Cetacean, under circumstances peculiarly favourable to observation, demands a more protracted notice. Having been familiar with several species of *Delphinidæ* in former Atlantic voyages, I had taken for granted that I should meet with some in this; and wishing to settle the question whether any of the true Dolphins spout, I had studied the Order a little before sailing; and, in particular, had made careful sketches of the form of the head in all the genera, that I might not depend on that treacherous guide, memory.

November 22d. — Lat. at noon 19° 1′ N., long. 45° 42′ W.; the trade wind blowing a most exhilarating breeze, with fine weather. Between three and four o'clock P.M., a herd of large Cetaceans appeared astern, trooping towards the ship. They soon came up and began to play around us, continuing to romp and frolic, in the manner of Dolphins, all the evening; and even long after nightfall they were still in company, being plainly visible by the light of the moon. During this long time, I had many opportunities of observing them. They frequently protruded their heads from the surface; and then, presently, the huge round back, with a small dorsal far behind, was seen. In going along beside the ship, one would occasionally turn on its back, displaying the white belly, and in this position swim a short distance. The muzzle was lengthened into a snout, but, as well as I could judge from many exposures, it tapered gradually without a furrow, and resembled that of *Delphinorhynchus*. As nearly as I could estimate from a position aloft, by comparison with the ship, their length was about thirty feet, or perhaps not quite so much. The body was elongated, black above, white beneath; the swimming paws appeared white, even on their upper surface, but surrounded by dark colour on the body; —this is remarkable. The lips and extremity of the muzzle appeared, when projected from the water, of a flesh colour. They usually expired with a rushing sound, the instant the blow-hole was exposed, but did not, as far as I observed, *spout*. Once, however, I noticed a little cloud of steam sailing away on the wind, from the spot where one had just disappeared; it exactly resembled that appearance which succeeds the spouting of the common Rorqual (which I have seen many times), but as my eye did not catch the animal itself, I cannot positively say that such was its origin on this occasion. The evenings being cool and refreshing after the burning days, and being generally fair, and now lighted by the moon,

> "——— pura nocturno renidet
> Luna mari, ———" *

we spend them on deck, as the pleasantest hours of the twenty-four. This evening, the wallowing and sporting of the Whales added a new interest; and at nearly eleven o'clock, we left them still in company.

November 23d. — On rising, we were surprised to find the Whales still attending us. I now had an opportunity of seeing the profile of one very distinctly, and of assuring myself that the form of the head was exactly that of the figured *Delphinorhynchus, no furrow being visible* between the forehead and the snout. One of the officers informed me that he had seen one of them *breach*, or leap clear out of the water. Soon after 8 A.M. they left us, having continued with us nearly seventeen hours, a period of extraordinary length, when we consider that the visits of frolicsome *Cetacea* to vessels rarely last more than half an hour or an hour. During the whole of this time, the ship had been running before a gallant breeze, and had proceeded nearly 120 English miles.

I have little doubt that the species was that very interesting and rare Cetacean known as the Toothless Whale of Havre, *Delphinorhynchus micropterus.* The small size of the dorsal, and its backward position, agreed well with the description of that species, and though these were nearly double the length of that

* Hor. Carm. ii. 5.

celebrated specimen, this incongruity is of little moment, since that was evidently a young one. If this was, indeed, the Havre Whale, the occurrence in associated numbers of a species, hitherto known only by a solitary specimen, possesses an interest which will be readily appreciated by naturalists; if, on the other hand, it was distinct, it is, perhaps, still more interesting, as it proves the existence of a gregarious Cetacean of large size in the Atlantic, which has hitherto escaped the observation of zoologists. The white hue of the flippers, isolated amidst the dark colour of the upper body, would seem to favour the latter conclusion.

I may add here that when we were off the west end of Porto Rico, I observed a shoal of Dolphins playing at a short distance; one of them in leaping fell in a perpendicular position, the tail downward, while the body was thrown into a double curve. I was thus enabled to see that the belly was of a bright rose-colour. Now this is the hue of the under parts of the other *Delphinorhynchus* (*D. rostratus*), which is about eight feet in length, and might well be mistaken, in the moment of leaping, for a true *Delphinus*. The coincidence is a curious one: especially as this species is nearly as rare as the former.

I had prepared a surface-net for towing after the ship in order to capture minute animals. It was thus made: — The towing line, a stout rope, was attached by a threefold bridle to a brass rod five feet in length, one foot of which was bent up at a right angle at each end. The three sides formed by this wire kept open the mouth of the net, which was about ten feet long, diminishing to a point. The size of the meshes also diminished until the extreme point was formed of gauze, and was not closed, but simply tied with a tape. Three or four hoops, sewed in transversely at intervals, kept the sides of the net from collapsing when in use. When employed, it was lowered over the quarter, gradually, so that the extremity might be carried out by the ship's way before the brass wire touched the water, to prevent entanglement. The wire being on three sides only of the mouth, would of course lie perpendicular, and the rope-bridle kept it transverse: corks at each upper angle of the mouth preserved it from sinking. The line was then payed out for ten or twenty fathoms, so as to pass the net beyond the dead water and eddy of the ship's wake. After awhile it was hauled in and examined. A bowl of water being ready, the gauze end was untied and turned inside out into the bowl, when the contents became disengaged by floating off, and after the net had been re-tied and lowered, were subjected to examination.

The booty taken by this apparatus, though for the most part minute, was not devoid of interest. A few of those delicate *Pteropoda*, the *Hyaleæ*, whose shells look as if they had been blown out of the thinnest glass, occurred; as also many specimens, more or less imperfect, of the *Cephalopodous* genus *Spirula*, with pearly septa; to some of these were attached several very minute but perfectly developed barnacles (*Lepas*), and many oblong and dark brown eggs, apparently of the same cirriped. It was interesting to see Barnacles no larger than rape-seed projecting from the mouth of the little *Spirula*, and throwing out their *cirri* with all the regularity and energy of the larger species. But the most curious thing was a small *Porpita*, a flat, circular Medusa of about one-sixth of an inch in diameter. It consisted of an annular disk, perfectly pellucid, inclosing an opaque central portion, occupying about two-thirds of the breadth, which seemed to be composed of fibres radiating from the centre. On the under surface this central part was flesh-coloured with a blue margin; and while alive, as it floated, I perceived on it several teat-like projections, which had a sluggish motion. On the upper side the pellucid membrane was continuous, and had a glistening appearance; the opaque part was here light brown. On being handled, the little animal gave out a milky fluid, which clouded the water.

I afterwards obtained several more of this minute *Porpita*, some of which were very beautiful; the margin being fringed with numerous oviducts (?) of varying length, which were transparent, but studded and terminated with knobs (germs?) of a deep blue colour. In the centre of the under surface was a fleshy sucker, and around it many teat-like projections, like the tentacles of an *Actinia;* these I saw thrown out and suddenly retracted irregularly, and independently of each other.* All these surface-

* As this seems hitherto undescribed, I propose for it the name of *Porpita* (*polybrachionea*) *minuta.* It differs greatly *in size* from those hitherto described, being only one-sixth of an inch in diameter, while Bosc compares *P. gigantea* to a 24 sous piece, and *P. glandifera* (figured, *of the natural size*, in the Phil. Trans. vol. xxiv. p. 2053),

swimming creatures occurred within a hundred and fifty miles of the northern tropic, after our crossing of which they ceased.

The common opinion that the motion of the Flying-fish through the air is nothing more than a vigorous *leap*, sanctioned as it is by many honoured names in science, I have long believed to be incorrect, having on several former occasions seen what appeared to me a distinct motion of the fins. To settle this point was one of the desiderata which I had particularly noted down on commencing this voyage; and the result has fully confirmed my belief. The observations as they occurred, I shall quote in the form in which they were recorded.

Nov. 20th. — Lat. 19° 24′ N., long. 41° 5′ W. Many Flying-fishes appeared in the course of the day; the first we had observed. They were the silvery species commonly seen in the Atlantic, of middling size, with clear wings; probably *Exocœtus volitans.* I now feel certain that these fishes have power to change their direction when in the air; more than one, which I saw to-day, *turned aside at nearly a right angle.*

Nov. 22d. — Flying-fishes leap from the sea every

is about half an inch wide; Guilding's *P. Linneana* is stated to have a diameter of eleven and a half lines, exclusive of the arms. It differs from *glandifera* also in having numerous tentacles on the disk, and likewise in colour, the disk being carnation, with a blue margin beneath, and brown above. The body of *Linneana* is described as of a cærulean hue. *P. nuda* and *P. appendiculata* of Bosc were evidently founded on imperfect specimens.

few minutes; several made courses distinctly angular, and some, I am quite sure, *rose and sunk in undulations.* To confirm my own observations, I requested a gentleman on board to notice this point; and he was quite certain of both these facts.

Nov. 24th. — I observed to-day a Flying-fish, after flying a very short distance, suddenly turn downward, abruptly and perpendicularly, as if alarmed, and enter the water. The action exactly resembled that of a bird.

Nov. 25th. — Several times I have observed in the flight of the *Exocœtus,* when near, *an occasional fluttering* of the pectorals. In general, these wing-fins appear motionless; but at the moment of rising to avoid the crest of a wave, there is a slight but rapid vibration of these organs, distinctly perceptible, if the fish be pretty close to the ship. I saw an *Exocœtus* to-day which was much larger than the species hitherto observed, with the pectorals wholly of a sooty black colour. Probably it is the *Exocœtus Noveboracensis* of Dekay. The others are still rather numerous, but do not rise in flocks.

Nov. 27th. — This morning multitudes of Flying-fishes rose, disturbed by the ship; a distinct species, differing from both the former in many points (Plate I. Fig. 1.). This species is much smaller; the pectorals are so delicately transparent as to be almost invisible; the upper parts are of a fine deep blue; their flight is swifter, more hurried, and continued for a shorter distance; they often merely shoot along at the surface, just cutting the water, and sometimes emerge for five or six feet only. The form appears

Plate. I.

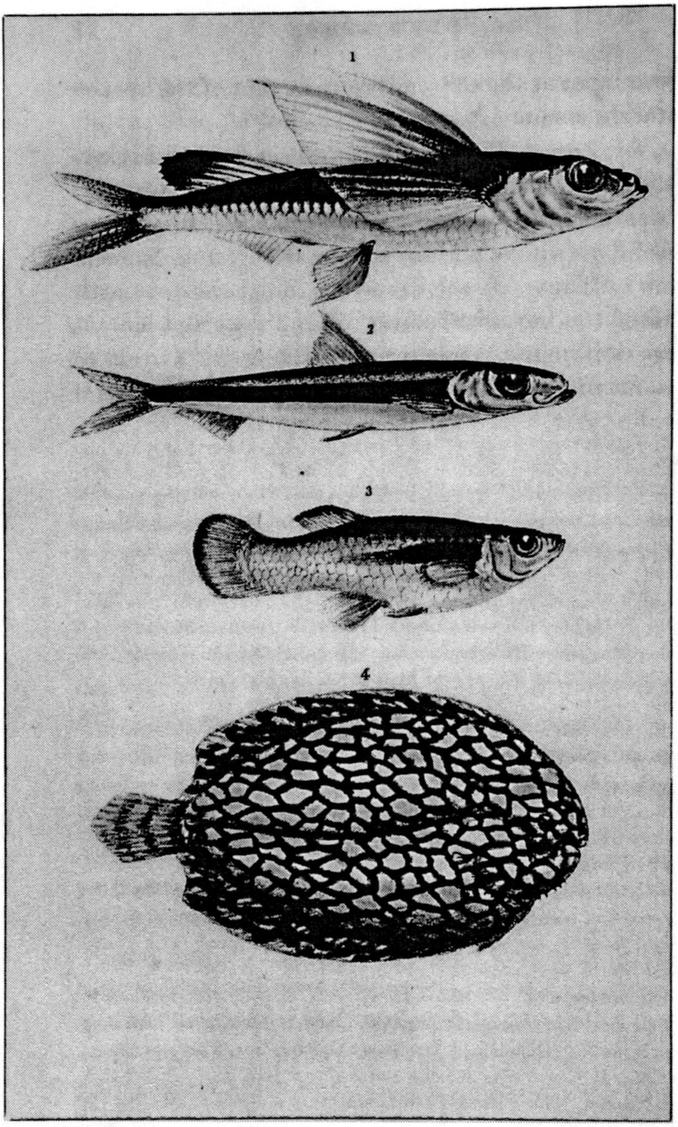

P.H. GOSSE DEL

M. & N. HANHART IMP.

1, EXOCŒTUS HILLIANUS, † N.S. 3, PŒCILIA MELAPLEURA, N.S.
2, CLUPEA LAMPROTÆNIA, N.S. 4, MONOCHIRUS INSCRIPTUS, N.S.

LONDON, LONGMAN & Co 1851.

more taper at the extremities than that of the species hitherto common.*

Nov. 28th.—Five of the little Blue-backed Flying-fishes flew on board early in the morning, by which I was enabled to secure specimens, and to make a careful drawing while the colours were yet undimmed. The brilliancy of the deep-blue on the upper parts, and of the burnished silver of the sides and belly is very striking. While passing Radonda, a rock to the north of Montserrat, multitudes of this species

* This is the "Volador" of Parra (pl. xv.), but certainly not the *Exocœtus mesogaster* of Cuv. et Val., which those learned zoologists have identified with Parra's *Volador*, after Bloch. They describe their *mesogaster* as having the pectorals of the same deep blue tint as the back, with a large transverse band; the ventrals of the same deep colour, with white at the base of the internal ray, and a little at the tip of the second external ray; the dorsal and caudal, grey; the anal white. Its length is seven inches French, or nearly seven and a half inches English. How importantly this differs from the species mentioned in the text will appear from the following description of the latter: —

Back deep blue; sides and belly pearly: pectorals perfectly hyaline and colourless, except a tinge of black on the basal half of the interspace between the first and second rays; no trace of any band; ventrals colourless; dorsal colourless for the basal half, the terminal moiety black; caudal white; anal colourless. The anterior part of the pectorals, and of the ventrals, runs out in a short point; the insertion of the latter is just midway between the extremity of the lower jaw and the fork of the caudal. The eye is large, the iris is blue, tinged with yellow anteriorly; the muzzle is tipped with black. Total length five inches.

Fin-ray formula. P. 13. — D. 12. — V. 7. — A. 13. — C. 19.

Believing that this species, though allied to *mesogaster*, is distinct from it, and as yet unnamed, I propose to honour it with the appellation of *Exoc. Hillianus*, after the name of Richard Hill, Esq., of Jamaica, a gentleman, whose scientific attainments are equalled only by his urbane manners, and truly estimable character.

rose on each side, flitting but a yard or so, and often only ruffling the surface. They were in small shoals, each containing about a dozen fishes.

Nov. 29th. — I noticed a Flying-fish curve its course so as to describe *more than half a circle.*

Nov. 30th. — Before breakfast the motion of the ship disturbed multitudes of Flying-fishes of the blue-backed species, which flew in shoals every moment from each side of the bows. Many Bonitos pursuing them were likewise leaping from the surface; at length a seaman from the martingale struck one of the latter with the grains, and secured it. It was a beautiful fish, remarkably plump and hard; the rainbow hues of its skin were most rich and lustrous, the pink predominating. Several black stripes running along each side marked it as the *Thynnus pelamys* of Cuv. et Val. We ate it for dinner, and found the flesh pleasant, but rather dry. In its stomach were six of the Blue-backed Flying-fishes, partly digested. Some large oceanic birds were likewise pursuing the little aëronauts, swooping down amidst the flitting shoals. As far as I remember, this is the only occasion in which the pursuit of the *Exocœtus*, by either winged or finned enemies, has ever occurred to my observation; I incline to think that it is much oftener described than seen.

Here end my notes of the Flying-fish; except that long afterwards I saw two or three of the Blue-backed species skimming the surface of Bluefields Bay. Besides the fact, which these notes appear to me to prove, that the *Exocœti* have the power of increasing, directing and terminating their progress while in the air, by the action of the pectorals — a true flight, those connected with the geographical distribution of the species are worthy of notice. The common Atlantic species, which I take to be *volitans*, suddenly appeared in considerable numbers in long. 41°, or just midway between Africa and the West Indies. They continued to be abundant until within a day's sail of Antigua, where they were replaced by the little Blue-backs, and were seen no more. This beautiful little species suddenly became quite as abundant as the former had been, and so continued until we neared Hayti. The great Black-winged kind appeared three days to windward of Antigua, and continued to appear, conjointly with the Blue-back, but rarely more than two or three in a day. The last seen was near the west end of Hayti.

One morning I rose at three to enjoy the delightful coolness of the scene by starlight. The watch were all seated on the booms beneath the shadow of the boats amidships, and the only living being visible on deck was the man at the wheel, and he did his easy work silent and motionless. The white sails were bellying out before the gallant trade-wind, and the rushing of the bows through the little waves, and the rustling of the water beneath the quarter, were the only sounds to be heard. The sky was almost cloudless; Orion, in glittering splendour, — "armatus auro Orion," — was nearly in the zenith; Ursus major had just risen from his ocean bath, no longer answering to the ancient description, —

> "Ἄρκτος ———
> Οἴη δ' ἄμμορός ἐστι λοετρῶν Ὠκεανοῖο," —

and was curving round the low Pole-star; the Moon and Jupiter had set. Presently Venus rose from the sea, enveloped in a slight haze, and looking, when a few degrees above the horizon, exactly like a light-house, but glowing like a torch as she rapidly mounted up the sky. About six o'clock I went aloft and sat in the maintop to see the sun rise free from distraction. It was a splendid sight: the gradual lighting up of the eastern quarter of the heavens, and the refulgent gilding of the few clouds that gathered there, were most gorgeous, and went on increasing in splendour every moment; till at length the sun leaped up "in his clearness," and irradiated the solitary ocean far and wide. The whole scene was indescribably beautiful; and though its elements are common to all parts of the earth, yet the effect was something new, and produced a peculiar emotion of delight; a sort of thrilling feeling, somewhat like that which the deep tones of a fine band of music will often excite.

There is not a more pleasant situation on board a ship than the maintop in fair weather. To take a book up for an hour's quiet reading, or to sit with our back against the topmast shrouds, and enjoy our own meditations; or to gaze upon the vast expanse of ocean all spread out around and beneath us; or to mark under the lifting sail the double hillock of foam tossed off by the vessel's prow breasting the waves; or to watch the curling of the dark eddies beneath the stern till they subside into the long wake, which, like a line of silver, we leave behind upon the sea,—all these are delightful to a passenger who has learned to mount a vessel's rigging. In the afternoon it is an especially pleasant post; for the broad shadow of the topsail begins to fall aft, and the elevated solitude is screened from the rays of the too fervid sun.

Night, too, had charms of its own. The pleasant breath of the steady trade-wind blew with a refreshing coolness over the level sea, breaking its surface into ripples and wavelets, that washed, with a whispering sound, along the counter of the ship. The ladies would linger on the quarter-deck till late; and cheerful conversation, innocent mirth, and mutual congratulations and anticipations connected with the land whose distance was diminishing with winged speed, beguiled the hours. The moon, "walking in brightness," poured down a flood of soft light on the ship and the wide sea around, putting out the stars above, but making amends by the thousand mimic ones that were momentarily forming below by the reflection of her silver face in the dancing, breaking wave-crests. Sometimes we watched the phosphorescent flashings of the sea, and the brilliant sparks that went and came among the curdled milkiness beneath the stern, stirred and whirled by the action of the rudder; or sometimes we would walk into the forepart of the ship, and see the same curious phenomena to still more advantage, where the bows dived into the sea, and threw off the luminous foam on either side. Or from the same spot we would gaze aloft, and admire the swelling canvass, partly white as snow in the glancing moonbeam, partly in deepest shadow; while each sail stretched and bellied out before the breeze, just as the poet has described it, —

> "—— tumidoque inflatur carbasus Austro."

Reluctantly the passengers one by one retire, and the meditative watch is left in undisputed possession of the silent deck.

At length all eyes were directed westward to look out for land: and presently (on the 28th of November) the little isle Desirada was detected on the horizon like a thin band of blue cloud. Antigua and Guadaloupe succeeded, both dim and distant; then Montserrat appeared ahead, and quickly grew distinct and palpable. It was afternoon, and the sun was sunk behind its heights as we neared it, so that we could see nothing but one dark blue ragged mass, tapering down at each extremity, and rising into irregular peaks in the centre. We rounded the north end, and saw the opposite side, where the beams of the slanting sun shone full upon the green woods and verdant cane-fields, displaying many white houses scattered about the hills. On the summits of the mountains many palm-trees elevated their noble feathery heads above the general mass of foliage. The central peaks were shrouded in a cap of cloud, and while we looked on the smiling scene, the mists began to roll down the mountain sides like the falling of a curtain; and soon the beautiful island was but a shapeless mass of grey haze.

The fresh breeze rapidly swept us along between Montserrat and another object of interest that now appeared. It was Radonda, a vast mass of barren rock, that rose steep and abrupt out of the sea before us, relieved against the horizon already glowing with the declining sun. As we neared it, I was struck with its resemblance to the Bass, as well in form and appearance as in the myriads of sea-birds that were congregated in clouds around it. They were settling down for the night; and strings of other birds, from all points of the compass, were seen on flagging wing wending their way to their island lodging, after their predatory seaward excursions of the day. Behind the rock in the northern horizon, was seen the conical form of Nevis, that lovely little gem of the Hesperidan archipelago, but blue and distant, and presently lost; for the sun had set, and the night, like a giant, strode quickly over the scene.

BLUE MOUNTAIN PEAK.

Dec. 4th.—My first sight of Jamaica was one that I never can forget. The approach to the land of one's destination after a long voyage is always a season of pleasurable excitement; a peculiar flutter of the spirits, a disposition to hilarity, and to mutual gratulation pervades more or less every one on board, from cabin passenger to the black cook in the caboose. Every one may be seen from time to time straining his eyes, in the direction from which the first dim blue indication may be expected; and those who dare to go aloft make free use of the advantage which that elevation affords them. I had arisen early, and at dawn of day had seen Cape Tiburon, the west point of Hayti, dark, distinct, and palpable, though distant sixty or seventy miles. The brightness of the east was behind it, and threw it out into strong and bold relief; but the moment the sun was visible above its outline, it faded away and became entirely invisible. Just before sunrise, or just after sunset, is a time much prized by mariners for discerning distant land.

During the forenoon the mountains of Jamaica were seen, and gradually grew more distinct as we neared the island. Yet the cloudiness of the day had prevented my having had any satisfactory view of it until evening. About sunset, I was standing forward, when one by my side said, "Look at the Peak!" I looked intently, directing my gaze to the neighbourhood of the horizon, where I supposed it was to be seen; but nothing but the dull white clouds met my eye. "Up there!" said my informant; and his finger pointed up into the sky; and there indeed was its noble head (perhaps elevated by refraction), a conical mass, darkly blue, above the dense bed of clouds that hung around its sides, and enveloped all beneath its towering elevation. Yet it is situated far inland, and was then full forty miles distant from our ship.

But night soon fell, and, as we were somewhat anxiously watching for the light on Point Morant, I had the pleasure of first seeing it from the main rigging. We were soon abreast of it, and as we passed on before an increasing breeze, that tempered the tropical heat with its refreshing breath, we saw the coast dark and high only a few miles off. Many lights were seen in the scattered cottages, and here and there a fire blazed up from the beach, or a torch in the hand of some fisherman was carried from place to place. My mind was full of Columbus, and of his feelings on that eventful night, when the coast of Guanahani lay spread out before him with its moving lights, and proud anticipations. With curiosity and hope, somewhat analogous (*parva componere magnis*), did I contemplate the tropical island before me, its romance heightened by the indefiniteness and obscurity in which it lay. I was on deck several times during the night, and during the intervals was still engaged, in dreams, in endeavouring to penetrate the darkness of the shore.

KINGSTON AND PORT ROYAL.

At day-break (Dec. 5th) we found ourselves off the mouth of Kingston Harbour, becalmed. How lovely was the scene! The sea was of crystal clearness, allowing the eye to penetrate far into its depth beneath the shadow of the ship's stern; while all around, sleeping in the perfect calm, it reflected the heavens like a plate of steel. No clouds broke the uniformity of that reflection; but all above was clear transparent blue, already suffusing with brightness in the seaward horizon. Before us stretched away on either hand the beauteous island, with its many mountain ridges, showing their purple summits in receding succession. But one object instantly arrested the eye, and long detained it; right over against us, apparently overhanging the city of Kingston, towered in its giant grandeur the Peak which I had seen the evening before. It seemed close at hand, from its vastness; though its uniform blue hue told on reflection its distance. Presently the sun rose, and purpled the conical head long before he shone on us, or even on the lower ranges of mountains. The well-known comparison by which Columbus is said to have given Queen Isabella an idea of Jamaica, — a sheet of paper crumpled up tightly in the hand, and then partially stretched out, — occurred to me, and I could not but admire its striking appropriateness. The oblique rays of the sun showed out the alternate hollows and ridges in vivid distinctness, and gradually raised thin white mists from the former, which soon grew into clouds, and spread themselves about the summits and sides of the higher mountains, until these were quite veiled, except where dark portions were here and there revealed through openings, like great rents in a garment.

The city of Kingston, with its steeples and white houses, was visible from the rigging, over the long low spit of sand called the Palisades, on the point of which we could see the collection of hovels known as Port Royal, with the masts of a line-of-battle ship towering above them.

By and by black specks appeared upon the shining sea as suddenly as if they had just emerged from its depths, each of which we understood to be the canoe of a fisherman, engaged in examining the wicker "pots" or traps, which had lain all night upon the bottom, or else in the more active employment of fishing with the hook and line. One of these tiny specks pulled out to us, and as she lay alongside, I observed with delight the brilliant colours and strange forms of the fishes that had just been caught; among which, however, was a young shark, some three or four feet long, and not yet dead. Three brawny negroes were on board, whose dark brown skins, plump and glossy, were only partially concealed by their ragged garments; yet one of these was the pilot, to whose skill and local knowledge we were to be indebted for a safe guidance among the kays, and through the coral channels, that guard the magnificent harbour of Kingston.

Not a breath yet stirred the sea; and though long gentle undulations rolled slowly in from the ocean, their surface was glassy and unbroken; and the reflections from the dark land were thrown up in broad masses, caught by one rounded swell after another, as the smooth side of each successively presented itself to the eye. It was a trial to patience to lie idly there, beautiful as was the scene; the heat was intense, though the season was December, almost too great, indeed, to allow of an emergence from the shadow of the awning into the unclouded beams, to examine the towing-net, which lay overboard. It had gathered little; masses of tangled gulf-weed and other *Algæ* had from time to time accumulated in it, some of the stems of which were covered with the pretty netted *Flustra*, so commonly found investing these marine plants. A few minute fishes, of the genera *Hemiramphus* and *Antennarius*, were entangled in the weed, and many little Swimming-crabs (*Lupa*), with the hindmost pair of feet dilated into flat rounded plates, and the shell produced on each side into a sharp spine, ran swiftly among their yellow mazes. Two specimens of a *Nereis* also occurred;

but the most curious creature was an *Annellide*, that inhabited a hollow stick, lined with a tough membrane, exactly in the manner of *Phryganea*.

It was noon before the welcome sea-breeze came in, and then it was so slight that we could scarcely feel its gentle breathing. It was sufficient, however, to impel us gradually nearer to the land, and thus to reveal the minuter beauties of the scene, whose grander features we had been admiring at a distance. Many little flat kays, as such islets are called, lay around, among which our tortuous course led us; scarcely more than the flat tops of coral rocks, almost level with the sea, on which sand and shells had been accumulated by the waves; yet pleasant to look on, because covered with low bushes of a refreshing greenness. On their snowy beaches, where the gentle ripple was sparkling, or perched on the irregular blocks of black rock that lay half covered with the tide, sat many Pelicans, preening their plumage, and dashing the water over their wings, or lazily resting after their morning's fishing excursion.* Some sat sleepily on the sea, their forms reflected from its bosom, inert and motionless, except for the alternate rise and fall which were produced by the undulation of the ground-swell.

At length, the peculiar harsh rattling of the chain cable rushing through the iron-lined hawse-hole, announced that the anchor was dropped, and at the

* "Tranquillo silet, immotâque attolitur undâ
Campus, et apricis statio gratissima mergis."
VIRG. *Æneid.* v. 127.

same moment the captain's gig was lowered to convey himself and the passengers across the beautiful broad harbour, to the thriving city at its head. But it was as a naturalist that I was there, and the sea-beaten shore promised me greater gratification than a visit to Kingston; hence I availed myself of the boat only to be put ashore at Port Royal Wharf, whence I speedily found my way to the long sandy beach of the Palisades.

It is true there was little of the luxuriance or beauty that we associate with tropical scenery, here. It is a low bank of sand nearly nine miles in length; but scarcely any where more than a few hundred yards in breadth, forming a natural breakwater that separates the broad lake-like harbour of Kingston from the Caribbean Sea. I found it barren enough; but it was all strange, and to feet which for nearly two months had not felt the firm earth, even a run along the beach was exhilarating. The graceful Cocoa-nut Palm sprang up in groups from the water's edge, waving its feathery fronds over the rippling waves that dashed about its fibrous foot. Great bushes of Prickly-pear and other *Cacti* were growing on the low summit of the bànk, covering large spaces of ground, with their impenetrable masses, presenting a formidable array of spines: as did also a species of *Acacia*, that grew in thickets and single trees. All along the line of high water lay heaps of sea-weeds drying in the sun, among which was particularly abundant a species of *Padina*, closely resembling the pretty "Peacock's tail" of our own shores, though less regularly beautiful. Sponges of various forms, and large Fan-corals, with the gelatinous flesh dried on the horny skeleton, were also thrown up on the higher beach; and I found in some abundance, a Coralline, of a soft consistence, and of a bright grass-green hue, each branch of which was terminated by a radiating tuft of slender filaments.

Shells were very scarce on the sea-beach; but on the harbour side many species were found in the crevices and pools of the low rocks, and just within the margin of the water. All were small, and few presented any facts worthy of being noticed; they were chiefly of the genera *Turbo, Phasianella, Planaxis, Buccinum, Vermetus* and *Fusus;* the bivalves *Ostrea, Anomia, Spondylus, Avicula, Arca, Cardium, Venus* and *Pholas.* Several specimens of a brilliant little *Chætodon* were swimming and darting about the narrow, but deep pools; they were not more than an inch in length, marked with alternate bands of black and golden yellow. In the vertical position in which they swim, with the eye of the observer looking down upon them, they appear to bear the slender proportions of ordinary fishes; and it is only by accident, as in turning, or on capturing one, that we detect the peculiar form, high and vertically flattened, of this curious genus.

THE HEMIRAMPHUS.

While lying off Port Royal, just within the mouth of Kingston Harbour, I had the pleasure of seeing a shoal of that curious and interesting fish, the Half-beak Garfish (*Hemiramphus*). A few single specimens had occurred in the towing-net, while approaching the island; but they were young, and not more than a few inches in length. But at this time about fifty were gathered, in the perfectly smooth water under the ship's stern, where they remained actively playing just beneath the surface. They were all of nearly the same size, about a foot in length. The negro-boys called them Ballahoo.* A few bits of bread which the steward had just shaken from the breakfast-cloth, were objects of great attention to the fishes; they clustered round each fragment, and bit at it every moment, not attempting, however, to take it in by opening the jaws; but each fish successively making a dart at it obliquely from beneath, and just touching its underpart with the point of the upper (and shorter) jaw. I could not perceive that their attacks, though pertinaciously continued, detached any sensible quantity, though some very small fragments were carried down a few inches below the surface, and then being let go rose again. Presently a Garfish (*Belone*) called Piper, about two feet in length, appeared among the Ballahoo, and was followed by another. I observed that the shoal of Half-beaks gave these strangers a wide berth, though they did not disperse. Probably this singularly formed fish habitually feeds on greasy substances, or minute animalcules that float near the surface, always taking its food from beneath. The Garfish and the Saury (*Scomberesox*) are both surface-swimming genera.

* Probably *H. apicalis* (Bennett), or *H. Brownii* (Cuv. and Val.).

ALLIGATOR POND.

Dec. 9th. — A few hours' run from Port Royal brought us to Alligator Pond. Most parts of the intermediate coast had a most sterile appearance, the hills being covered with small scrubby wood, unrelieved for miles by a single plantation. The little village of Carlisle displayed a pleasanter scene, where on the flat were seen fields of sugar-cane of a delicate light green hue waving in the breeze, windmills with their sails in revolution, the smoke ascending from the boiling-houses and other buildings of a sugar estate, and men and cattle actively performing the several duties of the culture or manufacture. Alligator Pond is the mouth of a valley which runs up between the May Day Mountains and the Santa Cruz Mountains. It consists of a few stores, with a wharf, a dwelling-house, and a few negro huts. I went on shore with an insect net. On the sandy beach, so loose and heavy that I could scarcely walk on it, the *Convolvulus pes capræ* was growing in profusion, covering its upper part with a carpet of verdure, and trailing its long stems in every direction over the beach, while its beautiful purple blossoms, mingled with the pink flowers of the *Canavalia rosea*, gracefully relieved the dark foliage. Here an active predaceous beetle (*Cicindela Guadalupensis*) was running and flying alternately with the wary agility common to the genus, so as to be difficult of capture.

The soil all around consists of the same heavy sand, into which the feet sink to the ankle at every step, which, added to the sun's rays direct and reflected, makes any exertion terribly fatiguing. Dragon-flies and Butterflies were numerous here; one of the first objects that attracted my attention was *Heliconia charitonia.* Its beauty and singularity of form, the great length and little breadth of the wings, the length and slenderness of the body, and the brilliant contrasts of colour, lemon-yellow and velvety black, together with the very peculiar flapping of the wings in flight, as if their length rendered them somewhat unwieldy, gave me a sensation of delighted surprise. It is one of the most easily captured of Butterflies, being slow of flight and fearless; it flutters heavily along over low herbage at the sides of roads, and by the sea-shore, rarely mounting as high as one's head, except when alarmed. *Colænis Delila*, *Agraulis vanillæ*, and the minute *Polyommatus Cassius*, were also abundant here. A large *Opuntia* was growing in great straggling beds, the broad spine-set articulations displaying both flower and fruit in profusion. Notwithstanding the pencils of fine barbed spines with which the latter is armed at intervals, well maintaining its title of *Prickly pear*, I was tempted by its plumpness and rich blooming colour to taste one. My lips were soon full of the spines, which are detached from their base with the slightest force, and left sticking in the flesh; yet the pulp and juice, both of which were of the richest crimson hue, were pleasantly sweet-acid, though somewhat insipid, and full of stony seeds. An excellent crimson dye, equal to true cochineal, has recently been obtained in the East Indies from the fruit of *Opuntia coccinellifer,* the Nopâl, on which the cochineal insect is fed; it would be worth while to institute experiments on the fruit of this species, which is so abundant in the most sterile places in the West Indies. While I was looking at a large Butterfly (*Callidryas Eubule*) that flitted about the expanded blossoms, and admiring the similitude of colour between the fly and the flower, both being entirely of a delicate sulphur-yellow, — a Humming-bird suddenly appeared probing one of the latter, but was gone before I could well observe his plumage By the ruby gleam that flashed from his throat, however, I conclude it to have been the Mango (*Lampornis porphyrurus*), the sides of whose gorget are crimson in some lights.

Among the joints of the tangled Prickly pear, many vertical spiders' webs were hung, some of which were of sufficient strength to offer considerable resistance to the hand. I looked at the tenant, and found it to be *Nephila clavipes,* a spider of exceeding beauty. It is of large size, being an inch and a half in length, exclusive of the legs, which extend over a space five inches in diameter. The body is lengthened, and studded with round white spots, each environed with a black border, on a rich greenish-brown ground, reminding one of the characteristic marking of the Tragopans among birds. The cephalo-thorax is shining black, its lustre half concealed by a clothing of short silvery down; the legs are very long, and have a remarkable appearance from having a bunch of black hair set around the extremity of the first and second joints, like the bristles of a bottle-brush; the third or short pair of legs, however, is destitute of these appendages. These peculiarities impart an interesting character to this fine spider, which seems also to have a wide geographical range, extending from Florida, and perhaps Georgia, to the Bermudas, all through the Antilles, and even over the continent of South America as far as Brazil. But to return to the individual which was now attracting my interested attention. Several young ones were scattered about the net of one of large size, each of which hung head downward on the threads in the manner of the adult, remote from her, and from its fellows. Hence it would seem that the young of this species learns its first fly-catching lessons on the parental web, and is not sent into the world to practise on its own account its net-weaving trade, until it has attained some size and strength. Most of the nets, however, in the Prickly-pear bush were occupied by Spiders (*Gasteracantha*) of smaller size, but equally curious; the abdomen having a shell-like hardness and polish, and being hexagonal in form, with the angles produced into sharp points. There seemed to be two species of these, some having the abdomen of a dull red, and others, the more common, of a porcelain whiteness.

I rode a few miles up the valley to dine with a gentleman at his coffee plantation. Looking into a deserted house not far from the shore, I was astonished to see that from every beam, rafter, and projection of the interior, hung multitudes of nests of a brown slender Wasp (*Polistes rubiginosa?*) in all stages of progress; some just begun; others as large as a teacup, and crowded with wasps; others old and abandoned. There must have been in that hut some hundreds of nests. Wasps of the same species were swarming about the spinous leaves of the Pinguin, a sort of wild pine-apple, of which fences are made. The road soon became interesting, being bounded by huge masses of white limestone singularly honeycombed with round holes of various sizes.* In these holes often were seen small shells, perfectly white like the stone itself, of the genera *Helix*, *Cyclostoma*, *Cylindrella*, &c.; and many were scattered about, some whole, and some in fragments. In many cases the shell accurately fitted the hole, and the stone bore evident marks of having been in a plastic condition when the shells were enveloped in it. I at first supposed that they were fossil, but I have since found that they are the common living species of the woods. This fact is interesting, as it seems o prove the comparatively recent formation of this honeycombed limestone, which forms so large a proportion of the rock in the central part of the island. Out of the hollows of the rock on either hand, their roots fast grasping the projections and twining round the sharp points between the holes, grew many tall trees of various kinds, interlaced with climbers, and hung with festoons of *lianes*, resembling long and twisted

* The appearance of this limestone is very singular. Sir H. De la Beche accounts for its structure, by suggesting that it is not homogeneous in substance; that some parts were originally more argillaceous than others; and that these becoming decomposed by the action of the atmosphere, left irregular cavities in the harder and more durable stone.

cords, thrown from one to another, or depending from the branches towards the ground. The great Keratto, or American aloe, as it is often called (*Agave kerato*), threw up its broad, fleshy, spinous leaves, and lofty flower-stalks from the crevices, in profusion; and numerous *Cacti*, both erect and trailing, helped to give a very peculiar character to the vegetation. I here saw epiphyte *Orchideæ* for the first time growing in native wildness. Some large bunches of crimson blossom caught my eye, far up above my head; and giving my horse to my negro guide, I climbed the rock to examine the plant. It was, as I had suspected, the elegant *Broughtonia sanguinea*, growing in profuse luxuriance in tufts, depending from the trunks of some Hog-plum trees (*Spondias graveolens*). On another tree, a species of *Bignonia*, I found masses of *Brasavola nodosa*, with long leaves resembling porcupine quills in form, and hanging racemes of elegant white flowers. I carefully detached all the specimens, greatly to the mystification of my sable companion; who, as I afterwards found, told his master how the "strange buckra had taken the trouble to get *parcel of bush!*"

The mansion of my kind friend afforded me one of the most magnificent prospects that I had ever beheld. From the wide balcony that surrounds the house, the eye roams in one direction over many miles of savannah, or open pasture country, with all the varieties of surface produced by rounded hills, valleys, and sheeted slopes, beautifully spotted with large clumps of bushes, imparting somewhat of the character of our English downs. In another direction the wooded sides of the limestone valley lay beneath us, a broad expanse of sea filling like a cup the hollow formed by the distant valley's mouth. But the most striking feature was an enormous mountain rising immediately in front of the house, covered to the summit with dark woods; so steep and towering that as I lay in bed in a lofty room I could but just see a little portion of sky in the upper corner of the window. The top of this mountain was the coffee-plantation, and would doubtless have repaid the toil of an ascent, if I had had time to accomplish it.

COLLECTING CHITONS.

Dec. 13th. — I walked a mile or two along the beach, towards the lofty cliffs called the White Horses. Two isolated masses of rock had attracted my attention from the ship, and I hoped to obtain some shells here. They were situated just within the tide, and the surf was dashing furiously over them. As I could see from the beach that shells were adhering to the very rugged surface, I stripped, and waded out and examined. After picking off a few species of *Phasianella*, which were very numerous, I was delighted with the sight of a *Chiton*, and presently, many others of several species, the largest about two inches in length. They were *Ch. marmoratus*, *assimilis*, *squamosus*, and *pectinatus*.

Chitons are rather difficult shells to procure. The force with which the broad muscular foot adheres to the rock is too great to allow them to be removed with the hand; and if they be touched without being detached, they instantly become alarmed, the shelly valves bend more downward, the leathery mantle clings to the surface, and the muscular adhesion of the foot becomes so great that no force can now detach the animal, short of that which would destroy the specimen. The approved mode of proceeding is as follows. Take an old rounded table-knife, with which the collector should be provided, in the left hand, so that the tip of the blade shall be close to one extremity of the *Chiton*, without touching it. Then, with the palm of the other hand, as a mallet, give a smart blow to the handle, the effect of which will be that the animal is pushed from its hold before it has time to be alarmed. In the present case, however, this device was impracticable; for, owing to the very carious surface of the rock, and the *Chiton's* generally filling a hollow, it was impossible to push them off. But I found that by suddenly inserting the dull point of a strong pocket-knife under the margin of the mantle, and giving a quick wrench, I could remove them with facility, and generally without cutting or tearing the edge. Out of more than fifty specimens which I essayed, I did not lose more than two or three. Yet the operation was not without difficulty, from the very sharp projections of the rock, and the force of the surf, which frequently dashed violently over my head, and once or twice knocked me down. I had to catch the momentary intervals of the waves, to dislodge my booty; and sometimes a sea coming at the moment I had done it, washed it from my grasp. The finest specimens were found on the seaward side, and on the under surface of projecting ledges, which made them more hard to be got at.

These Chitons I prepared in the following manner: —Having brought them on board, I threw them into a tub of cold fresh water to kill them. When dead, which was known by the absence of muscular rigidity, the soft fleshy parts were cut out, leaving only the shelly plates and the leathery margin. Some strips of board, just wide enough to receive the specimens, having been provided, I placed each Chiton lengthwise, exactly in the position it had assumed when alive on the rock, and with a thread, passed several times round it and the board, bound it firmly down. It thus dried in a natural form, instead of the condition in which we sometimes see these interesting shells in cabinets, coiled up like a millepede, from which, when once dried, it is difficult to relax them permanently.

A day or two afterwards, I obtained from the rocks near Black River *Chiton piceus* in abundance, a much larger and finer species than any of the former.

THE PAINTED SWIFT-FOOT.

Dec. 17*th.* — About the rocks in the vicinity of Black River, I noticed many specimens of this beautiful Crab (*Grapsus tenuicrustatus*). Its form is remarkably flat, as are also the legs, and particularly the thighs, which pack one on the other in a very curious manner: the hue is a chaste warm grey, marked with transverse zigzag black lines, somewhat like writing. It is difficult to capture, for it is very wary, and its swiftness of foot is amazing; it darts from side to side over the rough surfaces of the rocks, with a rapidity that is much more like flying than running; not in a continuous course, but fitfully and irregularly, now shooting hither, now thither, and remaining still for momentary intervals between, watching the intruder. What is remarkable is, that it does not matter whether the rock on which it runs so swiftly be horizontal or perpendicular, nor whether its back or its belly be uppermost; it shoots round the projecting ledges, and darts about on their under sides with as perfect security and ease, as on the broad flat top of the rock. Does not this indicate a delicate sense of touch in the tips of its shelly feet?

BLUEFIELDS BAY.

It was on the morning of the 19th of December that we entered the beautiful Bay of Bluefields. It had well nigh been a serious, if not fatal, acquaintance, which I made with it; for the ship here suddenly struck upon the reef that stretches off some distance from the shore. The wind was happily light, and by the good providence of God, an anchor carried out succeeded, after some hours of anxiety, in warping the ship off without much damage. When morning broke, it found us in the midst of the broad bay, that stretched in a wide semicircle before us, just off the open harbour, or rather roadstead, of Savanna-le-Mar. Under a gentle breeze we were running in, and I had opportunity to admire the lovely scene. The verdant Guinea-grass pieces and pastures of

Bluefields and the pens around, sloped up from the sea, studded with white houses that gleamed in the rising sun. Further to the eastward lay the park-like estate of Mount Edgecumbe, its greensward varied with groves and clumps of the graceful Pimento. Behind, rose the mighty rampart of the Bluefields Ridge, rising into one conical peak of half a mile in height, and others of less elevation, and jutting out into the bay in a bold promontory, covered, from the rounded summit to the very foot upon the sea-beach, with the dark and dense primeval forest. The town of Savanna-le-Mar, scarcely rising above the sea-level, could be recognised only by the clustered masts of the shipping at anchor; and from it stretched away, in a long needle-like point, the eastern extremity of the island, Cape Negril. Blue and distant, yet bold and well-defined in outline, rose above the flat country about Savanna-le-Mar, the Dolphin's Head, a single mountain, resembling in form a crouching lion, and reputed to be equal, if not superior, in altitude to Bluefields Peak. In the smooth water in-shore, that accurately reflected the outline of the land, long strings of Pelicans were alternately plunging after their prey, and sailing on heavy flagging wing; and far, far overhead, like black specks against the bright sky, a flock of Man-of-war birds were placidly floating; resting, if I may be allowed the term, in the lofty air, after their morning meal upon the flying-fish in the offing.

The water all over this beautiful bay is unusually transparent, so that in six fathoms the bottom, with every rock, patch of sand, or bunch of weed, was as distinct as if seen simply through a broad plate of glass. We ran in through a very narrow channel, the coral reef almost touching us on either side, as I saw plainly enough from a little way up the shrouds. The pilot had taken his station on the end of the bowsprit instead of the poop, that he might the better discern the slender passage, cut as it were through the coral rock, which we were threading; and very delicate work it was This was not the ordinary channel for large craft, but that was denied to us in the present case by the wind, except with the loss of a tide.

SAVANNA-LE-MAR.

The town of Savanna-le-Mar presents little which is attractive to a stranger, though its aspect is characteristic enough of West Indian manners. It stands very low; on the eastern side a grove of cocoa-nut palms, nearly a mile in length, fringes the white beach; and on the other the eye rests on nothing at all but a wall of sombre mangrove trees, growing actually out of the sea for miles. You climb the wharf, and are immediately in a broad and long straight street, that constitutes the town. There is no pavement but the sandy earth, ploughed into ruts by the waggons, some of which you may see with their long teams of oxen, bringing in puncheons of rum and hogsheads of sugar from the country. Right across, at irregular intervals, run great water-courses, dry sometimes to be sure, but in the rainy season, as now, liable every day to be flooded, each a foaming torrent of muddy water rushing and frothing into the great mangrove morass that environs the town. Most of the houses are shops, or *stores* as they are called in American fashion; each store, whatever the character of its merchandize, — shoes, drapery, "dry-goods," "hard-ware," spirits, tobacco, provisions, or what not, — fitted up in the same manner, with an open piazza in front, three or four yards wide, in which the various goods are exposed, and in which the owner may commonly be seen with a friend or customer, seated on chairs, the feet often on another chair (this too in American fashion) discussing the amenities of a cigar or a glass of "malt." Behind the piazza is the shop, with unglazed windows, through which communication freely takes place to the clerks and shopmen inside; this is fitted up with counters and shelves, rather more in English style. Above, the ceiling of the piazza being supported, on the street line, by one or two slender pillars, are the rooms of the dwelling-house, or else balconies; in either case furnished with jalousies, or strong Venetian blinds, which admit light and air from beneath, excluding the sun's rays; or can be entirely closed. Towards the upper end of the long street the shops cease, the houses become more elegant, each inclosed in a court or garden, often adorned with the beautiful or fragrant blossoming trees and plants of the island, or such as unite fruit with beauty and shade. Of the former the scarlet Cordia, the noble Agave, and the Oleander or South Sea Rose, both beautiful and odorous, are great favourites; and of the latter, the wide-spreading Genip and Tamarind, the tall Papaw, and the golden-fruited members of the *Citrus* genus, from the gigantic Shaddock to the diminutive Lime.

BLUEFIELDS.

A day or two sufficed to apprehend the beauties of Savanna-le-Mar, and I was anxious, besides, to commence my regular collecting. The ship's cutter was put at my service, and manned to row me across the Bay. Kind friends were waiting for me on the beach, and the hospitable roof of Bluefields soon received me, and became my home thenceforward as long as I remained in the island, a period of eighteen months.

In the prosperous days of Jamaica, Bluefields was a sugar estate; but is now, like many other beautiful properties, given up, almost entirely, to resume the original wildness of nature. The greater part is, therefore, what is called *ruinate*, the expressive term applied to land in such a neglected condition. About a dozen acres are kept open in pasture, among the grass of which grow many flowering weeds, such as the Mexican Horn-poppy (*Argemone*), the West Indian Vervain (*Stachytarpha*), Swallowworts (*Asclepiadæ*), small *Passifloræ*, and others; and about as much more is planted with the valuable, and always verdant, Guinea-grass (*Panicum jumentorum*), among the tussocks of which may generally be seen fluttering dozens of that pretty pink-winged moth, *Deiopeia bella.* But all around is covered with a dense and tangled mass of second-growth, chiefly logwood, interspersed with calabashes, breadnuts, and cotton trees, and with the usual fruit-trees of a plantation, the avocada-pear, the akee, a recent introduction from Africa, oranges and limes, cocoa-nuts, mangoes, guavas, papaws, sops, and custard apples. There is not a day in the year in which fruit from some or other of these may not be plucked. The fences consist of "dry-wall," that is, a low wall built up of loose stones without cement; which, when the stones are skilfully arranged, so as to bind each other, is sufficiently firm and durable. It is easy, however, to guess that such walls afford a prolific harbour for vermin; and, in truth, spiders and scorpions, great biting ants, centipedes, and all the kinds of lizards and snakes, find refuge in their crevices. Over these walls sprawl multitudes of creeping plants, covering their tops and sides with a festooned drapery of verdure and blossom; various kinds of *Cereus*, *Aristolochia*, *Aroideæ*, and beautiful *Convolvuli*, *Ipomeæ*, and *Echites*, are abundant; while at their bases spring up numberless bushes of *Lantana*, of several species, (always covered with their bunches of cheerful blossom, and clusters of berries), *Cleome pentaphylla*, *Solaneæ*, and many papilionaceous and other flowering plants, both herbaceous and shrubby; many of the latter overspread with the tangled leafless stems of Dodder (*Cuscuta Americana*), like a great yellow spider's web thrown over them. A beautiful rivulet runs through the estate in a winding course to the sea. The soil is a friable whitish marl; the highest elevation may be a hundred feet, sloping down gently to the shore.

THE SOUTHERN CROSS.

A few days after my arrival, I had set out some time before dawn, to go in a canoe round the neighbouring shore. As I walked along the road to Belmont, suddenly and unexpectedly the Southern Cross caught my eye, close to the meridian, and therefore erect. My first emotion was a gush of pleasure, at seeing what I had so often read of, and wished to see; my next, a feeling of disappointment at its effect. I had expected a much more splendid constellation; if I had not known its form from books, I do not think I should ever have been struck with it, or hardly have noticed any resemblance to a cross; the westernmost star is too near the beam, and too high, to be symmetrical with the opposite one; it is much inferior, too, in magnitude; and there is another star, quite supernumerary, a little below it; all which circumstances greatly detract from the effect of the constellation as a whole. Still there is a sort of classic association with it, which must always give it an interest to an European who looks upon it for the first time.

ASPECT OF THE FOREST.

It was pleasant to walk out the morning after my arrival at Bluefields, and survey my hunting-ground. It was not so much to collect specimens, nor to pursue any particular line of observation, as to take a general glance at Nature, and to delight my fancy with the richness and the novelty of the field in which I thus found myself. At one corner of the pasture a steep rocky hill rises abruptly, covered with pristine woods. The boughs of an immense fig-tree, which had been prostrated in a storm a few weeks before, enabled me to climb the ascent; but I was astonished at the difficulty of penetrating the forest. The number of tough withes, many of them fearfully spinous, that entwine about the trees and about each other; the long prickly *cacti,* too, that trail here and there; the lianes, that resemble ropes, or lines, or strings, according to their thickness, hanging down in loops, or loosely waving to and fro, — are wonderful: these last frequently extend from a lofty bough nearly to the ground, without a branch or leaf till near the extremity, where the cord commonly divides into three or four more slender ones. Some of the larger ones are woody, and are often seen tightly twisted together, like the strands of a cable. The bushes and smaller trees are sometimes very numerous and close, quite choking the ground, and preventing the view beyond a few yards in any direction. The oppressive heat, the insects, and often, as here, the loose stony character of the ground, render it impossible to go far into these woods.

Yet here was much to interest a stranger. The large trees, and many of the small ones, were studded with parasites, springing out of the bark of the trunk, from the angles of the forks, or from the upper surface of the great horizontal branches. Some of these were *Orchideous*, but the chief belonged to the same family as the pine-apple, the resemblance to which has been recognised in the appellation provincially conferred upon them, *Wild-pines*. They are found in all situations, on the tops of the mountains, in the plains, by the sea side, in the depths of the forest, in the open pasture; and on all sorts of trees, scarcely a tree being without some, large ones often carrying hundreds; nor do they seem to attach themselves, as far as I can discern, to dying or diseased trees, more than to such as are in perfect health. One species, called Old-man's-beard (*Tillandsia usneoides*), is very small and slender, but forms great matted bunches, with wiry stems, which often hang down like tufts of ragged hair. Others are highly ornamental. One, rather common, has a swollen base, much like the bulb of a lily (hence called *Tillandsia bulbosa*); and narrow, almost cylindrical, pointed leaves, of which those near the blossom are often crimson. A spike of flowers lengthened in form, and of crimson and purple hues, with singularly sheathing bracts, and projecting purple stamens, makes this a beautiful plant in the blossoming season. Another still more abundant kind, a species of *Vriesia*, throws out a long branching spike, of crimson and yellow flowers, imbricated or sheathed in a singularly compact manner. In the mountain-woods a crimson and purple species of *Æchmea* is very gorgeous; but the most imposing of all is the noble *Tillandsia lingulata;* the leaves of which are long, and shaped like those of the pine-apple, widened and, as usual, sheathing each other at their bases, and throwing out in July large flowers of a rich crimson hue and polished surface. The sheathing bases of the leaves form natural reservoirs for water; the rains and dews accumulating there in considerable quantities, and forming a resource for thousands of birds, and even for man himself, in the season of drought, and affording retreats, always cool and moist, for those reptiles that respire through the skin.

One of the largest trees on the hill of which I have been speaking was the Tropic Birch (*Bursera gummifera*), a tree not very common, but often attaining a giant bulk and height. It is less infested than many with these parasites, owing to the smoothness of its bark, which is glossy, and peels off in thin papery laminæ, like the northern tree after which it is named. A clear resinous substance exudes from the trunk, which is viscid, and appears to possess many of the qualities of turpentine.

THE SOLDIER CRAB.

Among the loose pebbles that formed the surface of this hill, wherever the huge outcropping masses of rock did not appear, and that added to the difficulties of penetrating the maze, were many large shells. The first of these that I saw was a large round brown Snail (*Helix Jamaicensis*), and as its exterior was very fresh, I thought I had a prize; but on taking it up, I saw the large claw of a Hermit-crab, red and tuberculous, closing up the orifice, which it exactly fitted. The negroes call these crabs Soldiers, perhaps from their red hue. Afterwards I saw other specimens, and some of a large *Trochus* (*Meleagris picus*), with pearly interior, as well as a capacious *Ampullaria*, inhabited by the same species. It crawls irregularly, but quickly; making its shell rattle against the pebbles as it proceeds; but if alarmed, it instantly withdraws into its house, and bringing its strong legs around its head in the form of a semicircle, claps its greater claw upon the whole, presenting, as I have said above, nothing but a hard, shelly, prickly, convex surface within the margin of the house, so accurately filling the aperture, and so strongly held down, that it is impossible to extract the animal alive without breaking the shell to fragments. Yet the wary Soldier is ready for fight; while I was holding one in my hand, the rogue protruded his claw, and seizing the skin of my palm, fairly took the piece out.

The species was the well-known *Cœnobita Diogenes*, and, as I afterwards found, abundantly common in the woods near the coast. I even found it numerous, inhabiting the shell of the same large *Helix*, far up on the side of the mountains, behind Bluefields, in the dryest and most rocky situations. Sir H. de la Beche found it inhabiting sea-shells, near the Rio Minho, ten miles from the sea. It is evident that the active little creatures must have crawled this whole distance, and as the *Helix* is found in a living state only in the woods, and, as far as I know, only in the mountains, it follows that the Soldiers must have travelled up the country in their sea-side shells, until they came to the region of the *Helix*, and there have changed their houses, and brought the latter down towards the sea in their return. Many of the *Helix* shells were so pearly in the interior, and so bright and fresh on the exterior, as to show that they could not have long lain exposed to the weather, since tenanted by the original proprietor. This suggests the inquiry, whether in any case the *Cœnobita* destroys the Snail to obtain his shell for a dwelling.

THE CALABASH AND ORCHIDS.

Just behind the pasture, among the fruit-trees, now all choked up with bushes and young wood, are many Calabash trees (*Crescentia cujete*), a tree commonly found around the homesteads and negro villages, and cherished for the sake of the large gourd-like fruit, the woody shell of which, divested of its pulp, makes admirable domestic vessels. The tree has a strange appearance, easily recognised when once it has been seen, from the peculiarity of the foliage, which does not form masses, but fringes the long branches; and as these are slender and straight, shooting out in all directions, and continually crossing each other, but never tortuous, the effect is much more curious than beautiful. The large oval gourds hang from the branches in all stages of maturity, together with the blossoms, which are large, and shaped like our Canterbury bell, but of a greenish hue, with dull purple lines. They are sessile, that is, without footstalks, and to add to their singular appearance, they sit as it were on the naked bark of the branches, and frequently on the bare trunk itself, without any leaves or bracts surrounding them.

No tree in the lowlands is so great a nursery of *Orchideæ* as the Calabash. On one of these I found large masses of ***Brasavola nodosa***, and on another a species of noble size. Great thick ovate leaves a foot and a half long, and four or five inches wide, without bulbs, formed immense bunches on several of the principal branches and their forks, from the axillæ of which sprang pendant flower-spikes eight feet or more in length, bearing, however, no flowers, but elliptical seed-capsules three inches in length. One of the tufts was throwing up a new blossom-shoot; and this I left untouched, in order to ascertain the species. I suspected it to be *Oncidium Carthaginense*, and this it afterwards proved to be. It came into full blossom about three months afterwards, in the beginning of April; at which time, masses of the same species growing on the trees at intervals all along the road between Bluefields and Savanna-le-Mar, burst simultaneously into flower. In a few weeks, however, the blossom had quite disappeared; and nothing was to be seen on any specimen but one or two maturing seed-vessels, till the following year. Notwithstanding its size, the blossom is not conspicuous for beauty; the loose, panicled character of the raceme, and the hue of the flowers (yellow, studded all over with red dots) detract from its effect. This species is essentially a lowland plant; for though it is found at a considerable elevation on the mountain-slopes, it is most abundant near the shore; it bears the open sun better than most *Orchideæ*.

On another Calabash, hanging over Bluefields

brook, was an Orchideous plant more curious, and much more rare; since I found it only on one or two other occasions all the time I was in Jamaica. It was *Angræcum funale;* entirely leafless at all seasons, and consisting of an intertwined congeries of contorted roots, long, slender, and cord-like, of a pale glaucous hue, except at the lengthening tips, which are bright yellow-green, and polished. The greater portion of these roots hang down in an irregular mass; but such as touch the bark of the tree grow to it, flattening themselves on that side which is in contact, and clinging so closely and firmly to the bark that it requires considerable force to detach them, the root often breaking rather than relinquishing its hold. This habit and mode of clinging is common to the roots of many *Orchideæ,* and is a curious provision for giving to the plants a secure hold in situations, such as the trunk of a smooth-barked tree, where it would have seemed impossible. This *Angræcum* throws out its flowers through the whole year, but not in great numbers, nor in spikes, but singly, or by twos or threes. The blossom is moderately large and pretty, the sepals and lateral petals pale green, the lip expanded and delicately white, with a long slender spur. The seed-vessel is slender, spindle-shaped, somewhat fluted, nearly four inches long, of a dull yellow-green when ripe.

While I was up in the Calabash tree, engaged in detaching the bunches of *Oncidium,* the beautiful Long-tailed Humming-bird (*Trochilus polytmus*) came shooting by with its two long velvet-black feathers fluttering like streamers behind it; and began to suck at the blossoms of the tree in which I was. Quite regardless of my presence, consciously secure in its power of wing, the lovely little gem hovered around the trunk, and threaded the branches, now probing here, now there, its cloudy wings on each side vibrating with a noise like that of a spinning-wheel, and its emerald breast for a moment flashing brilliantly in the sun's ray; then apparently black, all the light being absorbed; then, as it slightly turned, becoming a dark olive; then in an instant blazing forth again with emerald effulgence. Several times it came close to me, as I sat motionless with delight, and holding my breath for fear of alarming it, and driving it away; it seemed almost worth a voyage across the sea to behold so radiant a creature in all the wildness of its native freedom.

As I now, several years afterwards, here in the suburbs of London, copy these notes for the press, the impressions then produced on my mind as one novelty after another presented itself, — things that I had read of with eager desire to see, that had become encircled with halos of romance in my imagination, — come gushing upon my memory in all their fulness and freshness, like some sweet tune, that one has not heard for years, unexpectedly played. But how shall I transfer these impressions to my readers? I can name some of the prominent objects that helped to make up the picture, and by some short description, or a few well-selected epithets, may communicate a certain definiteness to those objects; but the picture itself, the thousand things that cannot be enumerated, birds, insects, flowers, trees, the tone of the whole, the sunlight, the suffused sky, the balmy atmosphere, the variety of the foliage, the massive light and shadow, the dark, deep openings in the forest, all new, rich, and strange; — not only new individually, but quite new and strange in character, quite unlike anything that I had seen before;—all this I cannot hope to convey. Nor can I hope to convey more than a very, very faint reflection of that delightful excitement with which I gazed around, bewildered and entranced, almost, with the variety of charming objects, all at once appealing for attention; the remembrance of which, protracted as it was through eighteen months' duration, with scarcely any abatement, has given in my habitual feelings, a kind of paradisaical association with lovely Jamaica.

BELMONT BEACH.

The great post-road of the southern side, after passing Bluefields, (supposing the traveller to be proceeding eastward), runs along the coast to Belmont, Mount Edgecumbe, &c., often at the very water's edge, and sometimes separated from the sea only by a narrow belt of woods. Close to Bluefields, the shore is a beach of white sand, not siliceous, but consisting almost wholly of coral, shells, *echini*, &c., bleached and pulverised by the long action of the weather. A small stream running through a fœtid morass crosses the road about half a mile from Bluefields, and has deposited a broad flat bank of mud at its mouth, which is uncovered at low water. At this time it is seen to be pierced with innumerable little holes; and hundreds of a tiny Calling-crab (*Gelasimus vocans*) are running over its surface, the males of which hold up their enormous claw in front, as if in defiance. At the approach of an intruder, every one hastens into his burrow, and in a moment the muddy bank, that was alive with the moving atoms, is perfectly still; except that a dull-coloured but agile beetle (*Cicindela Guadalupensis*) is flitting about and alighting upon it. The little Crabs are very swift and wary, so that it is difficult to capture them, except by making a sudden rush from a distance among them.

Beyond this *creek*, as the stream is called, a projecting point runs out into the sea, round which the coast is more rocky. A low cliff terminates the land, excavated in shallow caverns, from the roofs of which the Cave Swallows (*Hirundo pœciloma*) suspend their mud-formed nests; and great masses of honeycombed limestone lie in the sea at its base. After a while, the cliff becomes gradually obsolete, and the beach of coral sand reappears.

All along the beach at high-water-mark there lay, at the time of which I speak, an immense number of Sponges, exhibiting great variety of colour, form, and structure; they were mingled with Sea-weeds, and had been, as I was informed, thrown up by the sea in a very tremendous gale that had occurred a few months before my visit. I collected some hundreds of specimens of the different species, with little labour; and, as they had all dried with the gelatinous flesh enveloping the skeleton, they were in the very best condition.

A few yards from the beach the bottom is a white tenacious marl, covered with a dense but short coat of marine grass. On this lie in the shallow water many *Holothuriæ;* they are soft and flaccid when first taken out of their element, but after being held in the hand a few moments they become tense and stiff, and usually discharge a small stream of water from the extremity of the body. They do this, I find, even if held under water: it is doubtless the result of strong muscular contraction. They are sluggish, unattractive animals, of a dull olive tint. A species of Swimming-crab (*Lupa*) common in these shallows is very active and fierce, extending its open claws in a threatening manner when danger is near; but, if allowed to escape by swimming, it does so rapidly, bending up the claw of the side which happens to be foremost, and allowing the other to stretch out behind. Probably this is the arrangement in which these unwieldy members offer least resistance to the water in progression.

Raking with an insect-net among the weeds in these shallows, I captured a Sole of small size, prettily marked in a netted pattern of confluent lines.*

* It appears to be new, and may be thus described:—The Inscribed Solenette, *Monochirus inscriptus* (Plate I. Fig. 4.): Right pectoral composed of two minute filaments, left composed of a single one almost obsolete. Ventrals indistinct, brush-like, united to each other, and to the anal. Anal and dorsal united to the caudal. Form nearly oval, slightly tapering; caudal rounded; mouth and chin fringed with short bristles; mouth much decurved; lower eye close to rictus.

CORALS AND THEIR PARASITES.

Jan. 23rd. — Observing that what appeared to be rocks under water were really growing corals, I stripped and got in among them. They are of many kinds and of various colours; being covered with the round disks of the soft gelatinous animals aggregated so closely as to touch each other, giving a very slimy unpleasant feeling to the foot that treads on them, though with a shoe on; for I dared not trust myself with naked feet among the *Echini*, and other formidable creatures, not to mention the sharp points of the honeycombed rock. It was at the ebb of a spring-tide, the moon setting as I began my examination*; yet I found little variation in the height of the water, tide here being very small. Some of the corals (*Millepora complanata?*) grow in thin irregular perpendicular plates, joining each other at various angles, so as to form large honeycomb work, somewhat resembling the second stomach of an ox; others present thick flattened branches, covered with minute projecting mouths: these are of a bright fawn-

Lateral line arched at its commencement, and then curved slightly downward, rising again to the tail. Fin-rays, D. 46; A. 40; C. 12. Ground colour pale bluish-grey, covered with an irregular network of rich, deep brown, confluent lines; these marks, on the dorsal and anal, assume a longitudinal direction, but still irregular and confluent. Caudal pale brown, with indistinct dusky bands. Left side white. Length of specimen to extremity of tail, 2½ inches.

I scarcely know whether to call this a *Monochirus* or an *Achirus*; the under pectoral is certainly present, but only as a rudiment, while the upper is so minute as hardly to be anything more.

* High water at Bluefields is exactly at the southing of the moon.

colour, while alive; other close-pored masses, of a rounded form, are bright grass-green; and huge round brainstones (*Meandrina*) which are very numerous, are of a dull olive-brown hue. The first two kinds were easily broken, so that I detached large fragments without difficulty; but though touching them for this purpose did not sensibly affect the hand, the more tender skin of the thighs and legs was susceptible of a stinging influence from the slightest contact; and my leg, which was rudely scratched against one, presently swelled up into a large tumour, very painful. The water in some parts was up to my neck, and the rolling surge made it difficult to preserve my footing. All were slimy to the touch; but a very branched and flexible kind, growing in a tuft of numerous stems, springing from a common basal point, and waving gracefully in the roll of the sea,—was particularly slimy, and communicated to the hands more of the remarkably strong nauseous smell, which all living corals possess. Three or four living Sea-fans I took, and also some soft bunches of a plentiful Coralline.

After this I waded out to the reef which runs along parallel to the shore, at about a hundred yards' distance from it. The water here was knee-deep. Many small Corals were on the bottom, apparently alive, of different species, some of which were very pretty. On almost every specimen that we lifted there were marine animals, parasitically lodged in the interstices. Among them were two or three of a little *Sepia*, that adhered with exceeding tenacity to the coral, and contracted its arms so as to lie in the hollows, resisting all attempts at dislodgement; till suddenly, on a moment's respite, the creature would rear up its leprous form, like Satan before Ithuriel, and try to scud away. A species of *Aphrodite* was numerous, which on being handled thrust out bundles of white silky bristles *that adhered to the fingers;* their points had been visible before, just projecting from the sides. Star-fishes (*Ophiuræ*), of two or three species, entwined their snake-like tails in the crevices, and were difficult to get out; and when out, usually broke into fragments. One specimen dismembered itself the moment it left the water, before it had been touched.

A very singular creature was found on one piece of coral, which seems to approach the *Holothuriæ.* The body consists of a slender membranous bag of great length, of soft semi-transparent skin, filled with sea-water while immersed, and therefore plump and round; but on being lifted out the water escapes from some orifice, and the sac-like body becomes empty and flaccid. It is rather gaily coloured, being whitish on the under side, and yellow on the back, with broad transverse bands of dark brown. Around the mouth are set about fourteen broadly fringed tentacles, some of which in turn were continually bending inward towards the mouth. It was lively in its motions, twisting about in the water with considerable vivacity.

Besides these, some of the masses sheltered tiny fishes, brilliantly coloured, of the genus *Labrus;* a species of Sea-slug (*Aplysia*), and many other curious things, including some Nereidous *Annellida. Holo-*

thuriæ of various sizes and species lay on the bottom, as inert as usual.

On another occasion I found a large mass of soft coral-rock, lying in the water, dead. On being broken it was found to be perforated by cavities inhabited by two species of *Pholas*, two species ol *Sipunculus*, and some *Annellida*, which I judged to be of the genus *Syllis*. One of the *Sipunculi* had a long slender body inhabiting a membranous tube, and communicated with the surface of the stone, whence it protruded its tentacles in the form of a circular disk of great beauty; the individual tentacles radiating from the centre, and the whole disk marked with alternate concentric lines of light and dark hue. The other species, which was larger, when put into a vessel of sea-water, protruded by the evolution of the integument a long neck, bearing at the extremity a similar tentacular apparatus. I brought home several of these in a bottle of water, preserved pure by the presence of some sea-weeds, together with two handsome *Actiniæ* of a reddish-brown hue, with white warts. The *Annellida* were remarkable for the ease and agility which they now and then displayed in swimming, throwing their lengthened bodies into the most elegant serpentine curves.*

* The diversified animals seen by Captain Basil Hall on coral, in his voyage to Loo Choo, and described by him in a passage very interesting and often quoted, were certainly parasites analogous to those above mentioned, and not the worms that form the coral, as the Captain seems to have thought them. From his description I should take them to be *Ophiuræ*, *Holothuriæ*, *Sipunculi* (or perhaps some *Annellida*), naked *Mollusca*, and macrurous *Crustacea*;—the very forms which I have enumerated above.

MARKET-DAY.

Feb. 1*st.* — Having occasion to visit Savanna-le-Mar, I took the opportunity of crossing the Bay in a negro's canoe. Saturday is market-day; and many of the black and coloured people resort on that day to the towns from the country, some on horseback, some on foot carrying their own loads, others driving donkeys, and others by sea in canoes, with provisions for sale. The concourse, the gossip, and the excitement of the market present great charms to the negro's mind; I have known, repeatedly, a woman to carry on her head a huge tray of yams to Savanna-le-Mar, a journey of twelve miles, and return the same evening; when she had actually refused to sell her produce at Bluefields-house, close to her own door, for a price larger, *actually larger*, than she expected to get at the market.

At sunrise I walked down to the beach, and waited until the preparations were completed, much amused at the busy scene presented by the shore, usually so still and solitary. Three or four canoes lay half-launched at the water's edge, around which were congregated nearly a hundred persons; and more were continually arriving with trays or shallow baskets of provisions and fruit on their heads. Heaps of yams, cocoes, sweet potatoes, plantains, pumpkins, oranges, sugar-canes, and other produce, calabashes of water, bottles, &c., were lying about, which with much clamour were being deposited in the canoes. The jabber was immense; — a hundred negroes, many of them women, all talking at once make no small noise; and the white teeth were perpetually shining out in the sable faces, as the merry laugh — the negro's own laugh — rose continually. The figures of the women, many of them not ungraceful, though plump and muscular, were picturesque, clad in short gowns of showy colours, and wearing the peculiarly set handkerchief for a head-dress, in form of a turban, often also of bright hues, though in most cases white as snow. They moved about amidst the bustle, crowding up to the canoes to stow their ware; tucking up their frocks still higher as the depth of water increased, regardless of displaying their bronzed legs. At the edge of the water, on whose mirror-like surface the mounting sun began to pour torridly, the little children sat, sucking cane or oranges, while the elder ones played around them, helping to augment the noise.

As I was in no hurry, I quietly waited, enjoying the novel scene, until, at last, the canoe being loaded, I took my seat in the prow among the garden-stuff, and the men launched her into deep water, climbing in over her gunwales as she floated. They paddled swiftly along; and as we glided over the reefs and shallows, the bottom of the sea in many places was distinctly visible. On the snowy coral-sand lay hundreds of *Echini*, very conspicuous, like great black spots; while here and there a large Star-fish of bright orange hue varied the character of the spotting. The ripple of the canoe's motion prevented clear vision, notwithstanding the calmness of the air; but once or twice I caught sight of other objects still more interesting. On later occasions, however, I had better opportunities of admiring these large *Echini*, a species of *Cydaris*, with very long purple spines; and also of seeing what had escaped my notice at this time, the magnificent living Corals that form great bushes, if I may so say, on the bottom. They principally grow in large, irregularly-waved foliations, and in thick-set shrubs of cylindrical branches much ramified. After two hours' rowing we landed at Savanna-le-Mar, and I found myself in the midst of the market, held on the very beach; a scene somewhat similar in its character to the morning bustle on Bluefields beach, but much livelier and more like a rustic fair. The country people, as fast as they arrived, either by land or water, displayed their produce, in heaps, on white cloths spread out on the ground, and squatted down by them, or stood to chat and laugh with their acquaintances; while the townspeople were bringing their wares from their respective stores, and laying out cloths and linens, butter and pork, bread and buns, on standings and benches, or even in some cases on the bare sand.

The fishermen often see from their canoes large shells crawling on the bottom, and dive for them. The great rosy-mouthed Conch (*Strombus gigas*) is obtained in this way, and so are the more valuable Helmets (*Cassis*), and others. The noble *C. Madagascariensis* and *C. tuberosa*, the former called here the Heavy-back, the latter Turtle-back, with the large and beautiful *Tritonium variegatum*, are brought me by the fishermen for sale alive, and offered at about a shilling each. The animal of *Cassis* cannot be extracted in the ordinary way, owing to the narrowness of the aperture; the shell must therefore be buried in the earth for a few weeks, after which the flesh may be shaken out, carrying with it the long horny operculum.

In Helmet-shells buried for this purpose I found a *brachelytrous* beetle which enjoys a very wide geographical range. It is *Staphylinus* (*Creophilus*) *villosus*, which is so abundant in Newfoundland, as to be quite a pest, crawling about and devouring the drying codfish; it is there called the fish-fly. In Canada, and in Alabama (U. S.), I have also met with it, but rarely; and now I trace it to Jamaica. The *brachelytra*, however, are very scarce here, as are the carrion-eating beetles generally; their place is probably supplied by the Aura vulture. I only on one other occasion met with this fetid and disgusting beetle.

LUNAR RAINBOW.

Jan. 22d. — This evening I had the pleasure of seeing, for the first time, that rare meteor, a lunar rainbow. The arch was large and perfect; the southern limb was distinctly coloured, the northern just perceptibly; the middle portion was destitute of colour, displaying only white light.

THE LEAF OF LIFE.

About a hundred yards from Bluefields-gate there grows by the roadside a considerable bed of *Verea crenata*; a tall, rather handsome plant, with thick succulent leaves notched in a fashion which heralds term *engrailed*, leaving rounded segments. It is now in blossom, throwing up a spike of greenish flowers to the height of three feet. It is considered a great curiosity by the creoles, on account of its tenacity of the living principle, whence they call it the Leaf of Life, or the Life Plant. A single leaf is sometimes broken off, and suspended by a thread from the top of a window; when a number of delicate white rootlets are presently thrown out from the base, and the leaf forms a young plant. Some prefer to throw the leaf into a drawer or box, and to open it in a week or two, when the same results are found to have taken place. I have found that even half a leaf, cut across, will root; and it is impossible to dry the plant in an herbarium, without first killing it with a hot iron, or by boiling water. This property, in a greater or less degree, is common to many members of the Order *Crassulaceæ*, to which the *Verea* belongs. Perhaps it is not indigenous, but is a truant from some garden.

BLUEFIELDS MOUNTAIN.

Many days had not elapsed, after my arrival at Bluefields, before I determined to explore the summit of the lofty mountain that rises behind it, and bears the same name. A ride of four or five miles brought me to the brow of the mountain, the steepness of the ascent being mitigated by the practice (common on the precipitous mountains of the south side of Jamaica) of conducting the road in a zigzag line instead of a direct one. Owing to this circumstance, a much more varied, as well as more extensive, prospect is obtained of the country below. The opening of the bays and harbours that vary the sinuous coast, as we gradually mount higher and higher, is very interesting; and as we draw nearer the lofty summit, and look down on the variations of outline beneath our feet all marked as in a map, — Savanna-le-Mar, with a broad tract behind it, St. John's Point, and even the extreme western headlands, North and South Negril appearing just at hand, so that the eye looks into the bay between the last-named points, — and then, leaving the land, as we gaze out on the sparkling Caribbean Sea stretching away to the far, far distant horizon — the expanded prospect becomes indeed a noble one.

The side of the mountain up which the zigzag road winds, was formerly under cultivation; and many fruit-trees record the fact, amidst the maze of young shrubs that have sprung up since it has been thrown aside. Many a time when descending in the burning sun of noon, after a fatiguing morning's excursion, have I been refreshed by the delicious sour-sops or custard-apples detected by the keen eye of my faithful negro Sam within the "bush," and obtained by his dexterity. If these tropical fruits do not happen to be known to my reader, let him imagine, for the latter, a custard inclosed in a rough skin bag, most lusciously sweet; and for the former, nearly the same, but rather more fibrous, and of a delightful acidity. Both are as large as a child's head; the sour-sop often larger, kidney-shaped, and covered with flexible prickles.

Several of the primitive forest trees have been left on each side of the road for the comfort of their shadow. The common Cedar (*Cedrela odorata*), the Bastard Cedar (*Guazuma ulmifolia*), and the Fiddlewood (*Citharaxylon*), are numerous; a species of *Coccoloba*, called the mountain grape, and the Clammy cherry (*Cordia collococca*), covered with great bunches of scarlet berries, also occur, and two or three large Mahogany-trees. Near the summit the Mountain Mahoe (*Hibiscus tiliaceus*), a tall and spreading tree, displays its large gorgeous blossoms in the season, a mass of scarlet and yellow; and a considerable tract on either hand is covered with the elegant Bamboo, whose dead and dried leaves completely hide the ground, and preclude the intrusion of any other plant, where this gigantic reed has once obtained possession. Many kinds of flowering shrubs fringe the sides of the road; among which the most characteristic is a species of Heliotrope, possessing neither beauty nor fragrance. At the elevation of about (to speak roughly) 1500 feet, I found growing on the bare rock a terrestrial Orchid of much elegance, a species of *Bletia* (*B. verecunda?*) of a crimson hue, striped with yellow. It was not, however, at present in flower; but the tall grass-like leaves were beginning to wither, and young flower-shoots were already springing from the sides of the globose bulbs. Though very local, being confined to the space of a few yards, it had much increased in that spot; for I obtained upwards of three hundred bulbs, without including the very small ones, all closely packed together in masses that could be stripped bodily from the rock.

Just on the very brow of the mountain is a small plantation devoted chiefly to the Pimento or Allspice. At this time, the beginning of February, the groves were thickly covered with the green unripe berries, just fit to be gathered and dried for exportation; and accordingly a group of children were stationed around each tree, one of whom plucked the loaded twigs, from which the rest stripped the berries. Every part of this elegant tree is aromatic; the wood, the blossoms, the fruit, the leaves; but the seeds are especially hot in the mouth. The berries, if suffered to ripen, become black, pulpy, and sweet, retaining very little of their spicy character. Many birds eat them when in this condition; but even in the unripe state, as when I first visited these mountain-groves, they presented temptations to some. Flocks of green Parrots and Parroquets were shooting from tree to tree, screaming discordantly as they went, but, with the characteristic intelligence of the tribe, as quiet as mice the moment they had alighted, that no sound might betray their presence, so well concealed by the greenness of their plumage, while they luxuriously fed upon the aromatic berries. The planter, however, was warily marking them with his fowling-piece, and at the instant of my passing, he brought down a beautiful Yellow-billed Parrot (*Psittacus leucocephalus*), which he politely presented to me. The Solitaire (*Ptilogonys armillatus*), too, and the Glass-eye (*Merula Jamaicensis*), both of which, but especially the former, are most jealous recluses in common, haunting the dark recesses of the mountain forests, find in the green pimento a temptation too strong to be resisted.

In a piece of ground just reclaimed from the forest, cleared and burnt over, but not yet planted, full of blackened stumps and stones, I searched for land-shells. Under the stones were great numbers of the pretty pink *Cylindrella cylindrus*, with the whorls diagonally fluted in an elegant manner. Although multitudes were alive, I was surprised to observe that all were defective at the apex, which had evidently been broken off. But from some specimens which were less maimed than others, and from the discovery of one or two of the detached apices, I perceived that in a perfect state this shell would terminate in a slender elegant spire, colourless and pellucid; and at length I had the pleasure of finding a single specimen (the only one I ever obtained) in this perfect condition. Other shells were also found in some abundance, as *Helix sinuata*, *picturata*, and *soror; Cyclostoma xanthostoma* and *mirabile; Cylindrella Maugeri; Helicina pulchella*, *neritella*, *aureola*, &c., all under stones, inert and apparently torpid.

Some other interesting animals were found under stones in this locality, though not all at the same time. A curious little dark grey Oniscus, every segment of which is armed with two spines, was numerous; it has been described by my friend, Mr. Adam White, of the British Museum, under the name of *Acanthoniscus spiniger*. Some Earth-worms

(*Lumbricus*), very much like those of our gardens in appearance, but two or three times as large, and glossed with more vivid prismatic reflections, were also found. But what I considered the greatest curiosity was a small *Peripatus*, which I suspect to be different from the species found by the Rev. Landsdowne Guilding at St. Vincent's. It is a curious creature, and I certainly think rather allied to the *Annellida* than to the *Mollusca*. It is of a velvety appearance, of a blackish-brown hue, the tentacles tipped with white. From these latter organs there exudes, when the animal is touched, a thick glutinous substance, as adhesive as birdlime. It crawls about as fast as a caterpillar of a Bombyx, and much in the same manner. The skin repelled both water and spirit when immersed. When put into the latter, a considerable quantity of milky fluid was poured from the mouth. The animal was rather scarce, only five or six specimens having been discovered, all under the stones; one was twice as large as any of the others.

Immediately above the piece of ground in which these researches were made, the mountain rises into a conical peak of considerable elevation, though not the very loftiest. It is covered with the original forest, all in the rude luxuriant wildness that it bore in the days when the glories of these Hesperides first burst upon Europeans' astonished eyes. The gloom of the interior is so sombre, and its contrast with the effulgence of the sunshine without so great, as almost to deter one from penetrating it. I certainly felt a sort of undefined awe, as if I knew not what I might meet in those deep solitudes, where

> "——— The roof
> Of thickest covert was inwoven shade."

But when once we are within the shade, and the pupil of the eye, relieved of the broad glare, has had time to dilate a little, there is found to be abundant light for every purpose, softened and subdued as it is by being transmitted through the thousand green leaves that quiver and flash over head. The ground is very rocky; vast masses of rock lying irregularly heaped on one another, as if the fabled Titans had been attempting to scale heaven from thence. Comparatively few of the trees are large, the great majority showing trunks so slender as to suggest the idea of second growth; that is, of wood springing up spontaneously after the ground had been once cleared. But the great prostrate trunks, lying in all directions, so soft with decomposition as scarcely to bear their own weight, — long columnar masses of pap-like substance, woody fibre deprived of tenacity and saturated with water, the outer surface for several inches deep already turned to black mould, and bearing a dense crop of ferns and mosses, — proved convincingly that the hand of man had not disturbed the primal forest here; but that these giants had died a natural death where for ages they had reared their lofty heads. Many of these fallen trunks, and of the masses of rock, are covered with a dense and elegant sort of club-moss (*Lycopodium*); a graceful drapery concealing the decay of the former, and the rude angles and gaping crevices of the latter, like a thick and soft green carpet thrown loosely over them. The Broadleaf (*Terminalia*), a great timber tree, with leaves that remind one of those of our horse-chestnut in form, but immensely larger, and bearing edible nuts, like almonds, is a common tree in this region; and the edges of the woods are occupied by *Clidemia* and other forms of *Melastomaceous* shrubs, whose leaves, divided by a few longitudinal ribs, and crossed by transverse ones at right angles, have so very singular an appearance to an European observer. Among them stands conspicuous for beauty *Blakea trinervia*, whose expanded crimson flower has received the name of the Jamaica Rose; a tall shrub or small tree, with smooth but not glossy leaves of a lively green, supported by three nerve-ribs.

It was here that I first became acquainted with those elegant plants, perhaps more than almost any others peculiarly characteristic of tropical vegetation, the Tree-ferns. Many specimens were growing on this wooded hill, but all, I believe, of one species, *Alsophila aculeata*. I gazed upwards with peculiar delight at the minutely fretted fronds that formed a wide umbrella of open work between me and the sky. In one respect only was I disappointed; from descriptions and figures in books, I had expected to see the Tree-fern far loftier and more Palm-like than I found it. I do not think any specimen that I saw exceeded fifteen feet from the ground to the bases of the fronds; but then the expansion much exceeded my pre-conceptions. I should conjecture that some of these trees covered with their arching shade a circle of nearly twenty feet in diameter. There is no very close similarity between a Tree-fern and a Palm in appearance; there is a remarkable lightness and voluptuousness, if I may be allowed the expression, in the filagree work of the one, produced by the minute subdivision of the immense frond (of which every one may form a feeble notion from the commonest ferns of our heaths), that contrasts with the stiff, simply-pinnate or fan-shaped leaves of the Palms; yet these latter have a beauty and elegance of their own.

When the emotion produced by the first sight of these interesting plants had subsided, I still found much to admire in a more minute examination. The formidable prickles studding the knobbed bases of the fronds, that swelled out around the summit of the trunk, like the bulging branches of a candlestick; the elongated scars on the stem, that marked the position of the fallen fronds; and especially the basal part, that looked like a mass of intertwining wire, black and shining, as if running down with the concentrated moisture of those damp woods;—all were novel, curious, and pleasing.

URANIA SLOANUS.

Leaving, for the present, our description of this elevated region, which on a future occasion we may still further pursue, I will call my reader's attention to one of the most brilliantly lovely of animal forms; of which this spot forms one of the favourite localities. I speak of *Urania*, an insect which on account of its general form and appearance, its colours, and their arrangement in transverse bands, any one, at first sight, would pronounce to be a Butterfly of that typical division commonly known as Swallow-tailed; yet which is found, on anatomical investigation, to possess a very close affinity with Moths. Its habits partake of the same equivocal character, as observed in the Cuban species, *Ur. Fernandinæ*, so delightfully described by Mr. Macleay in the Zool. Trans., and in *U. Sloanus*, the species which inhabits Jamaica. The contrast produced by the rich velvet-black of the general ground, with the glittering emerald of the bands which cross the wings, is very fine; and, added to the broad notched band of ruddy gold, sparkling like powdered gems, that is so conspicuous on the posterior pair, entitles this insect to take its place in the very foremost rank of the most lovely class of animated beings.

The *Urania Sloanus* is not, like its Cuban relative, abundant on the sea-shore: nor does it manifest any predilection for the Sea-grape (*Coccoloba*); though this common tree fringes the beach in the immediate vicinity of Bluefields. A single specimen may be now and then seen in the winter months, flitting along over the sea-side bushes; but it was not until the middle of March that I saw it in any number. Just at that time the Avocado Pear (*Persea gratissima*) was bursting into blossom on the summit of Bluefields mountain, the drought having just yielded to the first vernal showers. Several trees of this species were growing at the spot I have described above; some (which had sprung up in an open space, long used as a pasture) well-grown and spreading; others (which had been spared in the recent clearing) slender, tall and branchless trunks, with a little tuft of foliage at the extremity. In an excursion to the mountain at this period, I observed *Uraniæ* fluttering around these trees, soon after sunrise; some half a dozen round each tree, and others very high in the air, dashing along in a headlong manner over the loftiest trees. They chiefly affected a very lofty isolated tree, and hence were with difficulty captured. Anxious to know at what time they began to be active, I ascended the mountain by starlight, cheered on my way by the rich gushes of melody from that worthy rival of the Nightingale, the Mocking-bird. I first saw *Urania* as the sun was rising.

But about three weeks later than this, viz. in the first week of April, I had much better opportunities of observing this magnificent insect at my very door. For it then began suddenly to be numerous at early day about the Pear-trees around Bluefields House.

Stationing myself under these trees in the morning twilight, I found that a little before sunrise the *Uraniæ* begin to appear. At first they come slowly, one by one; but about the time that the glowing sky over the mountain betokens the march of the advancing sun, they are fluttering in scores round the trees, sucking the blossoms. They are not here nearly so difficult of capture; on the first morning I secured in about an hour fifteen good specimens, besides several which I rejected; and I might have caught twice that number, but I abstained from molesting those which I saw to be imperfect. This is indeed the case with a large proportion of the specimens that one sees, the slender delicate tails of the wings being often broken off long before the downy surface is at all defaced. After a few mornings, though the numbers did not seem diminished, they were evidently grown more wary, and difficult to net. Except that a few individuals would occasionally hover round the Mango-trees that grew near, I have not found that the species affects, at least in numbers, any plant but the *Persea:* though many fruit-trees besides were blossoming at the same time.

These beautiful butterflies occasionally fly at a great elevation, far beyond that which has procured, for our finest British insect, the title of Purple High-flyer. I almost hesitate to say *how high*, yet I think I do not at all exceed the truth when I say I have seen them soar away over the open field to a height of 500 feet, when they were just visible as moving specks of black against the bright sky. They thus fully justify their name of *Urania* (heavenly). May not the human ψυχή draw an instructive moral from the heavenward soaring of its insect symbol?

When one alights, unless it is to suck the blossoms, it chooses a leaf, or other surface, that is nearly vertical, and instantly turns head-downward, and rests with the wings expanded in the plane of the body; the anterior pair, however, inclined backwards, so as to form an angle with each other, and partly covering the posterior ones. They chase each other about playfully; half a dozen or more sometimes joining in the gambols, when their wings glitter in the sun like the plumage of the Humming-birds.

Their manner of flight is much more that of a Moth than of a Butterfly. I may observe that the hooked form of the tips of the antennæ, which we see in cabinets, does not characterise the living insect; they are curved only in the process of drying.

As soon as the sun becomes hot, that is about eight or nine o'clock, they one by one retire; and scarcely a single individual is to be seen until after the rain, which at that season falls every afternoon, when they crowd around the Pear-trees, as numerous as in the morning. These semi-daily visits are continued for about a fortnight, while the Pear-trees continue in bloom; and then the insects gradually become scarce, and we see only an occasional straggler until the season returns.

To this scarcity, however, I found one remarkable exception. On the 18th of June, being on the lonely mountain road near Basin (or Beeston) Spring soon after sunrise, I was surprised by the sight of this beautiful butterfly, very abundant, though few other insects were abroad. They were flying low, and resting on the leaves of trees over the road, so as to be taken with ease. Nearly all were in the highest state of perfection, as if just emerged from the chrysalis. If this was the case, as I believe, these must have been the progeny of the brood which had been active in April; and as the abdomens of the females were at that period quite globose with eggs just ready for exclusion, we may consider the period that elapses from the deposition of the egg to the evolution of the imago to be about two months. I regret that I have nothing to say respecting the transformations of this species; no caterpillar at all resembling that of *U. Fernandinæ* ever occurred to me. Several of the eggs above mentioned I examined with a microscope; their form is that of a globe slightly flattened at the poles, marked with numerous ridges running perpendicularly, like meridians: their colour is yellowish-white.

LIZARDS.

One feature with which a stranger cannot fail to be struck on his arrival in the island, and which is essentially tropical, is the abundance of the Lizards that everywhere meet his eye. As soon as ever he sets foot on the beach, the rustlings among the dry leaves, and the dartings hither and thither among the spiny bushes that fringe the shore, arrest his attention*; and he sees on every hand the beautifully-coloured and meek-faced Ground Lizard (*Ameiva dorsalis*) scratching like a bird among the sand, or peering at him from beneath the shadow of a great leaf, or creeping stealthily along with its chin and belly upon the earth, or shooting over the turf with such a rapidity, that it seems to fly rather than run. By the road-sides, and in the open pastures, and in the provision-grounds of the negroes, still he sees this elegant and agile Lizard; and his prejudices against the reptile races must be inveterate indeed, if he can behold its gentle countenance, and timid but bright eyes, its chaste but beautiful hues, its

* "Nunc virides etiam occultant *spineta* lacertos."—VIRG.

graceful form and action, and its bird-like motions, with any other feeling than admiration.

As he walks along the roads and lanes that divide the properties, he will perceive at every turn the smooth and trim little figures of the Wood-slaves (*Mabouya agilis*) basking on the loose stones of the dry-walls; their glossy fish-like scales glistening in the sun with metallic brilliance. They lie as still as if asleep; but on the intruder's approach they are ready in a moment to dart into the crevices of the stones, and disappear until the danger is past.

If he looks into the outbuildings of the estates, the mill-house, or the boiling-house, or the cattle-sheds, a singular croaking sound above his head causes him to look up; and then he sees clinging to the rafters, or crawling sluggishly along with the back downward, three or four Lizards, of form, colour, and action very diverse from those he has seen before. It is the Gecko, or Croaking Lizard (*Thecadactylus lævis*), a nocturnal animal in its chief activity, but always to be seen in these places, or in hollow trees, even by day. Its appearance is repulsive, I allow, but its reputation for venom is libellous and groundless.

The stranger walks into the dwelling-house. Lizards, lizards, still meet his eye. The little Anoles (*A. iodurus*, *A. opalinus*, &c.) are chasing each other in and out between the jalousies, now stopping to protrude from the throat a broad disk of brilliant colour, crimson or orange, like the petal of a flower, then withdrawing it, and again displaying it in coquettish play. Then one leaps a yard or two through the air, and alights on the back of his playfellow; and both struggle and twist about in unimaginable contortions. Another is running up and down on the plastered wall, catching the ants as they roam in black lines over its whited surface; and another leaps from the top of some piece of furniture upon the back of the visitor's chair, and scampers nimbly along the collar of his coat. It jumps on the table; — can it be the same? An instant ago it was of the most beautiful golden green, except the base of the tail, which was of a soft, light, purple hue; now, as if changed by an enchanter's wand, it is of a sordid sooty brown all over, and becomes momentarily darker and darker, or mottled with dark and pale patches of a most unpleasing aspect. Presently, however, the mental emotion, whatever it was, anger, or fear, or dislike, has passed away, and the lovely green hue sparkles in the glancing sunlight as before.

He lifts the window-sash; and instantly there run out on the sill two or three minute Lizards of a new kind, allied to the Gecko, the common Pallette-tip (*Sphæriodactylus argus*). It is scarcely more than two inches long, more nimble than fleet in its movements, and not very attractive.

In the woods he would meet with other kinds. On the trunks of the trees he might frequently see the Venus (*Dactyloa Edwardsii*), as it is provincially called; a Lizard much like the Anoles of the houses, of a rich grass-green colour, with orange throat-disk, but much larger and fiercer: or in the eastern parts of the island the great Iguana (*Cyclura lophoma*), with its dorsal crest like the teeth of a saw running all down its back, might be seen lying out on the branches of the trees, or playing bo-peep from a hole in the trunk: or, in the swamps and morasses of Westmoreland, the yellow Galliwasp (*Celestus occiduus*), so much dreaded and abhorred, yet without reason, might be observed sitting idly in the mouth of its burrow, or feeding on the wild fruits and marshy plants that constitute its food.

SEA-URCHINS.

Feb. 21st. — I find lying on the sand and white chalky mud in shallow water along the shore at Belmont, some specimens of the large long-spined *Echinus* that I had noticed in crossing the Bay. On touching them, though cautiously, several of the very acute points entered my fingers, and I thought were broken off in the wounds; for something black remained there, and the part soon began to be tense and painful. But this substance afterwards proved to be merely a dark purple pigment, which came off from the spines whenever they were handled. The best mode of lifting the specimens, which at first seemed, on account of the points bristling in every direction, to be impracticable, was by putting the fingers under the animal; as the spines on the inferior side, that is, those surrounding the mouth, are short and comparatively blunt. The projection of the mouth is of a rich red-purple, and the spines and body purple-black. The spines, viewed through a microscope, display a beautiful structure; being each surrounded by regular and close-set rings of polished, pointed spiculæ, overlapping each other, most elaborate in their finish. The length and slenderness of the spines, combined with their brittleness, rendered it difficult to handle the animals; especially as, after they were removed from the water, they kept agitating these organs briskly, and each spine with a movement independent of the other spines. The sight of so many long needle-like points moving irregularly together, some oscillating to and fro, and others partially revolving on the curious ball and socket joints at the base, had a very singular appearance. Several, however, in spite of every precaution, were broken in removing the animals from the sea into a basket, and in taking them out at home.

A quarter of an hour's immersion in fresh water was sufficient to deprive them of life, and the soft viscera were easily removed by slightly enlarging the anal orifice in the centre of the upper surface. But now the problem was how to dry them, so that they might retain somewhat of their original beauty. For the spines, no longer sustained by muscular power, were now all fallen flat, and were lying across one another in confusion; and if the specimens were put to dry mouth downward, of course the spines would dry in such unnatural positions; and thus the whole of the elegance arising from their regular divergence would be lost. On the other hand, if they were set to rest in an inverted position, the weight would break the points of the spines, or distort them worse. I at first attempted to lay them, mouth upward, on two or three threads stretched across a box, but owing to the structure of the spines above described, the threads were caught by them every moment, and it was not possible to make them reach the root of the spines. This plan therefore would not do. Then I took a thread, and having tied a bit of stick transversely at one end, passed the other end, by means of a long needle, through the anal orifice, and brought it out at the mouth. The piece of stick retained the *Echinus* on the thread, suspended in the air, while the free end was tied to the branch of a tree. The long spines thus hung downwards, and presently assumed their natural divergence, in which position the wind soon dried them immoveably; and I thus had the pleasure of preserving the natural appearance of these fine *Echini* in great perfection.

BLUEFIELDS RIVER.

To have a permanent supply of clear, pure water is a very important advantage in a climate, where, during the long droughts, unmitigated by a single shower, the pastures become burned up, and the cattle languish for want of grass as well as water;—

> " ———————— nulla neque amnem
> Libavit quadrupes, nec graminis attigit herbam."
>
> VIRG.

At such periods, in many cases, the only resource is a filthy pond, whence the water has to be fetched several miles. The beasts are fed on the berries of the Bastard-cedar (*Guazuma ulmifolia*), which are very mucilaginous, and on the leaves of the Bread-nut (*Brosimum alicastrum*), great bundles of which are cut in the woods, and brought in every evening. No place is better off in this respect than Bluefields; a rivulet of the most cool and sparkling water running, with many meanderings, through its whole extent. In truth it is a romantic little stream. Here it dilates into broad but shallow pools; there, confined between narrow banks, it rushes like an arrow in a black and deep rapid; now it brawls among the rolling pebbles; then it pours foaming over a succession of round terraced rocks; and anon falls in a sheet over a little precipice, a Niagara in miniature. Here the ground is on a level with the water, and the brook flows beneath clumps of feathery bamboos, and luscious-fruited guava-trees: there the banks rise to high and steep walls, clothed with grass to the water's edge, where they are fringed with a luxuriant border of vegetation springing out of the stream. The broad, peltate leaves of the Cowfoot (*Piper umbellatum*), and those of the still broader Calalue (a species of *Caladium*), overshadow the water beneath; while among them peeps forth a lovely white blossom, resembling a star at the extremity of a long and slender tube. This, notwithstanding its treacherous beauty, is one of the most venomous of plants (*Isotoma longiflora*), commonly called Horse-poison, from its fatal effects on those animals, if they chance to eat it; causing their bodies to swell until they actually burst: even the juice of the leaves will raise bladders on the skin.

After having passed under a one-arched bridge, the rivulet is divided by a little islet, tenanted only, as far as I know, by a pair of Petcharies (*Tyrannus*

Plate II

F. H. GOSSE, DEL^T

M. & N. HANHART, IMP^T

BLUEFIELDS RIVER

LONDON, LONGMAN & C^O 1851.

caudifasciatus), that have built their nest in a shrub thereon, and make the air vocal with their early cries of O P — P P — P — Q, as they seem impatiently to call for the rising sun. Suddenly the brook plunges over a ridge of limestone rock into a secluded pool, hidden from view by overhanging trees, and by the pasture boundary-wall, which crosses the stream just at the waterfall. Thick matted masses of the Wall-Marigold, festooned with the clustered blossoms of a yellow Convolvulus, and the magnificent flowers of the Violet Hogmeat (*Ipomœa violacea*) cover the wall, and project far over the water on each side of the cascade, the spray from which keeps them ever verdant, and ever in bloom; while the living principle preserves them from becoming covered with the tuberculous incrustation of friable lime, which is thickly deposited upon every other object within reach of the spray. A tall Fiddlewood tree spreads its branches over the spot, gay at the time of which I speak (June), with its conspicuous bunches of crimson berries, relieving the dark green hue of its foliage.

To this little clear pool I usually resorted in the middle of the day to bathe; delighting to hold my head under the waterfall, and to receive its sudden coldness like an electric shock upon my back, until the stimulus ceased; and then to lie at full length in the shallow, just covered by the limpid element. How refreshing this is, those only can tell, who have felt the lassitude produced by a vertical sun; and how conducive to health, I can thankfully testify. Several pleasant observations have I made on the living things around me, while thus delightfully immersed by the half-hour together; and, as was indeed but fair, observations have been made on me. On one occasion I had been lying motionless for a long time, just beneath the surface, when a Vulture (*Cathartes aura*) that had marked me from a distance, began to descend in circles, swooping over me, nearer and nearer at every turn, until at length the shadow of his gaunt form swept close between my face and the light, and the rushing of his wide-spread wings fanned my body as he passed. It was evident that he had mistaken me for a drowned corpse; and probably it was the motion of my open eyes, as I followed his course, that told him all was not quite right, and kept him sailing round in low circles instead of alighting.

One day, about the middle of June, as I had just jumped over the wall, and was walking down the steep bank through the high Guinea-grass, with my towel under my arm, my attention was arrested by observing several little objects leaping up from the pool against the cascade. Presently perceiving that they were fishes, I returned to the house for my insect water-net, and proceeded to the margin to endeavour to capture one. I was not at once successful; my presence scared them: but soon one and another began to leap from the boiling pot beneath against the glassy wall above them. Most of them achieved only a foot or two; but I perceived one very nearly lodge at the top, though the height of the fall, from surface to surface, was not much short of six feet, and the little performers were scarcely more than half that number of inches. At length I succeeded in thrusting the net under a leaper, and into it he fell. I carried him to the house, and placed him in a large washing-basin of water, uninjured, though a trifle exhausted; when I found that I had captured a young individual of the Grey Mullet of our fresh waters. At first he could not maintain his equilibrium, but rolled over and over in his efforts to preserve his back uppermost. As he gained strength, however, he became more at ease; but was very impatient, swimming rapidly round and round the basin, projecting his open mouth. Sometimes he would swim rapidly in a perpendicular position, evidently striving to get out, but showing no tendency to leap; perhaps because there was not sufficient room in the basin to allow him to gain the required impetus. The energy, and muscular power, as well as the strength of instinct displayed by so young a fish (for it was only about three and a half inches long), not having attained more than one-sixth part of its adult dimensions, seemed to me remarkable. I continued to see the little creatures leaping at the cascade all through the summer.

In the holes and basins of this romantic little river, particularly at the foot of the many falls, large Mullet of the same species lurk, some as much as eighteen inches long. They are often taken by hand; the negroes plunging their arm into the holes beneath the bank, and feeling for the fish. They bite freely also at the hook, the most successful bait being a piece of a mellow Avocada pear, in the season. It is one of the most highly esteemed fishes for the table; and has considerable pretensions to beauty, for, every scale on the upper parts being bordered with black, an elegant reticulate appearance is given to the fish.*

Another fish of still more minute dimensions, but of remarkable beauty, is also numerous in the basin of the waterfall, and all along the stream. The negroes call it Tickiticky, a term formed like our word Stickleback (the initial *S* being omitted in negro pronunciation before another consonant), and alluding to the strong spine in the anal. All along the margins of the rivulet, particularly where a coarse grass grows in the shallows, and shoots its pointed blade-tips above the surface, the Tickiticky is numerous, herding together in little parties of half-a-dozen, or a dozen, of various sizes, but none exceeding two inches in length. They are active and amusing; and leap out of the water, if pursued, with great vigour. It is a beautiful little fish.†

* As this Mullet seems to be hitherto unrecognised, I shall describe it. *Mugil irretitus*, MIHI. Operculum smooth, silvery; pre-operculum scaled. Lips moderately thick. Fin-ray formula: — D. 4—8; A. 10; C. 17; P. 14; V. 6. Irides golden orange; upper part of the body pale olive; sides silvery; belly white. Each scale on the upper parts is bordered with black. Fins transparent; the first dorsal tinged with yellow; the rays of all irregularly marked with black, which, on the caudal, forms an indistinct blackish transverse band.

M. monticola of Bancroft is assigned to Jamaica in Griffith's Anim. Kingd.; no description is given, but the figure does not agree with the above, nor does the description of *M. albula* in MM. Cuv. and Val.'s Poissons.

† The Tickiticky is an undescribed species of *Pœcilia*. with the following characters. *Pœcilia melapleura* (Μέλας, black; πλευρά, the side). Fin-rays, D. 11; A. 10; C. 28; P. 13; V. 6. The caudal

Some interesting *Crustacea* are also found in Bluefields rivulet. I obtained in some numbers a new *Atya* which has been since described by Mr. Newport under the name of *occidentalis,* and a new *Palæmon,* named in the List of the Crustacea in the British Museum, *P. Procles.* Besides these, a Prawn of large size is common, provincially known as the Crayfish (*P. Jamaicensis*). Specimens are often found seven inches in length, independent of the claws, one of which is sometimes enormously developed, while the other remains of the usual size. They are frequently seen from the banks, darting with the swiftness of an arrow, tail foremost, from the shelter of one stone to that of another; and, as their flesh is esteemed, they are much sought after. The most common mode of capturing them is by a small fish-pot, or lengthened basket made on the principle of a wire mouse-trap, admitting the entrance, but refusing the exit of the prawn. I procured a good many specimens of these and other *Crustacea,* by means of this device.

In the hollows of the limestone-rock beneath the waterfall, and at either side of it, subject to a continual dripping, where the calcareous matter with

has an upward direction. There are thirty-one scales in a straight line, from the gill-aperture to the base of the caudal; and three more *upon the caudal;* there are nine in transverse series at the vent.

The irides are green, brilliantly iridescent. The body almost pellucid; olive above; the sides pearly, with rich amethystine reflections; green and golden hues prevail on the cheeks and gill covers; a black band, rising from the upper side of the base of the pectoral, runs along the side about one third of the length: the dorsal and caudal are edged with blackish. Length to two inches. (Plate I. fig. 3.)

which the river is impregnated forms little irregular pillars and groins of coarse stalactite, play a number of small reddish-brown Crabs, about three-fourths of an inch in diameter, a species of *Sesarma*, allied to *S. reticulata*. They are slow, deliberate, and circumspect in their general motions, creeping on a horizontal or perpendicular surface with equal facility; but, on being alarmed, they dart hither and thither with great swiftness, shooting into the dark irregular recesses of the stone, so that it is not easy to take hold of them.

A RIDE TO CONTENT.

March 1*st.* — About fifteen miles to windward of Bluefields, there is a little cottage, most singularly and romantically perched on a mass of bare rock on the steep mountain-side. The coffee property on which it stands is called by the favourite appellation (in Jamaica) of "Content;" and here, in the society of some valued friends whose residence it is, I passed many contented days. Higher engagements than those connected with Natural History called me into the vicinity every other week, and this cottage was always my home on such occasions. Setting out from Bluefields in the cool of the newly-opened morning, I could attain a considerable portion of my journey, before the torrid rays of the sun began to pour down upon the road; the gigantic shadow of the mountain-ridge, rising up in a lofty barrier between it and the glowing east. The shadow that intercepts the rays of the declining sun is a grateful relief; but the coolness of a spot that remains enwrapped in deep shadow in the fore part of the day, where the dew of the night lies collected in the hollow leaves, and glitters on the blades of grass, long after all moisture has been dissipated around, is peculiarly refreshing. Four miles of a very rocky road, bounded, on the upper side at least, by a belt of high shrubs, that much remind us of the tall, luxuriant, uncut hedges that we see in many parts of England, lead through the estates of Belmont and Shafton; the former apportioned out in small negro allotments, and cultivated in gardens; the latter a "pen," that is, a grazing farm. The hedgerow-like shrubs are gay with the beautiful scarlet blossoms of the Morass-bark (*Malvaviscus arboreus*), in particular abundance. The rich vegetation of the negro gardens, surrounding their cottages, many of which are neat and pretty, forms a pleasing interchange with the broad sunny pastures, studded with clumps of the peculiarly dense and massive Mango-trees, under the shade of whose dark foliage, the cattle are congregated, and peacefully ruminating. The purple leaves of the Dragon's blood grow in thick tufts out of the crevices of the loose walls, on the top of which many withes and cactaceous plants creep in loose confusion; and beside them spring up luxuriantly Annotto bushes, the fruit of which is so much used to give a heightened colour to cocoa and chocolate.

Presently we open the wide pastures of Robin's River, through the midst of which runs the little stream that bears the same name. The sun is by this time high, and pours down slantingly on our heads and into our scorched faces the torrid fury of his effulgence, unmitigated by a cloud: —

> "———————— πέπτατο δ' αὐγὴ
> Ἠελίου ὀξεῖα, νέφος δ' οὐ φαίνετο πάσης
> Γαίης, οὐδὴ ὀρέων."—(*Il.* xvii. 371.)

amidst which the brawling, murmuring sound of the running brook, and the sight of its sparkling, dashing, crystal waters, have an indescribably refreshing effect upon the senses. The wild scream of the Kildeer Plover is suddenly heard, and up springs a flock of these birds, which wheel in swift flight around the traveller's head, and alight close to their first station. In the rushy shallows of the stream the stately form of the Snowy Gaulin is seen, deliberately wading hither and thither; or watching, motionless and silent, for his aquatic prey. Plump Peadoves, with large liquid gentle eyes, walk about on the turf beneath the pimento trees, picking up the fallen fruit, or the seeds of papilionaceous weeds; and now and then their reiterated cooing, a very soft and mournful sound, comes from the bordering woods, falling gently and soothingly on the ear. In the farthest corner of the pasture, a great bed of that truly gorgeous plant, the Barbadoes Pride, or Flower-fence (*Ponciana pulcherrima*), glowing with scarlet and orange, indicates the commencement of a road that leads through the gloomy forest to Basin Spring, near the summit of the Bluefields Ridge. But this we leave on the left; and pursue our road beneath the orange-trees, that have been planted in bordering rows along each side of the highway. Beauty, fragrance, and fruit, combine in the citron family; and wherever they will grow they are sure to be favourites. In Jamaica they are planted in profusion, and in the winter months refresh the eye and the palate of the wayfarer with their golden fruit.

We are again in shadow. The open pastures afford no shelter from the sun, except that of their scattered trees; but now we enter a romantic road of delightful coolness and shade. The sea is at our back; and as we turn from time to time on the road, now rapidly ascending, we get more and more extended views of it, over the intervening woods, sparkling and flashing as its clear blue surface is broken up by the awakening sea-breeze. In the woods on our right hand, whose tall mass casts the high road into shade, the Trumpet-tree (*Cecropia peltata*) is particularly abundant, giving a remarkable character to the scene. It shoots up a slender jointed stem to the height of forty or fifty feet, from the summit of which a few digitate leaves, resembling gigantic horse-chestnut leaves on long footstalks, radiate horizontally, with a very palm-like aspect. The stem is hollow between the joints, and composed of light porous wood; the bark when wounded bleeds a milky fluid, which is said to make good india-rubber. Hundreds of these curious trees are seen in these woods, and on the mountain-side before us, conspicuous a long way off, amidst the dense surrounding foliage of the other forest-trees.

The opposite side of the road is bounded for some distance by a bank fifteen or twenty feet high, out of which the highway has been scarped. Several shallow caverns have been excavated in the gravelly sides, from the roofs of which many ferns and climbers hang in wild grace and beauty. The Cave-Swallow shoots in and out, pursuing the minute insects that dance in the air; and the forest around rings with the voices of many birds. A little way farther, among many other flowers, a beautiful little *Ipomea*, resembling the delicate *Quamoclit*, but pure white, with digitate leaves, covers the low wall-fences with its long graceful twining stems and star-like blossoms. That fine butterfly *Papilio Pelaus*, may very commonly be seen flitting about this shaded lane; with its low, irregular, not very rapid flight, it dances along from bush to bush, and from flower to flower, rifling them as it goes; now and then resting on a leaf to suck while it vibrates strongly its half-erected wings, in a manner peculiar to itself.

After awhile we break away from the woods, and open a scene of exquisite beauty, but of a totally different character. It is the noble sugar-estate of Peter's Vale; now, however, like so many other fine estates, thrown out of sugar-cultivation, and its buildings and offices hastening to swift decay. It is a long and spacious valley, bounded by wooded hills, which almost everywhere assume a rounded or conical form, clothed principally with clumps and groves of the dark green Pimento. Yet in one quarter the native forest covers the shaggy sides of the mountains, which slope up till they merge into the ridge of Bluefields; and, in another direction, over the intervening hills, the mountain of Grand Vale rises to view, hazy and blue in the distance, but with the

Bamboo-walk cutting a conspicuous zigzag line upon its dark mass. The ride through this beautiful natural amphitheatre, which extends for three or four miles, is for the most part over the short soft turf of expanded pastures, which sustain hundreds of cattle, and are studded with many noble and useful trees. Among these the Star-apple (*Chrysophyllum cainito*) is especially worthy of notice; the golden tint of the under surface of its leaves, moved with the breeze, perpetually interchanging with the deep glossy green of their upper sides.

Such open pastures as these, intermingled with fields where the valuable, always verdant, Guinea grass grows in thick tussocks, are the favourite resorts of the Tichicro, a prettily-marked ground sparrow, who calls "*tichicro! chi-chi-tichicro*" in a loud tone from his grassy cover, or sits on a large stone and warbles forth a pretty melody, as the traveller is passing. Among the orange and pimento trees, the Wild Canary hops and twitters; a colony of imported strangers, it is said, which have lost in song, what they have gained in richness of colouring. Thousands of little *Teriades* and *Polyommati* (minute yellow and blue Butterflies) with white *Pierides* and tawny *Danaides*, flit over the pasture in the bright sun, paying perpetual homage to the blossoms of *Asclepias*, *Argemone*, *Stachytarpha*, *Cassia*, and many other weeds that spring up amidst the pasture grasses.

To this beautiful estate succeeds another very similar in character, but more uneven in surface, called Grand Vale. It is still cultivated as a sugar estate; and here only, within a circuit of many miles of Bluefields, may yet be seen the luxuriant beauty of the majestic Sugar-cane, the busy scenes of industry of which it is the subject, and all the varied processes by which it is converted into sugar and rum. The hum of many voices, the cheerful song, the merry horse-laugh, the shrill notes of the women and children, with the creaking of ungreased wheels,—all tell pleasantly of industry and happiness, in a country where certainly a stranger is apt to be painfully struck with the prevalence of silence and neglect, and of that sort of decay which consists in the too successful efforts of wild nature to reconquer from man the possessions which he had once wrested from her sway.

At length the quiet smiling valley of Peter's Vale, and the busy laughing one of Grand Vale, are both at our back, and we enter the Cotta-wood, a dense but low coppice, and begin to ascend, by a narrow path, the steep mountain-side. The rock projects in many parts in huge tabular shelves, forming rude steps, up and down which it is terrific to ride, though I have done both, trusting to the surefootedness of the horses bred in the mountains, which are used to these precipitous paths. Long, tough, spinous stems trail in every direction through these woods, tangling them beyond all description, and making it a most laborious and painful task to penetrate them. Birds abound in them, especially various species of terrestrial Pigeons, the White-belly, the Partridge, and the Ground-Dove: and many fine and curious insects, I have exclusively found here. At length we suddenly break through the bushes, and find ourselves on a

Plate. III.

P.H. GOSSE DEL^T

M & N. HANHART IMP^T

VIEW FROM CONTENT

LONDON, LONGMAN & C^O 1851.

good broad road, that leads from Black River to Hampstead and over the Luana-mountains; and here, just round a corner of the zigzag line, is perched the little cottage of Content.

Why this situation was chosen for the house, I can hardly imagine, unless the wide-extended prospect afforded the inducement, or else the contiguity to the high road. The thin stratum of sloping earth, that originally supported the forest-trees, has been quite cleared from a small area, leaving only the naked furrowed rock, which has partly been built up with masonry on the lower side, to form a site sufficiently level for the *barbican*, on which the coffee, pimento, &c. are dried in the sun. Yet a few hundred yards within the forest, on the same plantation, a spacious and fertile level dale exists, which would seem to have offered a spot far more eligible for habitation.

Above, below, and around, is the primeval forest, scarcely interrupted by the small and widely scattered clearings that here and there occur. From so singular a position, — the tops of the trees immediately beneath the little space that surrounds the dwelling scarcely reaching to the level of its base, — the eye commands a magnificent prospect, embracing the indented coast from the bold promontory of Pedro Bluff on the east as far as the park-like slopes of Mount Edgecumbe on the west; ranging over the sombre intervening forest, with its cultivated openings, and resting on the broad savannahs and flooded meadows that surround Black River; this town with its bay and shipping in the distance, and the course of the river itself visible at intervals, winding like a silver thread through the dark morass.

The rocky hill-side below the house is thickly studded with fruit-trees, principally consisting of the luscious and juicy Sweet-sop (*Anona squamosa*), but mingled with Oranges and Limes, all greatly multiplied by self-plantation. Some trees grow among them apparently for ornament; the brilliant scarlet blossoms of a Coral-tree (*Erythrina*), and its equally showy seeds, display themselves at the different seasons here; and just below the window is a fine Moringa or Horse-radish tree (*Moringa pterygosperma*), where we may see the various species of Humming-birds buzzing round and round, like bees, all day long, and at all seasons of the year, within a yard or two of our face.

ENTOMOLOGY.

I had left England with high expectations of the richness of the West Indian entomology: large and gaily-coloured beetles, I supposed, would be crawling on almost every shrub, gorgeous butterflies be filling the air, moths be swarming about the forest-edges at night, and caterpillars be beaten from every bush. These expectations were far from being realised; a few species of butterflies, chiefly *Pieris*, *Callidryas*, *Terias*, *Heliconia charitonia*, *Argynnis passifloræ*, and *A. Delila*, *Cystineura mardania*, and one or two *Nymphalidæ* and *Lycænadæ*, are indeed common enough at all times, and in almost all situations; others are abundant at a particular season or locality; but in general butterflies are to be obtained only casually. Moths are still more rare: I had provided myself with bull's-eye lanterns, and repeatedly took them out after nightfall, carefully searching the banks and hedges by the sides of roads, the margins of woods, &c., but never, in this way, took a single specimen. At some seasons, however, as in December, and more particularly June, on rainy nights, hundreds of little *Noctuadæ, Pyralidæ, Geometradæ, Tineadæ,* &c. fly in at the open windows, and speckle the ceiling, or flutter around the glass-shades with which the candles are protected from the draughts. A good many small beetles, and other insects, also fly in on such occasions, and several interesting species I have taken in this way which I never saw at any other time. But in general beetles and the other orders are extremely scarce, and especially Diptera; I have often been astonished at the paucity of these, as compared with their abundance in Canada, the Southern United States, and other localities (in which I have collected) during the hot weather. One may often walk a mile,—I do not mean in the depth of the forest, but in situations comparatively open, beneath an unclouded sun,—and not see more than a dozen *specimens* of all orders. Nor is the beating of bushes productive of insects and their larvæ, as I have found it in North America. In Canada I have shaken off perhaps twenty species of lepidopterous larvæ in the course of an hour or two on an autumnal morning; but I think I have seen scarcely more than half that number in Jamaica during a year and a half's collecting.

To this scarcity of insects, however, there are two or three local and seasonal exceptions. The high-road, passing just behind and above the cottage of Content, climbs the mountain in the zigzag direction so frequently adopted in Jamaica, to diminish the steepness of the ascent; and it is a mile or two of this road that forms the most remarkable exception to the general scarcity of insects that I have noticed. During the month of June the shrubs and trees that border the road (which is cut through the forest) are alive with insects of all orders, but particularly Coleoptera; many species of *Longicornes*, *Lampyridæ*, *Buprestidæ*, *Cassididæ*, *Chrysomelidæ*, &c., occur by hundreds on the twigs and leaves; and the air is alive with butterflies, Hymenoptera and Diptera. I cannot at all tell why this abundance exists; it is very local; beyond a certain point, the road, the forest, seem to be unchanged, but the insects have ceased: it is very temporary also; it suddenly commences about the end of May, and by the middle of July scarcely a dozen beetles are seen where there were thousands. I might have supposed it a casual thing, if I had had but one season's experience; but in 1846 it was the same as in 1845, the same abundance at precisely the same season, and with the same local limits. It is worthy of record, that at the same time and place the leaves of the trees were studded with shelled Mollusca, of the genera *Helix*, *Cylindrella*, *Helicina*, *Cyclostoma*, &c., as I never saw them elsewhere.

It is not improbable that some peculiarities in the geological or the botanical character of this region would account for what I have mentioned; but of this I can give little information. The Mahoe (*Hibiscus tiliaceus*), the Bastard Cedar (*Guazuma ulmifolia*), the Mammee Sapota (*Lucuma mammosa*), the Locust (*Hymenæa coubaril*), and the Trumpet-tree (*Cecropia peltata*), are some of the forest-trees, with others called Burn-wood and Down-tree, of which I know not the systematic appellation, unless the latter be the *Ochroma lagopus* of botanists. But there is one tree which grows numerously in that locality, which I suppose to have some influence on the Lepidoptera and Hymenoptera; it is provincially called the Potatoe-wood; it is at that time covered with blossoms, which, though they grow in thick racemes, offer nothing pleasing to the sight or the scent. But these form the centres of attraction to the insects I have named; *Pierides* and *Theclæ* in particular flutter around the summits in considerable numbers, and swarms of small beetles and flies. The *Bauhinia* displays its elegant blossoms, and in one corner a large patch of *Cassia* attracts *Papiliones* and *Coliades;* but in general there is an almost total lack of the flowering herbaceous vegetation that fringes the roads in most other places. It is remarkable also that the trees in these woods are nearly, if not quite, destitute of epiphyte *Orchidaceæ,* which are so abundant on Bluefields Mountain at a similar elevation, that hardly a tree is without one or more specimens. But in other respects the character of the vegetation in the two regions differs greatly.

This district I habitually visited every alternate week, very frequently spending eight or ten days at a time with my worthy friends at Content. Probably two thirds at least of my collection of insects were the result of my labours here. Yet I never found insects abundant except at the season named above. The elevation of the region may be assumed (I speak only from my own estimate) as ranging from 1500 to 2000 feet above the sea.

I may add, that during the period of insect-abundance on the Hampstead road, a large number of species were taken by flying in at the open windows of Content cottage by night. Many valuable specimens occurred in this way, not only of the crepuscular and nocturnal Lepidoptera, but of other Orders, in considerable variety. *Curculionidæ*, *Longicornes*, and *Lampyridæ* were very numerous. I am inclined to think that a far greater number of insects are active by night than by day.

The other exceptions to the general paucity of insects were principally in Westmoreland. In going from Bluefields to Savanna-le-Mar, the road for some miles borders the sea-shore, which at first is a sandy beach, but soon rises to a shelving, rubbly sort of cliff, at the top of which the highway passes. The first portion, extending to about a mile from Bluefields, is called Sabito Bottom; the soil here is a heavy sand, mixed with shingle, doubtless washed up by the surf in strong gales; large masses of the Maritime Lily (*Pancratium*) spring up on each side of the path; a narrow belt of single trees, chiefly of the Sea-side Grape (*Coccoloba*), on the left hand, overhang both the road and the sea-beach, and on the right a dark and fetid morass is hidden by great bushes of the black-withe. This would seem an unpromising place for a collector, and yet it forms one of the signal exceptions I have mentioned to the general paucity of insects. Many magnificent butterflies frequent this bottom, as *Aganisthos Orion*, *Charaxes Cadmus*, *Charaxes Astyanax*, *Papilio Pelaus*, *P. Cresphontes*, *P. Polydamas*, *P. Marcellinus*, and other *Papilionidæ*, besides more common Lepidoptera. And when we get up the hill, — where the trees are Manchioneel, Cedar (*Cedrela*), Mahogany, Bully-tree (*Achras*), Logwood, &c., with the fragrant Wild Coffee (*Tetramerium odoratissimum*), the Papaw, the Trumpet-tree (*Cecropia*), the beautiful Spanish Jasmines (*Plumeria alba et rubra*), and the rocks are hung with festoons of *Portlandia grandiflora*, gay with their noble tubular blossoms, — we find insects very numerous. Many species of *Pieris*, *Callidryas*, *Terias*; of *Nymphalidæ*; *Heliconia charitonia*; of *Lycænadæ*; of *Hesperiadæ*; and not a few of other Orders, are at most seasons, abundant here. A large portion of my insect spoils was collected in this locality.

Pursuing the same sea-side road, but in an opposite direction from Bluefields, we come to the estate of Belmont. It is very sandy, close to the sea, and on the same level with Sabito Bottom; yet it possesses some peculiarities both in botany and entomology. Prickly Acacias of several species border the road, intermingled profusely with the formidable Pinguin (*Bromelia pinguin*). The fences are logwood hedges, over which trail many beautiful creepers, as different kinds of *Ipomœa*, and the lovely

Clitoria Plumieri; and Passion-flowers throw their feeble stems and entwine their tendrils among the shrubs and herbaceous plants that fringe the road-sides. Some small *Melitææ*, *Cystineura Mardania*, and *Charaxes Astyanax;* some pretty low-flying *Glaucopidæ* and *Pyralidæ*, haunt these lanes, and a few rare Coleoptera have been taken from the shrubs.

A few rods' length of the high-road at the brow of Bluefields Mountain, along the edge of the dark wood where grow the tree-ferns already described, has also produced several fine insects. Here, and in the neighbouring parts that have been once under cultivation but are now "ruinate," bushes of numberless kinds have sprung up, many of which are in blossom at all seasons. Though the flowers of most of these are individually small and inconspicuous, yet from their profusion they present an attraction to Hymenopterous and Lepidopterous insects; and such a wilderness of vegetation is usually more or less productive to the entomologist. In this particular locality I have usually found butterflies pretty numerous, principally *Nymphalidæ* and *Hesperiadæ*, and those of sorts rarely found in the lowlands; but from the tangled character of the "bush," and from the height of the blossomed summits about which they hover, they are less readily obtained than observed.

FIRE-FLIES.

> "How beautiful is night!
> A dewy freshness fills the silent air;
> No mist obscures, nor cloud, nor speck, nor stain,
> Breaks the serene of heaven;
> In full-orbed glory yonder moon divine
> Rolls through the dark blue depths."
>
> *Thalaba*, i. 1.

A noble description, though expressed in few words; but fully to enter into its beauty, we must have seen the glorious nights of the tropics. There the "dewy freshness" receives a heightening charm by its contrast with the burning day: there the sky has, indeed, a "dark blue depth," into which the eye seems to go onward and onward interminably; and gains an idea of the illimitable expanse, the infinity of space, that our paler skies fail to convey. And surely nowhere else does the moon appear invested in such "full-orbed glory," or pour down such a flood of brilliant lustre on forest and sea, as from those "serene," transparent heavens.

A scene particularly lovely is presented by one of those little glades, or open spaces of greensward, in the midst of the woods, that are not uncommon in Jamaica, — when seen under the light of a brilliant vertical moon. The forest, like a bounding wall, rises all around, within which everything is hidden under the blackest gloom. The thick leathery leaves of many of the trees, however, reflect from their glossy surfaces the moon's light, and, moved by the gentle land-wind, throw off the flashes in all directions. And, after the daily shower in the rainy season, the drops, depending from every twig, and lying in globules in the hollows of the leaves, both reflect and refract the beam, like thousands of diamonds.

I have said that the interior of the forest is veiled in the deepest gloom, concealing every object; but this is true only of such objects as depend for their visibility on external light. The very depth of the darkness only makes more perceptible some objects there, which shine by their own proper radiance. Here and there, all around, among the trees and shrubs, little lights are flitting along a few feet above the ground, which the beholder can scarcely persuade himself are not candles borne about by some human inhabitants of the forest. These are Fire-flies, species of the same family (*Lampyridæ*) as the Glow-worm of our own summer evenings, but in many instances far exceeding it in lustre. There are other lights, however, which surpass the brightest of these; a red glare dashes by with headlong rapidity along the grassy edges of the woods, now concealed, then flaming out again, which we at once see to be of a superior character to the sparks of the woods. This also is the torch of an insect (*Pyrophorus noctilucus*), to which I shall give the English appellation of Glow-fly a beetle of the family *Elateridæ*. To each of these families I shall devote a few remarks.

The *Lampyridæ* are, in Jamaica, far more abundant than *Pyrophorus noctilucus*.* At all times, their

* Mr. Sells' statement, that "the splendid luminous spectacle in Jamaica is produced exclusively by the *Elateridæ*, the light of the

sparks, of various degrees of intensity, according to the size of the species, are to be seen, fitfully gleaming by scores about the margins of woods, and in open and cultivated places. About fourteen species have occurred to me, all luminous. *Photuris versicolor*, a large species with drab-coloured elytra, I found abroad soon after my arrival, in December. One flying around the house in the evening, I was struck with its swift and headlong flight and nearly permanent luminosity, which was much more brilliant than that of any species which I had at that time seen.

A large *Pygolampis*, which I have called *P. xanthophotis*, I did not meet with until May, when one flew into the house at Bluefields in the evening; and two nights afterwards I observed it rather numerous on the very sea-beach at Sabito. It was conspicuous for the intensity of its light, much exceeding that of *Photuris versicolor*. Sometimes it is only the last segment but two that shows luminosity; but when excited, the whole hinder part of the abdomen is lighted up with a dazzling glare.

It is in the woods of St. Elizabeth's, in the month of June, that I have seen the *Lampyridæ* in their glory; and particularly along the road leading up the mountain from Shrewsbury to Content, where it is cut through the tall forest, which overhangs it on each side, making it sombre even by day, and casting

glow-worm which is *occasionally* seen being no more comparable to that of the *Elater* than a dying oil-lamp to a jet of pure gas" (*Tr. Ent. Soc.* I. xlvi.), I can by no means confirm.

an impenetrable gloom over the scene by night. The darkness here, however, and especially at one point, a little dell, which is most obscure, is studded thick with Fire-flies of various species, among which the two large ones above-named are conspicuous. I have delighted to watch and study their habits in this lonely spot, while the strange sounds, snorings, screeches, and ringings, of nocturnal reptiles and insects, sounds unheard by day, were coming up from every part of the deep forest around, giving an almost unearthly character to the scene.

Pygolampis xanthophotis I have observed only in flight: its light is of a rich orange-colour when seen abroad, but when viewed in the light of a candle appears yellow. It is not of so deep a tint as the abdominal light of *Pyrophorus noctilucus.* It is intermittent.

Photuris versicolor is noticeable by its frequent resting on a twig or leaf in the woods, when it will gradually increase the intensity of its light till it glows like a torch; then it gradually fades to a spark, and becomes quite extinct; it thus remains unseen for some time, but in about a minute, or, it may be, two, it will begin to appear, and gradually increase to its former blaze; then fade again; strongly reminding the beholder of a revolving light at sea. The light of this species is of a brilliant green hue. I have seen a passing *Pyg. xanthophotis*, attracted by the glow of a stationary *Phot. versicolor*, fly up and play around it; when the intermingling of the green and orange rays had a charming appearance, like the two lights of *Pyrophorus noctilucus*, of which I shall presently speak.

The smaller species have, some yellow, some green light: I have noticed only these two colours in the luminosity of such *Lampyridæ* as I have observed.

Pygolampis xanthophotis, when held in the fingers, will frequently illuminate the antepenultimate segment of the abdomen, over which the light plays fitfully, sometimes momentarily clouded, more or less, but generally saturated, as it were, with most brilliant effulgence. This species occasionally comes in at open windows at night, but much more rarely than *Photuris versicolor* and the smaller kinds, a dozen or more of which may be seen almost every night, especially at Content, crawling up the walls or flitting around the room and beneath the ceiling.

Southey's well-known lines on the Fire-flies of tropical America, are, with one or two trivial exceptions, as correct as they are poetical:—

> "Sorrowing we beheld
> The night come on; but soon did night display
> More wonders than it veiled: innumerous tribes
> From the wood-cover swarmed, and darkness made
> Their beauties visible; one while they streamed
> A bright blue radiance upon flowers which closed
> Their gorgeous colours from the eye of day;
> Now, motionless and dark, eluded search,
> Self-shrouded; and anon, starring the sky,
> Rose like a shower of fire." *Madoc*, I. v.

I would substitute "*green* radiance" for "blue;" I have never seen a blue gleam from a Fire-fly, either in the Antilles, or in continental America. The phrase "starring the sky" seems rather at variance with the lowly habits of the genus: I would not say

I have never seen one against the sky, but, generally speaking, their flight takes a range of only a few feet or yards from the ground. The term "swarmed," too, must be understood as expressing only their numbers, which are often very great; not any association, like the swarming of gnats or bees, for the *Lampyridæ* are essentially solitary in their habits.

I will now speak of our other luminous insect, the Glow-fly (*Pyrophorus noctilucus*). From February to the middle of summer this beetle is common in the lowlands, and at moderate elevations. Lacordaire's account of the luminosity of this *Elater* (known to me, however, only by the citation in Kirby and Spence's Introd. to Ent. ii. 333., 6th edit.) differs so greatly from the phænomena presented by our Jamaica specimens, that I cannot help concluding that he has described an allied but very distinct species, and I feel justified therefore in recording what I have myself observed. The light from the two oval tubercles on the dorsal surface of the thorax is very visible even in broad daylight. When the insect is undisturbed, these spots are generally quite opake, of a dull white hue; but, on being handled, they ignite, not suddenly but gradually, the centre of each tubercle first showing a point of light, which in a moment spreads to the circumference, and increases in intensity till it blazes with a lustre almost dazzling. The colour of the thoracic light is a rich yellow-green. In a dark room, *pitch-dark*, this insect gives so much illumination as to cast a definite shadow of any object on the opposite wall, and when held two inches from a book the whole line may be read *without moving it.* The under part of the thorax has a singular appearance when the tubercles are fully lighted up; for the horny coat of skin being somewhat pellucid, displays the light within redly and dimly, as if the whole thorax were red-hot, particularly at the edges, immediately beneath the tubercles. When left alone, the insect soon relapses into stillness, and the tubercles presently fade into darkness, either total, or redeemed only by a spark scarcely perceptible.

I had been familiar with this Glow-fly for some weeks, and had made the above observations on it, without being aware that it possessed any other source of light than the thoracic tubercles. I had indeed remarked that when flying at liberty the light which it diffused was of a *rich ruddy glow*, and yet these individual insects, if captured and held in the hand, showed only *green* light. I much wondered at this, but knew not how to account for it, until a friend explained it, illustrating his remarks by experiment. On the *ventral surface*, when the abdomen is extended, there is seen, between its first segment and the metathorax, an oval transverse space, covered with thin membrane, which glows with orange-coloured light; totally concealed, however, when the abdomen is relaxed, by the overlapping of the metathorax. When the insect is placed on its back it throws itself into the air like other *Elaters;* but if it be made to repeat this many times it appears to become weary, and endeavours to raise itself by bending the head and the abdomen back, so as to rest on the extremities, in hope to *roll* over. It is when thus recurved that the abdominal light suddenly appears, the oval space being uncovered. When held in the hand, the same effect is produced by forcibly bending back the abdomen with the fingers; but this is not very easy of accomplishment, on account of the resistance of the closed elytra; but if these be held open with one hand, and the abdomen recurved with the other, it is readily shown. As the open space, then, can be exposed only when the elytra are expanded, the reason is manifest why the red light is never displayed by the insect when walking or resting: the green thoracic light on the other hand *may* be displayed at any time; it is however very rarely shown during flight. On one occasion two or three Glow-flies, having entered the sitting-room in the evening, gave out the red light most brilliantly as they flew round near the ceiling, the spectators being beneath them; one of these, being alarmed by my efforts to capture it, gave out the thoracic light also very brightly; and the mingling of the green and red light in the evolutions of flight produced an effect indescribably beautiful.

That the thoracic light is subject to the will of the insect is indubitable; but whether the same can be predicated of the abdominal light I am not assured. During flight it is every second intermitted, as far as the observer can detect; but its appearance or disappearance may depend upon whether the dorsal or ventral surface is presented to the eye. This is when, soon after dark, the insect is sweeping in rapid, headlong, irregular curves over the fields or along the edges of the forest; when the appearance resembles that of a stick with the end on fire (but not in flame) carried or whirled along by one running swiftly, quenched suddenly after a course of a dozen yards, to appear again at a similar distance. When slowly flying over the grass, the progress of one may often be traced by the red glare on the ground beneath; a space of about a yard square being brightly illuminated, when no light at all reaches the spectator's eye from the body of the insect.

Whether any light would appear pervading the abdomen if the segments were stretched, I cannot positively say, for I have not in my journal any note on this point. I think not, however; for in my repeated handlings of these insects and experiments on their abdomens, I could scarcely have avoided extending the segments, even unintentionally; but I am quite certain I never saw any light except in the one ventral and the two thoracic spots. If one be trodden on, a mass of mixed light remains for some minutes among the fragments. The story told by Peter Martyr of these *Elaters* having been hunted for, to eat the mosquitoes is sufficiently amusing; of course it is not right to *contradict* a statement because one has never verified it, but I may be permitted to observe that I utterly disbelieve it That they might afford a substitute for candles in performing household operations that required no great exactness, is certainly true, provided they were constantly carried in the fingers; but if put under a glass or allowed liberty in a room, as I have abundantly proved, they very quickly conceal their light. I have found too, that one kept beneath a glass would display very little light the next evening, even under the excitement of being handled, and on the following night would be irrecoverably dark; this may have resulted from the lack of food or of exercise, not I think from the lack of air or of moisture.

Peter Martyr asserts that the natives of Hispaniola, at the time of the discovery, were in the habit of tying one of these Glow-flies to each of their great toes, when they journeyed by night through the woods; a thing not at all improbable. The two insects would throw a considerable light around the traveller's steps; and if they should withhold their luminosity, might easily be replaced by others freshly caught. On this custom Southey, in the beautiful Poem already quoted, has founded a pretty incident. When Coatel was guiding Madoc through the cavern,—

> "She beckoned, and descended, and drew out
> From underneath her vest, a cage, or net
> It rather might be called, so fine the twigs
> Which knit it, where, confined, two fire-flies gave
> Their lustre." *Madoc*, II. xvii.

Of the earlier stages of any of these light-bearing insects I have been able to procure little information. About the middle of May a larva of an Elateridous beetle was brought to me which was luminous; in the dark the whole insect was pellucid, but the divisions of the segments showed distinct light, blue and pale, not very vivid. It was impatient of being handled, and bit fiercely at the hand, but ineffectually. I suspect that it was the larva of the Glow-fly: the specimen is now in the British Museum. And at Content, in the latter part of July, I found in fresh-turned earth a larva of a *Lampyris*, small and lengthened: the abdomen, like that of the European Glow-worm, was furnished with a retractile brush of divergent filaments, ordinarily concealed; but having no lens with me I could not examine it particularly.

Mr. Hill has favoured me with the following interesting speculations on the phosphorescence of these insects; particularly on the *Pyrophorus noctilucus*, which he indicates by the term "Fire-fly."

"Humboldt states that the larva of the Fire-fly feeds on the roots of the sugar-cane, and proves destructive to that plant in the West India Islands. This remark was no doubt made on information derived from Spanish planters; and relied on, because consistent with the known habits of the larvæ of European *Elateridæ*,—particularly the well-known wire-worm, which devours the roots of vegetables, and does considerable damage to corn-fields. No one can have looked upon a stretch of canes in some rich and teeming soil in one of our serene nights, and seen the numerous luminous insects shooting athwart the gloom like meteors, or spangling the wide landscape as with a thousand stars, without being struck with the relation which subsists between the prevalence of *phosphorescent* insects, and the growth of a plant, like the sugar-cane, which depends on the presence of an unusual degree of *phosphates* in the soil. The fact is, that the peculiar economy of these insects with respect to their phosphorescence is carried on by the aid of vegetable food in which phosphorus is elaborated. Vegetables fashion elementary or mineral matter; and when fashioned those matters pass ready formed into the bodies of animals; — animals change one portion of them, and store up another in their tissues: — they engender heat, and elicit force in consuming that which vegetables have produced and slowly accumulated. This is the relation between the luminous insect, and the soil charged with phosphates. What the plant reduces, the insect appropriates and consumes; — plants decompose carbonic acid to seize upon its carbon, and they decompose water to seize upon its hydrogen; animals burn carbon to form carbonic acid, and they act on hydrogen to form water. The Fire-fly, in its economy of life, burns the phosphorus, absorbed from the plants that nourished it, to give forth light. The phosphorus in a state of combustion unites with the oxygen of the air, and when we experimentalise this process of combustion, in order to trace the parts severally played by vegetables and by animals in the economy of nature, we find that phosphorus, when it unites with the oxygen of the air, produces a solid acid, which falls down in the included air like flakes of snow, and in this way it again combines with the soil.

"It is certain that the Fire-fly feeds upon the sugar-cane; and should the larva do so likewise, as it is *xylophagous*, this insect must be added to those that do mischief to the planter; considering the abundant swarms which nightly, at certain seasons, illuminate the cane-fields. When Mr. Lees, from the Bahamas, carried the living Fire-fly to England, he took sugar-canes to sea with him, on which the beetles fed. They readily broke away the wood to obtain the saccharine matter; and after his stock of canes was consumed, they ate brown sugar; and were kept alive the whole of their voyage from June to the middle of September. (Zool. Journal, vol. iii.) Of the two kinds of luminous beetles which we possess, the *Elater noctilucus*, with the large phosphorescent tubercle on either side of the thorax, produces the wire-worm of the corn-fields, — the *Lampyris*, which we call the *Blinker*, has not been traced through its transformations. It is usually found on the trunks of trees, and in a state of inactivity; — during the day it clings to their bark or is concealed in their fissures."

CYDIMON.

Two species of the genus *Cydimon*, the one *C. lunus* of Cramer, the other considered by my late lamented friend, Edward Doubleday, Esq., as new, though closely resembling the former, I have occasionally found in the summer months around Content, but very sparingly. The genus is interesting, as being, like *Urania*, one of those dubious forms which connect the Butterflies with the Sphinges. In habit, as well as in form and coloration, *Cydimon* is more allied to the Moths than *Urania* is; for, as far as my observation goes, I believe it to be exclusively crepuscular, if not nocturnal, in its activity. The few specimens that I have taken, have all been found resting on the walls and in the angles of houses, in the morning, just in the manner of the *Geometradæ*, having flown in at the open windows between sunset and sunrise.

One evening near the end of June, I was returning to Content from a visit to the Kepp. The sun had already gone down into the wide Caribbean Sea that spread out before and almost beneath me; and the little peak-like clouds, that always appear like so many islets on the horizon after his departure, were beginning to lose the brilliancy of their golden borders, when I entered a part of the road where the beauteous scene was suddenly shut out by the lofty woods towering on either side. The trees nearly met overhead, so as almost to exclude the little daylight that yet lingered in the sky, when my attention was attracted to what at first appeared an *Urania* in the air; but which I presently perceived, by the broad-tailed processes of the wings, to be a *Cydimon.* Its manners were singular, and unlike those of any Moth or Butterfly that I had ever seen; so that I drew up awhile to watch it. It hovered in one spot high in the air, immediately over the road, at an elevation of perhaps twenty-five feet, just in the manner of a *Syrphus* or of some Bees, the front margin of the wings strongly and rapidly tremulous; at intervals of a few seconds, it gave a single stroke with the wings, and darted laterally to a distance of several feet. Here it would become instantly stationary as before; and so proceeded, alternately hovering and shooting to and fro, without leaving the open space between the trees, as long as I remained.

TRAP-DOOR SPIDERS.

In digging their mountain-gardens the negroes often expose the curious subterranean nests of the Trap-door Spider (*Cteniza nidulans*), many of which are brought to me. This Spider makes its tubular dwelling in soft earth, frequently choosing cultivated ground, on account, doubtless, of this quality; each nest is cylindrical, or nearly so, from four to ten inches deep, and about one inch in diameter; the bottom is rounded; and the top, which is at the surface of the soil, is closed very accurately with a circular lid. They are not all equally finished, some being much more compact, and having the lid more closely fitted than others. Some have irregular bulgings, and ragged laminated off-sets on the outer surface; but all are smooth and silky on the inside. This smoothness, however, does not preclude little irregularities or unevennesses of surface, nor is it glossy; its appearance rather resembles that of paper which has been wetted and dried; it is always of a reddish-buff hue, but the outside is stained of the colour of the surrounding earth. The mouth of the tube, and the parts near it, are very strong; the walls here often having a thickness of from $\frac{1}{8}$th to $\frac{1}{4}$th of an inch; but the lower parts are much thinner. The lid is continuous with the tube for about a third of its circumference, and this part may be called the hinge, though it presents no structure peculiar to itself; it is simply bent at a right angle, as is manifest if a nest be cut longitudinally through with scissors, the incision passing through the midst of the lid.

The mode of construction I judge, from examination of many nests, to be this. The Spider digs a cylindrical hole in the moist earth, with her jointed fangs or mandibles, carrying out the fragments as they are dislodged. When the excavation has proceeded a little way, she begins to spin the lining which forms the dwelling. I conclude thus, because nests are occasionally found a few inches in length, with the lid and upper part perfect, but without any bottom, these being evidently in course of formation. I suppose that she weaves her silk, at first, in unconnected patches, against the earthy sides, perhaps where the mould is liable to fall in; and thus I account for the loose rough laminæ of silk that are always found projecting from the outer surface. These are overlaid with other patches more and more extensive, until the whole interior walls are covered; after which the silk is spun evenly and continuously all round the interior, in successive layers of very dense texture, though thin. Under the microscope, with a power of 220 diameters, these layers are resolved into threads laid across each other and intertwined in a very irregular manner; some are simple, varying from $\frac{1}{7000}$th to $\frac{1}{2000}$th of an inch in diameter, and others are compound, several threads, in one part separate, being united into one of greater thickness which cannot then be resolved. No pellets of earth are ever interwoven with silk to form the outer layers of the walls, though the adhesive nature of the silk, when freshly spun, causes fragments of earth to remain attached to the surface. The mouth of the tube is commonly dilated a little, so as to form a slightly recurved brim or lip; and the lid is sometimes a little convex internally, so as to fall more accurately into the mouth and close it. The thickening of the hinge by additional layers is, I think, accidental only, as, out of many specimens that I have examined, only one or two had such a structure. In the neatest examples, the lid is of equal thickness throughout its extent, agreeing also with the walls for the first few inches of their depth.

One of peculiar compactness now before me I have slit open longitudinally with a pair of scissors in the manner spoken of above. The thickness of the substance is in no place greater than $\frac{1}{16}$th of an inch, which is very regularly maintained throughout the lid and upper parts. The appearance at the cut edge closely resembles that of millboard so divided; the layers of which it is composed being very numerous and compact, especially towards the interior side, where they can scarcely be distinguished even with a lens. In this specimen there is what I cannot find in any of the others that I have examined. A row of minute holes, such as might be made by a very fine needle, are pierced around the free edge of the lid, and a double row of similar ones just within the margin of the tube. There are about fifteen or sixteen punctures in each series; and they penetrate through the whole substance, the light being clearly seen through each hole. Now, what is the object of these orifices? I do not think, as I have somewhere seen suggested, that they are intended to afford a hold for the Spider's claws, when she would keep her door shut against the efforts of an enemy; for what would be the use of having them in the tube, *close to the lid,* so close that not the eighth of an inch intervenes between the series of the lid and that of the tube, when the former is tightly closed? I would suggest whether they may not be air-holes; for so tight is the fitting of the lid, and so compact the texture of the material, that I should suppose the interior would be impermeable to air, but for this contrivance. And as those in the horizontal lid might possibly be closed by minute particles of earth rolling on it, the second row around the edge of the perpendicular tube, just at the surface of the ground, would still be available in such a contingency. They may admit also an appreciable amount of light. On viewing such a structure, with its beautiful contrivances for security and comfort, we may well say with the Psalmist, "O Lord, how manifold are Thy works! in wisdom hast Thou made them all."

The Spider that inhabits this nest is black, with the thorax of an exceedingly lustrous polish; its abdomen is full and round, its legs very short. It retreats on alarm to the bottom of its tube, whence it is very reluctant to be dragged; and when exposed, seems inert and helpless. Yet it is much dreaded, its bite being reputed to cause tumefaction, and painful fever.

TERMITES, OR DUCK-ANTS.

In going only a few rods into the woods, a stranger cannot help observing many of the trees encompassed with large black, round masses, often as big as a hogshead, adhering to the trunks or the branches. Curious to know the nature of so singular an appendage, he is told that these protuberant masses are the nests of Duck-ants or *Termites;* and on examination he finds that they are composed apparently of an earthy substance, comminuted very fine, and made into a sort of paste with animal gluten. The outside is friable, and much resembles, except in colour, that description of pastry technically called "short," as does the whole of that portion which is newly constructed; but the interior has a hardness and firmness superior to that of wood. The whole mass is composed of numberless passages of the diameter of a man's little finger, separated by walls about one-tenth of an inch thick, and running without any obvious regularity or design. On the surface-walls being broken in, out swarm hundreds of the inhabitants, most of them active little fellows with black heads, but many, rather larger and plumper, with *yellow* heads are scattered among the crowd. I presume the former to be the labourers, and the latter the soldiers. A single covered gallery is invariably found to extend from the edifice along the trunk of the tree to the earth. Frequently I have observed in the morning a similar covered gallery, as thick as one's thumb, extending all across the high-road daily travelled by vehicles of various kinds; whence I conclude that the construction of such a passage was the work of the preceding night.

After a time, from some cause or other, the great nests are deserted; but continue to maintain their integrity for an indefinite period. In this state the

Yellow-bellied Parroquet (*Conurus flaviventer*) chooses to build her nest in them, excavating a chamber with her powerful beak. Snakes and their eggs are often found in them, and the eggs of Lizards also I once found in one several eggs of the small Pallette-tip Gecko (*Sphæriodactylus argus*). It is now an excellent fuel, burning readily, with a flame and a glowing brightness, little inferior to coal. As no ammoniacal smell proceeds from it, and as it consumes into a clear white ash, I conjecture that the substance is of vegetable origin. It is sought after, in those districts and seasons in which the Musquitoes make a more than endurable pest, in order to be burned in a chafing-dish; as it gives out a good deal of smoke, which is the only weapon that those formidable, though minute, warriors fear. The smoke clears the house of the insect-hosts in a few seconds; and is much preferable to that of wood, because far less painful to the eyes.

The *Termites* do not often enter dwelling-houses; but sometimes they do penetrate the floors, and devour whatever lies in their way, encrusting the residue with their galleries, which they invariably make as they go along. Some spare bedding that had lain in one corner of my bed-room for some weeks, tied up in a blanket, I found, on removing it, much injured in this way; the blanket being devoured in long meandering lines, and so defiled with the crustaceous deposit as to be irreparably spoiled.

In the spring a swarm of the winged males and females often enters the house, to the great annoyance of the inmates. The insects flock to the candle-shades, and spread themselves over the table, and especially crowd upon books, papers, or any other white materials. Their four wings, which are ample, are of a dull brown colour, possessing little transparency. Presently we are surprised to see several dismembered wings lying about; and, looking at the insects more carefully, we perceive that they are casting these organs voluntarily, which seem to be annoying encumbrances rather than serviceable parts of their bodies. They are seen to throw the wings very far forward towards the head, and then suddenly jerk them back; these motions soon dissever their connexion with the thorax, which appears to be very slight.

RIDE TO KILMARNOCK.

March 3rd. — Rode from Content to Hampstead, and thence, along the summit of the Luana mountains, to Kilmarnock; returning through the Bamboo walk of Grand Vale. Near Hampstead, by the side of the road, half-hidden among the luxuriant herbage, was a large mass of stone which I alighted to examine. Beneath the ledges, adhering to their under surface, were many specimens of a pretty white shell, *Cylindrella gracilis.* The delicate pointed spire seems less deciduous in this than in other species, for most of the individuals were perfect.

At the summit of the mountain are two elevated peaks of a rounded form, about half a mile distant from each other; on the very apex of each is a dwelling-house, the one of which is named Highgate, and the other Hampstead, doubtless in allusion to the somewhat similarly situated villages in the neighbourhood of London. It is spoken of as something rather wonderful in the country round, that in these elevated dwellings the apartments are furnished with *fire-places.*

The scenery about this part is singularly romantic. Large, round hills, almost hemispherical in their contour, rise out of the valleys in great number, apparently without any order; yet so regular in their form that they seem as if cast up by art. The road often winds round the sides of these, and opens delightful and ever changing panoramas. The valleys and plains beneath, smiling in verdure, and studded over with clumps of ornamental trees, now hidden and interrupted by these conical hills, now breaking into view between them, strike the passenger with ever fresh delight; and the various hills themselves, half in the glowing sunlight, half in deep shadow, changing their relative places as he moves on, have a wonderfully beautiful effect, totally unlike anything I have ever seen elsewhere. We look across a deep grassy vale from the hill-side around which we are winding, and see another similar mound, with the narrow line of road passing across its rounded side in like manner. We are told that we shall presently pass along that line: it looks almost within gunshot, but we wind on, and lose sight of the opposite hill, and it is perhaps half an hour before we arrive at it, having made many tortuous courses and opened many new scenes in the intermediate space. The summits of many of these hills have been planted with clumps of bamboos; and these verdant, plumose crowns, expanding their arched heads like gigantic tufts of ostrich feathers against the brilliant sky, finish the picture.

After travelling some miles through an open country, we entered the forest at Kilmarnock, where some new provision gardens had recently been reclaimed and planted. The grace and luxuriance of the yam-festoons, and other cultivated plants already shooting up, nearly concealed the hideousness of the blackened stumps that thickly studded the ground. Here I saw *Brasavola nodosa* in full flower, the spikes of white blossom large and massy; and from a huge fallen tree by the side of the road, I obtained many bulbs of a species of *Maxillaria*. The woods were now high and dense, and presented much of the same character as those on the Bluefields Ridge. Presently we began to descend, and soon opened the beautiful sugar estate of Grand Vale, with its bright green cane-fields and pastures spread out below us as on a map. The mill-house with its curious conical roof; the boiling and trash-houses, and other offices; and, at a little distance, the "great house," beside a beautiful sheet of water; all could be traced from our elevated position; while the groups of busy labourers, the teams of working oxen, and the pasturing cattle, moved about like ants over the sunlit scene. The dark forest bounded the estate on every side, heightening by its sombre contrast the cheerfulness of the variegated inclosure; and beyond this stretched away the boundless sea, sleeping in silvery beauty beneath the noontide sun.

The number of dead and bleached shells that lay on the road hereabout, induced me to search for living ones; and I spent an hour or two in the examination. Nor was the search unsuccessful: beneath the loose stones on the hill-side, a little within the forest, were several species in considerable abundance, some of which I had not at that time met with. Among the most conspicuous was *Helix Jamaicensis*, which I now saw alive for the first time; the surface of the foot was, in most of the specimens, infested by a minute species of mite in great numbers. This fine shell was still rare. A large flattened species of much value (*Helix Spengleriana*) occurred more numerously; *Cylindrella sanguinea* and *cylindrus* were abundant; and several species of *Achatina* were found sparingly, such as the lovely *A. Philippiana*, *A. venusta*, *A. Phillipsii*, and others.

The zigzag avenue of Bamboos afforded a grateful shadow as we descended the mountain; and the discovery of a nest of the wild Guinea-fowl, with twenty eggs, was an incident which enlivened a peculiarly toilsome part of the journey, the passage through a long but narrow water-course, now dry, filled with masses of loose slippery stone, almost impassible for a horse. In the midst of a thick tuft of grass, within the wood beside this rocky path, the Guinea-hen had deposited her numerous eggs. At length we arrived at the bottom of the Cotta wood, and were again sheltered from the burning sun till we arrived at Content.

THE WATER-WITHE.

March 5th. — From Shrewsbury I went a little way into the woods to see the Coulter-Spring, a stream so named. A walk of half a mile through the tall and dark forest brought me to a wild rocky defile, in the bottom of which the stream ran. In the rains this is a roaring, impetuous torrent, and must be wildly magnificent; at this time it was romantic enough, though in another way. The water was dwindled by the parching weather, until it no longer formed a stream, but lay in calm, glassy pools, bounded by the huge, angular masses of black rock that lay in confused disorder in the ravine. Tiny threads of water trickled from one reservoir to another, and produced a tinkling music, sufficiently audible in the deep silence of the woods. The lofty trees that shot up their straight branchless stems all around, were reflected in the dark pools with perfect outline; not a bird, not an insect was visible; the obscurity, the stillness, and the silence gave a gloomy awe to the scene, and I felt a sort of relief at again breaking out into the sunny fields of Shrewsbury. In this obscure glen the friend who acted as my guide pointed out the Water-Withe (*Vitis Indica*), a valuable plant, for the resource it affords to thirsty travellers. A long twisted stem, much like that of the common Grape-Vine, and about as thick as one's wrist, was hanging down from one tree to another; with a stroke of his heavy knife he cut this in two, and putting one extremity to my mouth, bade me drink. A copious supply of fluid instantly flowed from it, which I should not have distinguished from pure cold water. A junk of a yard long, it is said, will yield a pint; and lives have been saved by the seasonable supply of this plant, when travellers have lost their way in the woods, and have been fainting with thirst.

THE KEPP.

March 8th. — A friend having business at Painstown, about eight miles distant from Content, I accompanied him. We rode through a lovely mountain country, chiefly laid out in *pens* or grazing farms, well studded with trees, and broken by tracts of forest. In the neighbourhood of Highgate, on the side of a conical hill, covered with huge masses of limestone and with small rubble, and crowned with a tuft of Bamboo, I found some shells, especially the pretty little new species, which Dr. Pfeiffer has done me the honour to name *Bulimus Gossei,* but which seems to be rather a *Cylindrella.* This was in great abundance beneath the loose stones.

In these mountain estates there are no streams; and the resource of the inhabitants is to dig large ponds in the hollows, into which the rain-water collects in the wet season. Owing to the long continued drought, these ponds were now very low, some dry, and others reduced to a small space of water in the centre of a large area of parched and cracking mud. At the ooze which margined the water, I was interested to observe the honey-bees thronging to drink in great numbers; I have seen the same thing, however, in other places, lately. The ants, also, are very thirsty, crowding to every vessel of water in the house; forming long, serpentine, black lines up the sides, around the edges, and down the concavities of the basins and ewers in our bed-rooms, and black circles around the water's edge. The orange-groves were in blossom, and delighted the senses of sight and smelling with their beauty and profuse fragrance.

After winding round many of the curious hemispherical hills that I have already mentioned, we came to the Kepp, the estate of George Marcy, Esq., to whom I had an introduction. This property is particularly beautiful: its surface is very varied, presenting continual changes of scenery. The broad swells and slopes of pasture, verdant in spite of the drought, are enriched by a great variety of fine trees, standing singly and in clumps, all of them useful, and many of noble and imposing beauty. The negro huts of the estate, embosomed, as usual, in luxuriant groves of deepest green, and in small gardens of provisions; — the white mansion crowning a swelling hill;—the dark belts of surrounding forest, admitting, here and there, peeps at the smiling fields of neighbouring estates; — impart a peculiar character of loveliness to the scene. One lofty tree was nearly covered with the dark-green foliage, and magnificent white trumpet-blossoms of a climbing plant, probably a species of *Bignonia*, that had spread itself over the branches like a mantle, and hung down in the richest profusion. On the fruit-trees of the estate, many species of epiphyte *Orchideæ* were numerous; we had obtained a noble mass of *Epidendrum ciliare*, from an Avocada Pear near Highgate; there were forty-two of the fine lengthened bulbs in the cluster, but the singular fringed blossom was not displayed. On the trees in the Kepp, this species occurred again; *Epid. fragrans* was in blossom, most beautiful and sweet-scented; *Brasavola nodosa* was also in flower:—other species displayed only the leaves and pseudo-bulbs.

A slight depression in the ground, gradually deepening, led us into a most romantic little glen. One side of it was formed by a very steep slope, covered with trees, and difficult to be climbed; the other side was an enormous sheet of limestone, greatly overhanging, from the moist surface of which, slender lianes were hanging, and many interesting ferns threw out their gracefully curling fronds. The bottom of the dell was damp and tenacious, and supported a rank vegetation; largely consisting of the dangerous Dumb-cane (*Caladium seguinum*), a tall kind of *Arum*, so virulently acrid, that the juice of any part, incautiously applied to the mouth, causes the tongue to swell so as to take away the power of speech, and produces burning torments of long duration. It is said to have been one of the modes of torture employed by cruel masters in the dark days of slavery. Amidst these fine-looking, but deleterious Arums, I found a liliaceous plant with ovate-leaves, the bulbs of which I took; these subsequently blossomed in my possession, and proved a fine species of *Pancratium*, delightfully fragrant. The smooth overhanging rock I found bristling with hundreds of minute slender shells, all of the same species (*Cylindrella subula*), alive, but shrunk into the interior of their dwellings, which were attached by their mouths to the surface of the rock, at various degrees of inclination from the perpendicular. The great majority were decollated, as is customary with this genus; but some half-dozen occurred in a perfect condition, the shell running off into a delicate taper spire with an acute point. The summit, however, was excessively fragile, so that it was almost impossible to secure the specimens, without reducing them to the ordinary condition of amputation.

The gentleman on whom my friend had intended to call was lying on what proved to be the bed of death, too ill to receive us; we therefore returned without accomplishing the main object of our journey, but in other respects much gratified.

NIGHTSHADE AND MISSELTOE.

Immediately behind Bluefields, extending over the plain at the foot of the mountain, and considerably up its steep side, is the pen of Pinnock Shafton. A small inclosure, into which the numerous cattle may be driven for examination when required, contains a few objects of interest. The loose stone walls that surround *it*, as well as the other fields, are half-hidden by a vegetation whose wild luxuriance conceals or adorns their naked unsightliness. Long Cacti, with cylindrical or angular stems, especially *Cereus grandiflorus* and *C. triangularis*, trail irregularly over them to a surprising length, and hang about in loose festoons, mingling with the contorted stalks and broad noble leaves of Arums, the Five-finger (*A. auritum*), and other species. Especially characteristic of these walls are the plants called Nightshade, species of *Echites*, with handsome yellow or white flowers, somewhat resembling, at first sight, those of a Convolvulus. These are elegant slender climbers, which blossom profusely at most seasons of the year, and, being very common on fences, along with the noble Hogmeat (*Ipomœa violacea*), which exactly represents a Major Convolvulus of three or four times the usual dimensions, materially contribute to the gaiety of the fields in this island. The seed vessels of the *Echites* have a curious appearance; they are long, slender, and slightly spindle-shaped, always disposed in pairs at the end of a stalk, from which they diverge either at a very wide angle, or in opposite directions at right angles to the foot-stalk, like the head of the letter T. These plants have been named Nightshade because of their poisonous character; they are not botanically allied to the true Nightshades: they belong, however, to an equally virulent order, *Apocyneæ*. An instance of the deadly qualities of these beautiful plants was named to me. Two men, not long ago, were found lying dead in the road: — the only clue to the cause of their decease was that a bottle of rum was found with them, from which they had been drinking, and which they had stopped with a plug made of the leaves of the *Echites*, growing in its usual abundance in the spot, instead of a cork. The alcohol had probably extracted the poisonous properties of the leaves, and had become fatally impregnated with them.

Here grow several bushes of the true Aloe (*A. socotrina*); whose serrate, fleshy leaves, and tall spikes of tubular flowers, red, yellow and green, are well-known. Though cultivated in some of the Antilles, it does not appear to be much known in Jamaica, at least in the parts with which I am familiar; this being the only locality in which I remember to have seen it.

But what interests me most in this place is a flourishing Misseltoe, or God-bush, as the negroes call it. It is growing on a Soursop (*Anona muricata*), a tree which it principally affects, overspreading every branch, and effectually, though gradually, killing its supporter. The seeds are viscous, and are to be seen sticking *on the leaves and twigs*, as well as on the trunk; in every instance rooting and shooting where they adhere; so that hundreds, perhaps I might say thousands, of young plants, in various stages of forwardness, may be seen on this Soursop, springing up from the surface of the leaves, three or four on one leaf, and that on both the inferior and superior faces. This I take to be a somewhat unusual phenomenon. It would be curious to ascertain what becomes of these young plants when the leaves on which they grow, fall and decay. Do they become terrestrial, or perish? Certainly I saw no Misseltoe plants growing beneath the tree.

BLUEFIELDS RIDGE.

The road which leads from the shore to the summit of Bluefields mountain, branches off right and left at a point just beyond the Tree-ferns, formerly described; or rather abuts upon another road running at right angles to it along the ridge. We will first pursue the right-hand branch, which, at the distance of about three miles, leads us to Rotherwood, a coffee plantation. The intervening space, for the most part, consists of the primitive woods; and the road is nothing more than a narrow rocky bridle-path, just wide enough for two persons to pass each other. The negroes of the lowlands have little gardens or provision-grounds embosomed in these woods, where they rear various kinds of produce; on Fridays they resort hither, early in the morning, to perform the little tillage that is required, and to collect the fruits and roots which, on the next day, they carry to the market at Savanna le Mar. With the exception of the Friday in every week, however, when the woods re-echo the voices of men, women, and children, and the sounds of the axe and the hoe, this path is almost an unbroken solitude, where the voices of the mountain birds by day, and of the strange reptiles that are vocal by night, and the sighing of the breeze in the upper foliage of the forest, are the only sounds that break the solemn silence. The vegetation here is totally different from that of the lowlands; more luxuriant, more close and matted, the parasitical plants are more numerous and varied, the air is more humid, and the whole scene is much nearer that described by South American travellers, as characterising the interior of the forest in that region. That giant of the lowlands the Ceiba or Cotton-tree (*Eriodendron anfractuosum*) hardly reaches these elevated woods; but its place is supplied by scarcely less bulky Fig-trees, whose hoary trunks and broad horizontal limbs are a perfect nursery of *Orchideæ* and *Bromeliaceæ;* and magnificent Santa-Marias (*Calophyllum*), Broad-leafs (*Terminalia*), Parrot-berries (*Sloanea*), and other lofty trees, tower up to an enormous pre-eminence above their fellows. In parts which have once been cleared, and since neglected, — according to that law by which a primitive forest, when once cut down, is succeeded by a spontaneous growth of a totally different kind of wood, — dense thickets of a species of *Piper*, called (from the propensity of this tribe to form thickened nodes, like those of grasses, at regular intervals on its trunk and branches) Jointer or Jointwood, grow in large tracts to the exclusion of everything else. In these Jointer thickets the Green Tody, Green Sparrow, or Robin Redbreast of the colonists (*Todus viridis*), is particularly abundant; a lovely little bird, with the upper parts emerald green, the belly pale yellow, tinged with rosy, and the throat and gorget deep rich crimson-plush. It sits with the utmost fearlessness on the low twigs that jut out into the road, almost brushing our faces as we pass; or flits about on feeble wing, pursuing flies, with a soft plaintive squeak.* In other places

* In The Birds of Jamaica, p. 77., Mr. Hill has described the eggs of this little bird as "grey, brown-spotted." He has since had

the trees are tall, slender, and somewhat open in growth, but the edge of the woods is formidable with cutting Sedges, and spinous *Solaneæ*, relieved by beautiful tufts of *Cannæ*. Hereabouts *Orchideæ* are very numerous; almost every tree, from the size of one's arm and upwards, being studded with masses of pseudo-bulbs of various species, enwrapping the trunks with their matted roots, and throwing out their fleshy leaves and spikes of fantastic blossom. *Epidendrum ciliare*, *Maxillaria Barringtoniæ*, *Brassia caudata*, and other epiphyte kinds, grow here in great luxuriance. Farther on, the road is bordered by shrubby *Gesneraceæ*, of which the curious *Rhytidophyllum tomentosum*, with its woolly spindle-shaped leaves, and branches of pale green flowers, and another species called Glass-eye berry, from its forming the principal food of the Thrush of that name, are pre-eminently abundant. Here and there the immense leaves of the wild plantain (*Heliconia*), of the most delicate green hue, wave out from the more ignoble bush; and, looking in, we see a clump of these magnificent plants, and catch sight of their massive branches of blossom, enveloped in great boat-like bracts of richest scarlet. Looking over the forest from an eminence, the eye is attracted by the elegant Mountain-pride (*Spathelia simplex*) rising here and there above the mass; a beautiful tree, whose straight slender stem, terminated by a crown

an opportunity of seeing other specimens, and of finding that this was an error. The eggs are *pure ivory-white, unspotted.* The spots that deceived him, were probably stains produced by the earth in which they had lain.

of large pinnate leaves radiating from the centre and arching outwards, bears a remarkable resemblance to the habit of the Palms. Beautiful it is at all times; but it is in the flowering season that its name of Mountain-pride is felt to be appropriate; when its summit is crowned by a dense pyramid of lively pink flowers, several feet in diameter and in height. In districts where this tree is numerous, the effect is said to be most magnificent.

Two fine species of real Palms are found in these lofty woods, though not in great numbers. The one is the Long Thatch, a species of *Cocos,* whose long pinnate fronds are used for thatching the houses of the negro peasants; the other is the Mountain Cabbage (*Areca oleracea*), one of the very noblest of this kingly race of plants. It shoots up its verdant tuft of feathery fronds to an enormous elevation, some specimens even to the height (as is credibly asserted) of two hundred feet. To think of a tree as high as the Monument of London, with a slender branchless stem, as straight as an arrow, perfectly cylindrical, yet not more than a foot in diameter! The immense spike of blossom that projects in the early autumn from the base of the crown, arching gracefully downwards, is a fine object. I have seen, at such times, the earth beneath the tree, for a space of many square yards, quite white with the scattered pollen, as if a light snow shower had fallen. Bees, Beetles, Flies, and other insects, throng around it in this season, attracted by the nectariferous bloom, and themselves forming an attraction for numerous Swallows, which, darting by on rapid wing, snatch their selected victims as they pass, and, wheeling round, return again and again to the prey.

But if I were asked to name the most prominent character of the vegetation on these lofty peaks, I should designate it a region of Ferns. Scores of species, and thousands of individuals, fringe the sides of the path with their graceful fronds, and almost choke the way. If we sit down on the grey-spurred root of a tree, the great fronds of *Phlebodium aureum*, so elegantly pinnate, arch over our heads, and spread widely on each side; while the eye is pleased with its massive twisted leaf-bases, covered with golden hair that shines like silk, and with the brown, delicate, thread-like roots, that cling to the bark of the tree spur, meandering over it like a spider's web. The large triangular pinnæ of *Adiantum macrophyllum* overlapping each other, and gradually diminishing, have a very striking appearance; and many of the minuter kinds growing in the hollows of the stones, and beneath the roots of the trees, display a grace and beauty peculiarly their own. Within the gloom of the forest other forms are seen in luxuriant profusion. The trees are loaded with them: many of the terrestrial kinds spring in feathery tufts from the crevices of the bark, and curve gracefully towards the ground; others fringe the horizontal limbs, and conceal the forks; and others, perhaps the most curious of all, as *Phlebodium lycopodoides*, *Ph. vaccinifolium*, *Polypodium acrostichoides*, &c., crawl up the trunks of the tall trees, from the earth to their summits; their lengthened slender stems clinging fast to the bark, fringed, throughout all their irregular windings, with their small oval or oblong leaf-like fronds. The sides of the bare rocks, and the surfaces of the large loose stones, that lie in the woods, half concealed by bushes, are sprawled over by similar caulescent and clinging species of the great Fern tribe, which is estimated to constitute one ninth part of the whole vegetation of Jamaica.

I will mention but one more member of this tribe, a Tree-fern of peculiar beauty, that I found growing in some abundance in a spot of more than usual gloom and grandeur, far on towards Rotherwood. The species was, I believe, *Cyathea arborea,* taller and more graceful than the *Alsophila* of the mountain-brow. The slender stems, each marked with its oval, scale-like scars, and throwing out from its summit its swelling cluster of leaf-bases so compact and so regular as to look like the elegantly fluted knob of some cast-iron pillar, again constricted before they spread abroad in a wide umbrella of finely cut foliage, —had an imposing effect here in the rather open woods, surrounded by the naked irregular trunks, moss-grown and studded with parasites, of the tall trees that towered up, and interwove their branches far above their heads, shutting out the sun, and almost the light.

In a very dense, and nearly impenetrable, part of the bush that borders this lane, I found, about the end of February, some fine plants, in full flower, of that noble and magnificent terrestrial Orchid, *Phajus Tankervilliæ.** The flower-spikes, which rose to a

* Though the extremity of the labellum in the plants above mentioned, was of a deeper and more purple tint than in the pub-

height of about thirty inches, sprang from a mass of pseudo-bulbs that had attached themselves to the surface of a huge stone. I was much struck with the presence of so rich and elegant a flower in so gloomy, obscure, and tangled a place, uncared for and unknown. For years and years it may have successively displayed its gorgeous beauty, without the eye of man having ever rested on it; and for ages more it might have continued to blush unseen, but for the curiosity of a prying naturalist.

This lonely road was one of my most favourite resorts, and will ever be associated with my most delightful recollections of Jamaica. I am sure I cannot, by any attempt to describe the scene, or by any enumeration of its more prominent constituents, give any adequate notion of the peculiar charm that belonged to it. The gradual ascent higher and higher up the mountain, with the commanding view thus obtained, widening and spreading beneath, imparted an exhilaration to the spirits that probably had something to do with it. Then the leaving behind, and far below, the habitations of man, to plunge into the deep and wild solitudes of Nature; and the thought that pro-

lished figures of this species, the similarity is so close as to warrant their identification. Yet *Phajus Tankervilliæ* is a Chinese plant, and the genus has never before been recognised in the New World. The suggestion occurs that it may have been introduced, and have been increased by cultivation, and that the seeds may have been scattered by the winds, or carried by birds over the island. The lone humid forest at the summit of Bluefields Peak, does not certainly seem a very likely locality in which to find a large and beautiful flower escaped from a garden. I may add that just a year afterwards I met with it again in blossom in the neighbouring woods.

bably few Europeans, at least few who could at all appreciate their beauty, had ever trodden them, were pleasant accessories; not to mention the delightful coolness of a temperature which, perhaps, never exceeds 70°, after the oppressing heat of the lowlands. I used frequently to set out two hours before dawn, and ride leisurely up, by the light of the moon or of the twinkling, quivering stars, listening to the rich melodies poured forth by dozens of Mocking birds from the fruit trees and groves of the lower hills, and arrive at the mountain-brow about sunrise. There I would leave my horse, and, throwing the bridle over his neck, allow him to graze on a little open pasture until my return, while I pursued on foot the track which I have been describing. Here I often spent many hours at a time, engaged in collecting specimens of zoology and botany, in making observations, or in calmly drinking-in the quiet enjoyment that the scene afforded. Very few birds occur in this region; but such as are found are of peculiar interest. The wary Jabbering Crow (*Corvus Jamaicensis*) gives utterance to his strange, uncouth articulations, as he calls to his fellow from the topmost branches of the very loftiest trees, or sails along, on labouring wing, from one such post of observation to another, taking good care to keep out of harm's way. This, too, is the region of the Solitaire (*Ptilogonys armillatus*); that mysterious recluse, whose rich flute-like tones fall slowly upon the ravished ear, like the measured notes of a psalm. One, and another, and another, take up the strain, till the mellow notes come one by one from all parts of the dark surrounding woods, at different distances, and at considerable intervals; the effect of which, so unlike the ordinary voice of a bird, is startling until you are acquainted with the performers, and always charming when you are. You listen with the more interest, because the bird itself is very shy, and consequently very seldom seen; ordinarily keeping with jealous suspicion in the most dense and sombre parts of these mountain woods.

But there is one bird which is very abundant here. As the Ferns are eminently characteristic of the botany of this lofty elevation, so is the lovely Long-tailed Humming-bird (*Trochilus polytmus*) of its ornithology. The velvet crest, and emerald gorget, and long streaming tail-plumes of this lustrous living gem, flit, and flutter, and hover about this shady lane all day long, and all the year round; but it is especially numerous in the spring, when scores, and even hundreds, may be seen rifling the perpetually-blossoming shrubs that are its denizens. To sit on a fallen log in the cool shadow, surrounded by beauty and fragrance, listening to the broken hymns of the Solitaires, and watching the Humming-birds that sip fearlessly around your head, and ever and anon come and peep close under the brim of your broad Panama hat,—as if to say, "Who are you that come intruding into our peculiar domain?" — this is delightful. There is nothing to mar the charm of the situation; no wild beasts in the forest behind glaring at you, and ready to make their fatal spring; no deadly reptiles coiled beneath your seat, or swinging from the branches of the neighbouring trees; for Jamaica possesses, as far as I know, neither the one nor the other; scorpions, and centipedes, and red spiders, are far too rarely seen, unless searched for, to be any objects of dread; and even the petty insect annoyances, which commonly make the woods rather intolerable, are wanting here. You may sit for hours without having your clothes and your hair full of little stinging ants, or without disturbing a city of those formidable warriors, the great-jawed Corromantee Ants (*Odontomachus*); and even the Musquitoes, except on rare occasions, are so little troublesome, as to be scarcely more worth minding than the gnats in England in an autumnal evening.

> "Hìc ver assiduum, atque alienis mensibus æstas.
> At rabidæ tigres absunt, et sæva leonum
> Semina; nec miseros fallunt aconita legentes;
> Nec rapit immensos orbes per humum, neque tanto
> Squameus in spiram tractu se colligit anguis."
>
> VIRG. *Georg.* ii. 149.

Above this secluded road on the seaward side, shutting out the view in that direction, except in little peeps here and there through the valleys, — rise the peaks of the ridge. Up the highest of these I climbed, with no small toil, one day in March, and stood on the loftiest point in the western part of Jamaica. I was well rewarded for the difficulty and labour of the steep ascent, by the extensive prospect. I counted six ranges of mountains to the eastward, each beyond the other, besides Bluefields Ridge, on which I stood; the most distant of which was in all probability that of the Manchester Mountains, near the centre of the island. A few insects commemorated my visit: *Passalus interstitialis*, a curious flattened

Beetle of large size, somewhat allied to our Stag-beetle, was found in some numbers beneath the decaying bark of a fallen tree; as also some specimens of a large and noble *Forficula*, and of a small *Osorius*. A very lovely little Hawkmoth (*Empyreuma lichas*), of two colours, greenish-black, and rich crimson, was sitting beneath a loose stone, very perfect, and apparently just out of pupa. Attached to the under surfaces of contiguous stones were several thin cocoons of yellow silk, all empty; very probably belonging to this species. This pretty moth is not uncommon in the lower hills, and even by the sea side, occasionally flying into dwelling-houses at night: when seen by day it is always in dark woods.

THE VENUS LIZARD.

The first time I met with that fine Iguaniform Lizard, *Dactyloa Edwardsii*, called Venus* by the negroes, and sometimes Green Guana, was on this ridge. And though I afterwards found it in lower situations, as on the moderate elevations of the St. Elizabeth's hills, and even in the pimento groves of Mount Edgecumbe that border the sea shore, it seems rather to affect the higher mountain-woods. The mode in which I formed an acquaintance with

* This name has probably no allusion to the Goddess of beauty. Mr. Hill writes me as follows: "The brilliant green Lizard you speak of is usually called the Green Venus. *Venus* in this case I take to be an Indian word; for I found it, — in a district of St. Domingo (Yasica) in which all the rivers had Indian names, — as the same of one of the streams. We have an Indian name for another of our Lizards in *Iguana*."

Plate. IV.

P. H. GOSSE DEL[T]　　　　M & N. HANHART IMP[T]

CHILABOTHRUS INORNATUS ⅓ N.S.　　　　DACTYLOA EDWARDSII, ½ N.S.

LONDON, LONGMAN & C[o] 1851.

the species may be worthy of being related. One day in February, having ascended the ridge with a companion, my attention was arrested by a Lizard about a foot long, and of a lively green colour, on the trunk of a small tree, head downward, intently watching our motions as we stood near. My young friend suggested the possibility of capturing it by slipping a noose over its head, while its attention was engaged by whistling. I laughingly proceeded to try the spell; and having made a noose of small twine, which I tied to the end of a switch, I gently walked towards him, whistling a lively tune. To my astonishment he allowed me to slip the noose over his head, merely glancing his bright eye at the string as it passed. I jerked the switch; the music ceased; and the green-coated forester was sprawling in the air, dangling, greatly to his annoyance, at the end of my string. He was very savage, biting at every thing near; presently his colour began to change from green to blackish, till it was of an uniform bluish black with darker bands on the body *, and a brownish black on the tail: the only trace of green

* As published descriptions of tropical reptiles are commonly made from specimens in spirits, with the colours and even the forms greatly altered, I may be excused for the following note made on the species during life. The length of adult males varies considerably, from 13 to 18 inches, but I have invariably found the difference to lie in the tail, the body being always about 5 inches, the tail varying from 7 to 12 inches. The colour, when the animal is not excited, is brilliant yellow-green; the hind half of the tail pale grey; the goitre orange, set with longitudinal rows of separated pale-yellow scales; the edges of the eyelids white. In some specimens the body, tail, and limbs are marked with transverse bands of a bluer green, those on the trunk more slender and oblique. (See Plate IV.)

was just around the eyes. I carefully secured, without injuring him, and brought him home in the collecting-basket, into which I had no sooner put him, than he fiercely seized a piece of linen in his teeth, and would not let it go for several hours. I transferred him to a wired cage, linen and all; and at length he suddenly let go his hold, and flew wildly about the cage, biting at anything presented to him. At night I observed him vividly green as at first; a token, as I presumed, that he had in some measure recovered his equanimity.

The next day he continued very fierce. I hung the cage out in the sun; two or three times in the course of the day I observed him green; but for the most part he was black. The changes were rather quickly accomplished.

After he had been in my possession about four days, I observed him one morning sloughing his skin; the delicate epidermis, loosened from the body and legs, looked like a garment of thin white muslin, split irregularly down the legs and toes, and separated from that of the tail, on which the integument yet adhered unbroken. Throughout the day the loosened skin hung about the animal, though more and more loosely. He had not abated a whit of his fierceness; leaping at a stick pointed at him, and seizing it forcibly with his teeth.

Another individual, caught in the same locality, and by the same device, I introduced into the cage of the former, who did not offer any molestation to the intruder. After they had remained in my possession, the one about six weeks, the other about four, they both died, almost on the same day, and both in the process of sloughing. In this operation the skin appears to be first separated from the head; for in one of these, it was perfectly loose from the whole head, and was removable in one piece, but to the neck and entire body it still adhered by organic union. I suspect that the sloughing of the skin is, at least sometimes, the result of universal excitement. All that I have taken alive, and caged (amounting to many individuals), after most violent behaviour at first, soon sloughed; usually the very next day.

The food of this Lizard appears to include both vegetable and animal substances. I was never able to induce one to eat in captivity; but the dissection of several has given me this result. Thus in one I found hard seeds and farinaceous substance; in another the fragments of a brilliant *Curculionidous* beetle, and other insects. I once observed a large one on the summit of the mountain, deliberately eating the ripe Glass-eye berries, munching half of one away at a mouthful.

It would require no great warmth of imagination to identify these sunny, spicy, pomiferous groves with the golden-fruited gardens of the Hesperides, and this fierce, sinister, saw-crested Lizard, with the watchful dragon that guarded them. If I had had the naming of him, I would have called him Ladon.

THE GRAVE-DIGGER.

On the earthen floor of the building, formerly used as the boiling-house on Bluefields estate, but now dilapidated and partially unroofed, where twine-like roots depend from the rafters, and elegant ferns spring out of the crevices of the crumbling walls, a good many large wasp-like Flies may be observed in the hottest part of the day, briskly flying to and fro. It is a species of Sphex, closely allied to *S. ichneumonea*, but with the abdomen wholly rufous. On closer examination we discover numerous holes entering diagonally into the dry and dusty ground, into which some of these bright-coloured flies are crawling, and from which others are emerging. From some of the holes proceeds a shrill, but intermitted, buzzing; and if we watch one of these we perceive the Sphex at work therein. At first we cannot see what she is doing, for she crawls in head-foremost, and in a second or two comes out tail-foremost, recedes a few inches, and then advances again; again emerges in the same manner, and again enters; and continues thus to crawl backward and forward with bustling activity, and with much flirting of the purple wings. She is almost white with dust, and is evidently very busy, if we can but comprehend her motions.

On stooping down and bringing our face very near the scene of labour, we discover by narrow watching, that she is digging the hole; and hence the negro children have given her the appropriate title of grave-digger. Every time that she comes forth, she brings a load of the powdery earth, much larger than her head, tightly held between the shanks of her two fore feet, her breast, and her chin, and this she drops an inch or two from the cave's mouth. Sometimes she brings a stone still larger, and this is grasped in the jaws, and dragged to the distance of four or five inches, for fear it should roll in again. I have seen her bring two stones together, one grasped beneath the chin, the other in the jaws. Each time she has dropped the load, she never fails, as she advances, to keep the road clear by scraping with the fore shanks, throwing the dust behind her. But for this, the earth brought out would soon accumulate in a heap, and roll back. If a dry leaf or small stick happen to drop against the mouth of the hole, she seizes it with her curved jaws, and carries it to a safe distance.

I observed one filling up a hole. No doubt she had deposited her egg at the bottom, and stored sufficient provision (caterpillars or spiders disabled but not killed, according to the custom of these interesting insects) to last the young grub when hatched, until its maturity,—"haud ignara ac non incauta futuri." With her tail towards the hole, she scraped back a little heap of dust; then turned, and with her head moved it about, that it might fall to the bottom. Then she turned again, and did the like, repeating this procedure several times in succession. At length no more earth would go down, for the hole was full; she then rammed it two or three times with her head, and flew away, leaving still, however, the situation of the orifice obvious enough.

These insects work very fast in the soft dusty earth, for they are indefatigable in their exertions. The Bee is the recognised symbol of industry, but the labour of the bee is play compared with the efforts of the grave-digging Sphex.

WAX COCCUS.

March 28th. — Visited Basin-Spring, a romantic little fountain high up on the side of Bluefields mountain. A sugar estate was here formerly, but now the greater part is mere wild forest. On the trunk of a slender Lancewood tree, I found many small sub-conical bodies closely adhering; they were of a pale yellow hue, smooth and shining. They were evidently insects allied to those which produce cochineal, lac, and wax. The bodies of these specimens, when applied to the flame of a candle, I found to melt into a sort of wax.*

INVERARY.

The left branch of the mountain road has in many respects the same character of vegetation as the one already described. It is, however, more frequented, and is therefore less romantic. Considerable tracts of Bamboo occur, and the road side is bordered with large beds of *Urena lobata*, a pretty Malvaceous flower, with a profusion of pink blossoms. Within the woods is situated a curious limestone cavern of several chambers, which I have described elsewhere.† Farther to the westward, the forest becomes particularly gloomy, and more humid than any other part that I have seen; hence Ferns and Parasites are

* These have been since described by my friend Adam White, Esq., F.L.S., in the Annals of Nat. Hist., under the name of *Ceroplastes Jamaicensis.*

† Birds of Jamaica, p. 103.

growing in great luxuriance. Handsome *Begoniæ* are numerous, some of which are semi-parasitical, clinging to the trunks of the larger trees, and climbing with a profusion of stems and foliage to the height of twenty feet or more. This is not the usual habit of *Begoniæ*. Within the edge of the woods, which is always much more dense and bushy than the interior, a terrestrial Orchid, which I have no means of identifying, with large irregular tubers, and ample plaited leaves much like those of *Bletia*, grows to the height of three feet. The appearance of the plant is striking, but the blossom, which appears in June, has little beauty.

A tangled path through the woods of what was formerly a coffee property named Bognie, now deserted and allowed to fall into neglect and ruin, leads to a little retired plantation called Inverary, embosomed in the lofty primitive forest. Its elevation is but little inferior to that of Bluefields Peak, and the prospect from it, if not quite so extensive, is more rich and varied. In one direction it takes in Montego Bay on the north side of the island; in another it looks over the capes of Negril and the expanse of blue sea far beyond; to the eastward it extends to the parish of Vere, over four ranges of mountains.

A very noble species of wild Pine (*Æchmea*) was in blossom at the time I visited the place, about the end of January; a magnificent spike of densely set flowers, crimson and purple, shooting out from the tuft of broad sheathing leaves, parasitically growing on the trunks of the great trees. *Epidendrum nutans* was also in flower, and very abundant.

A small garden, in which English vegetables were cultivated, was interesting, and its owner was not a little proud to show it. Cabbages, carrots, parsley, thyme, and sage, were healthy and thriving; and an apple tree displayed a blossom or two. It is curious that our common Plantain (*Plantago major*) is as abundant by the way sides, and in the clearings, on these mountain-summits as it is in our meadows at home. It has doubtless been accidentally introduced, and its seeds dispersed by birds. The negroes call it English Plantain, to distinguish it from the stately Musaceous plant that bears the same name. In like manner Water-cress, apparently identical with ours, grows abundantly in Bluefields rivulet, and forms an agreeable salad all the year round.

Near Inverary there is a deep gully, or narrow rocky glen, between almost perpendicular rocks. The bushes and trees that shoot out of the clefts, and the profusion of long lianes and climbing plants that sprawl over the surface of the precipices, make it very interesting. I endeavoured to go through it; but after advancing some distance, was compelled to retrace my steps, on account of the excessive difficulty presented by the enormous masses of rugged stone heaped one on another along the bottom in the most Cyclopean confusion, as if thrown there by an earthquake.

MOUNTAIN GARDENS.

The provision-grounds of the negroes present an interesting object to a stranger. As I have said, many of the black peasantry inhabiting the lowlands, have their gardens on the summits of the mountain (where the soil is more productive), as well as around their dwellings. The mode of proceeding is as follows. The negro having chosen his spot, in the untouched forest, hires it from the owner of the land, at a certain yearly rent per acre. He then cuts down the timber, piles the logs, and in a dry time sets fire to the piles, much in the same manner as in the United States and Canada: this is called "burning-over" the piece. It now presents a very unprepossessing aspect; the large charred and blackened stumps stand as thick as tombstones in a churchyard; the bare ground is strewn with half calcined stones, unrelieved by a green leaf; and great heaps of ashes lie here and there with fragments of burned wood, the only remains of the giant trunks that once reared their verdant crowns to the skies.

If we visit it in a few months, however, how different is the scene! A large portion of the ground is seen to be occupied by that indispensable West Indian root, the Coco (*Colocasia esculenta*) *, whose

* The term "Coco" is applied in the West Indies to three very distinct kinds of vegetable productions cultivated for the food of man. The first is the Cocoa, which is prepared from the seeds of the Chocolate tree, and forms a well-known beverage. The second is the equally familiar Cocoa-nut, the fruit of a Palm; the third is the thick farinaceous root of an Arum, mentioned in the text.

broad leaves expand themselves, and touching each other form a surface of the most beautiful green hue, which conceals the earth and every thing upon it. The English reader may form a very correct idea of this useful plant from the Cuckoo-pint, or "Lords and ladies" of our hedges, only magnifying the leaves to two feet in diameter, and all the rest of the plant in proportion. From the midst of this sea of great green leaves rise many young Plantain trees in rows, already putting forth the great spike of blossom, which will soon be thickly studded with whorls of close-set fruit. The leaves, five or six feet in length, and a foot wide, are noble objects when entire; especially as they are then of a very brilliant light green hue; but each plant rarely can show more than a single leaf in this condition, the action of the wind soon tearing them up into lateral strips, in the direction of the transverse veins.

In another part of the ground we are reminded of the hop-fields of Kent, or the vineyards of France; for the graceful Yam, a plant not inferior in beauty to either, twines its slender stems up tall poles, and stretches from one to another, making wild natural arbours; while various sorts of pumpkins and melons trail over the ground at their feet. Perhaps a little patch of Sugar-cane occupies one corner; a few bushes of the Castor-oil plant, or of the Cassava, another; with two or three Cotton-trees, not the lowland giant of that name, but the Malvaceous shrub that throws out its snowy bunches of genuine cotton, capable of being manufactured into calico; — but a small tract, carefully cultivated and kept free from weeds, we shall see exclusively devoted to those useful and closely allied plants, the Arrow-root and Ginger; each consisting of succulent green shoots formed by the sheathing leaves, and the former displaying handsome heads of scaly flowers. The rootstocks of Ginger remain long in the ground after cultivation has ceased, and continue to increase and to throw up their verdant shoots. We often used to dig them up in the neglected bush of second growth on Bluefields ridge.

After a few years, the first energy of the virgin soil being somewhat diminished, the ground is thrown up, and allowed to resume its native wildness. Another plot is then selected from the forest, and rented in like manner; the same process of clearing and cultivation is pursued as before, and after a few years this also is relinquished; no attempt being ever made to maintain the fertility of the soil by manure.

No house is attached to these gardens, their owners dwelling, as already said, around Bluefields; a slight hut of logs, however, is sometimes erected, "a lodge in a garden of cucumbers," as a shelter to creep into during the brief, but deluging, torrents that descend in the afternoons of the rainy season; and the floor is strewn with twigs of trees, or "*trash*," that is, the dried leaves of the Plantain, as a rude couch on which the negro may take his customary siesta.

The whole, however, would be incomplete, at least in the opinion of those old negroes, few now in number, who are of African birth, without one plant of little beauty, and of no use, except the imaginary one for which it is planted. It is the Horse-eye bean (*Dolichos ensiformis*) bearing large swordlike legumes, with white oval beans. An old Guinea woman, whom I asked about the matter, after much evasion of the question, and apparent reluctance, at length mysteriously told me that — "if any somebody look into de groun', him make dat him no cut him eye;" that is, the plant will prevent any one from casting an evil eye into the ground. This dread of the evil or "cutting" eye, is a very prevalent form of superstition among this superstitious race, as it seems to have been among the ignorant of all nations, in all ages.

VOICES OF EARLY BIRDS.

April 29th. — I rose some hours before the sun, and proceeded to the Peaks of Bluefields. Passing through the wooded pastures and grass-pieces of Pinnock-Shafton, I was interested in the voices of "earliest birds." While as yet no indication of day appeared over the dark mountain, no ruddy tinge streamed along the east; while Venus was blazing like a lamp, and shedding as much light as a young moon, as she climbed up the clear dark heaven among her fellow-stars; — the Piramidigs or Night-jars were unusually vociferous, and careering in great numbers; they flew low, as I could perceive by listening to their sounds, but were utterly indistinguishable to the sight from the darkness of the sky across which they flitted in their angular traverses. Presently the Flat-bill uttered his plaintive wail, occasionally relieved by a note rather less mournful.

When the advancing light began to break over the black and frowning peaks, and Venus waned, the Peadove commenced, from the neighbouring woods, her five-fold coo, hollow and moaning. Then the Petchary cackled his three or four rapid notes; and from a distant wooded hill, as yet shrouded in darkness, proceeded the rich, mellow, but broken song of the Hopping Dick. Now the whole east was ruddy, and the rugged points and trees on the summit of the mountain-ridge, interrupting the flood of crimson light, produced the singularly beautiful phenomenon of a series of rose-coloured beams, diverging from the eastern quarter, and spreading like an expanded fan aeross the whole arch of heaven, each ray dilating as it advanced. Then Mocking-birds all around broke into song, pouring forth their rich gushes and powerful bursts of melody, filling the ear, and overpowering all the other varied voices, which now helped to swell the morning concert of awakening birds.

PHŒNIX PARK.

Many of the opulent merchants of Savanna le Mar have pleasant country seats, a few miles out of town. At one of these, the residence of a kind friend, I frequently spent a few days; though the neighbourhood was not peculiarly favourable to my pursuits. As Jamaica houses are commonly built on one principle, I will briefly describe it. The furnished part of the house is all on the same level, forming what we should call the first floor, the whole of the ground-floor being devoted to store-rooms and cellars.

An arched passage open at each end leads through the house, beneath the dwelling apartments, from the road in front to the yard behind.

A flight of stone steps, with iron balustrades, on which run beautiful twining and creeping plants, such as the lovely crimson Quamoclit, the wax-like *Hoya carnosa* and others, leads the visitor up to the front door, and he is immediately ushered into a spacious hall, of the form of a cross, extending the whole length and breadth of the house. This large hall is characteristic of all Jamaica houses; it forms the principal sitting room; and, from its shape, admits the cooling breeze to sweep through it, whenever there is a breath of air. The two square areas formed by one side of the cross are filled by bedrooms; but with these exceptions the whole of the sides and ends of the hall are either occupied by windows, or open, and furnished with jalousies, a broad sort of transverse Venetian blinds, which freely admit the air while they exclude the glare of light, which in this country is scarcely less distressing than the heat. This large and cool apartment is furnished with sofas, ottomans, tables, chairs, &c., not differing from ours; but there is no fireplace, nor any carpet. Instead of the latter the floor is made of the most beautiful of the native woods, in the selection of which much taste is often displayed, as also in the arrangement, so that the various colours of the wood may harmonise or contrast well with each other. Mahogany, greenheart, breadnut, and blood-heart are among the trees whose timber is employed for floors. Great hardness is an indispensable requisite in the wood used, and capability of receiving a high polish, which is given and maintained with great labour. Scarcely anything surprises an European more than to tread on floors as beautifully polished as the finest tables of our drawing rooms. The mode in which the gloss is daily renewed is curious: if the visitor should peep out of his bedroom about dawn of day, he would see some half a dozen sable handmaids on their knees in the middle of the floor, with a great tray full of sour oranges cut in halves. Each maid takes a half-orange, and rubs the floor with it until its juice is exhausted; it is then thrown aside, and the process is continued with another. When the whole floor has been thus rubbed with orange-juice, it is vigorously scrubbed with the half of a cocoa-nut husk, the rough fibres of which, acting as a stiff brush, soon impart such a reflective power to the hard wood, as would put Day and Martin into ecstacies. After the last touch is given, it is amusing to see the precautions taken by the waiting maids to avoid dimming its beauty. The preparation for breakfast, and various other duties, performed by servants with bare feet, would seem to make it impossible that the floor should remain untarnished, but it does; and it is thus managed. The girl takes two pieces of linen cloth, and sets one foot upon each; then with her great toe and its next neighbour, she grasps a pinch of the cloth (for the negroes' toes are almost as effective as fingers), and thus scuffles about the floor; practice enabling them to do this with facility, without their feet ever coming into contact with the wood.

A small garden partially surrounds the house, filled with ornamental trees and shrubs, some native and some exotic. The beautiful *Nerium*, called the South-sea rose, is prominent among the latter, as also the gorgeous *Hibiscus rosa-sinensis*, and the Spanish Jasmine, as fragant as it is elegant; and the scarlet Belladonna lily, and many others add to the gaiety of the parterre. In the garden and around it are several curious trees. The noble Malay apple (*Eugenia malaccensis*) or, as it is here called, the Otaheite Cashew, there erects its conical head, covered either with its beautiful flowers, like clusters of crimson tassels, or with its close, luxuriant richly-green foliage. The Sago Palm, likewise transported from farthest India, has found here a climate and a soil congenial to its growth, and presents a singular object in its stiff bristling leaves, radiating in all directions, and its heart covered with a brown woolly or mealy substance. Immediately before the door is a large arborescent *Euphorbia*, probably *E. grandidens*, a native of South Africa, with rather inconspicuous flowers, but sure to attract the attention of a stranger by its long angular leafless branches, set with spines, like long *Cacti* growing from the trunk of an ordinary tree. A row of Shaddock trees, hung in the season with their golden fruits, as large as a child's head, combines beauty, fragrance, and utility; while Cashew trees, Mangoes, Custard-Apples, Sops and Guavas, all valuable fruits, but too common to need description, form groves around the mansion.

Having thus introduced my friend's dwelling to

the reader, I may proceed to make him acquainted with some of the inmates. Scores of Humming-birds hover from day to day around the blossoms of the trees, sucking from flower to flower upon the wing, just as the Hawkmoths do in our English gardens in the summer twilight, or Bees in the sunshine. A Hawkmoth with its long sucker exserted, and plunged into the corolla of a jasmine or a honeysuckle, and with its wings vibrating so rapidly and powerfully as to produce a humming noise, forms a capital representative of a tropical Humming-bird, if the imagination of the observer will only supply the green and gold of the plumage, and the gem-like play of flashing colours on the crest and throat. I have, however, elsewhere described in detail the manœuvres of these elegant creatures around this very Malay Apple-tree*, and shall therefore dismiss them for denizens of the air of a very different character.

THE CHESTNUT MASTIFF-BAT.

Between the ceiling and the shingles of the roof certain Bats find a lodging, emerging at nightfall from a small hole beneath the eaves. Soon after sunset we hear the scrambling of little claws along the plaster, gradually tending towards the point where the hole is situated. At length, just as the stars begin to come out, one by one, in the sky, one of the boldest peeps forth his sombre face, and plunges down into the air, rising with expanding

* Vide 'Birds of Jamaica,' p. 92.

wings at the distance of a foot or two; and sails away to procure his insect-supper. Another immediately follows the first, and at his heels is another, and another, until perhaps half a dozen are out. Then there is a pause for some little time, during which more scrambling is heard, and presently another leader emerges with a "tail" of three or four more in Indian file; — and so on till about fifteen have come forth, — the whole household. I judge that they crawl along one after another in a straight line to the outlet, in parties.

After having witnessed the exit of the party on one or two evenings, and been assured by my friends that the same performance was nightly repeated, I determined to essay the capture of some specimens to ascertain the species. The hole in the eaves was within reach of the top of the steps, where I stationed myself with an insect net about sunset, to the no small interest of the children. We presently heard one slowly scuffling along upon the plaster, making a smothered chirping sound: this enabled us to trace his progress by the ear, and when he seemed to be near the hole, I held up the gauze net in front. The sight of this, however, evidently deterred the Bat, for he remained at the mouth chirping, but without attempting a flight. At length I removed the net to the distance of a foot or two, holding it in act to strike. He now ventured forth, thinking the coast clear, but the projecting eaves compelled him to fly downward, and I was alert enough to receive him into the bag of the net. It proved a *Molossus*, of chestnut-coloured fur, and rather large size; the *M. rufus* of Geoff., as I believe.

Several other specimens I obtained in the same way, on the same and other evenings; but the residue evidently became increasingly suspicious, and loath to leave the hole while I was watching. Wishing to examine one alive more carefully, I allowed it to remain in the bag of the gauze-net all night, simply giving the net a twist over the rod. In the morning, however, to my great chagrin, I found the Bat almost entirely devoured by Rats, which are here very bold and voracious. However, I had other opportunities of studying the habits of the species in captivity.

When handled, its impatience of confinement is manifested by a continuous screeching, not very loud, but excessively harsh and shrill. The ears are commonly so pendent as completely to cover the eyes; but they are occasionally contracted, so as to expose the eyes, especially if the face be touched. The mouth also is then opened widely and threateningly, and a sufficiently grim armature of teeth developed. The volar membranes are narrow in proportion to their length. There is a gland on the throat with an orifice capable of admitting a small straw; it emits a very rank odour. This orifice is manifest in the female no less than in the male, but I could not detect any odour from it in the former, even though irritated by the insertion of the head of a pin. The duct does not penetrate farther than the gland itself.

A female which I kept a few days was very sullen. I offered her cockroaches, a *noctua*, and a longicorn beetle, but she would not eat: she would seize and bite them spitefully, but presently dropped them. She would not drink when held in the hand, and the muzzle presented to water; but if it were allowed to fall in drops upon her face, she would suck them in with a motion of chewing, with apparent satisfaction. There was much of mastiff-manners, as well as physiognomy, in her; she often bared her teeth by contracting the sides of the lips, and watched with open mouth to seize any object presented, which she then held with surprising force and pertinacity.

I incline to think that *some* Bats at least are crepuscular, and not nocturnal. The family assured me that after these Mastiff-bats had emerged a few hours, they invariably returned into the hole again; aud they several times directed my attention to them when returning. On looking out, I indeed saw that they were flying up towards the hole in the eaves, but, on account of the darkness, I could not be sure whether they entered. One moonlit evening, however, looking through the window, between eight and nine o'clock, I distinctly saw one, after flying up and hovering a minute or two, and then coursing round again, twice or thrice in succession, — I distinctly saw it enter the hole, whence it emerged no more, at least while I continued to look at it. I asked Sam's opinion, without intimating to him what I thought. He without hesitation affirmed that they do not fly all night, stating that when he had been abroad through the whole night, grinding cane, as was often the case, he had noticed Bats numerous a little after the evening, and again a little before the morning twilight. And when fishing on the reef, or a short distance from the shore, he had often seen them in company with Piramidigs (*Chordeiles Virginianus*) at the same hours, but not during the middle part of the night. Yet I have observed, when the moon is at the full, small Bats (*Phyllostomes*, I suspect), flitting round the house as late as ten o'clock, or even later.

All Bats are called by the negroes Rat-bats, probably to distinguish them from Butterflies, to which they give the name of Bats. A little sable urchin came to me one day, and asked me in evident sincerity, "Do you want to buy any Bats, Massa?" Then opening his hand he displayed two of our most common butterflies, squeezed flat, rubbed clean, and one of them deprived of two of its wings! what price he intended to charge, I did not ascertain.

AMPULLARIA'S EGGS.

April 5th. — In the little rapid stream called Sweet River, I found several specimens of an *Ampullaria* alive, and many groups of the eggs of this mollusk. They are laid, for the most part, in a double row, attached by a glutinous substance to the stalks and leaves of plants overhanging the water, but not immersed. The eggs are oval, shelly, pure white, and nearly as large as sweet peas. In many I found the shell and operculum of the young animal, perfectly formed; in others only a white cream, or curdy substance.

LUCIFUGOUS INSECTS.

April 24th. — Under a stone by the roadside at Sabito Bottom, I found a Centipede performing the duties of a mother. It was a blue species, about three inches in length; it was lying in the form of a bow, the head and the tail curved forwards toward each other, almost on its back, the curved body embracing some ten or fifteen eggs, which slightly cohered. The eggs are oval, about as large as hemp-seed; pale-yellowish, pellucid, and under a lens perfectly glossy, with a membranous integument. The parent on being disturbed, darted away among the stones, leaving the eggs, so that I did not capture her. I brought home the eggs, and having taken out a few for preservation, placed the rest carefully on moist earth in a phial, hoping to rear them. They soon, however, became covered with mould, and decomposition destroyed them. The mother's care is perhaps indispensable, as in the case of ants, regulating the admission of heat and moisture to them, according to circumstances.

Scolopendridæ of this size and under, are common enough beneath stones; but the larger species are very rarely met with. Scorpions also, of two species, one with slender claws (perhaps *Tityus griseus* of Koch), the other with these organs much thickened (probably an *Atreus*), are also common under stones; they are inert when uncovered, and almost helpless; and are not at all dreaded.* In the same situations is found

* Respecting a curious organ peculiar to these animals, of which

a little spider with round glossy abdomen, black, with three scarlet spots near the extremity (probably a *Theridium*); this is held in great terror as highly venomous. About this time I find them carrying yellow membranous egg-bags, globular, thin, and *empty*. When uncovered they are unwilling to run, keeping close to the under surface of the stone that has served them for shelter and concealment.

A much larger Spider, of the genus *Lycosa*, inhabits houses and heaps of stones; in both of these rather dissimilar situations, it retains the same habits: the female carries constantly about an egg-bag of buff-coloured silk, perfectly circular, rather more than an inch in diameter, but flat, being about two lines in thickness, with a narrow thin edge all round. The spider crawls about with this nest beneath her body, her feet spreading around its edges; and nothing will induce her to part with it.

That curious flattened Arachnidan that looks like a Scorpion without a tail (*Phrynus reniformis*), with the second pair of feet so produced and attenuated as to resemble lengthened threads, is not uncommon

the use is so little understood, Mr. Hill offers the following suggestion. "The comb-like appendage under the corslet of the Scorpion, so little noticed by naturalists, and when noticed, never determined as to the part it acts in the animal's economy, seems to me very obviously to relate to the deficient tactile powers of these *Arachnida*. I take it to be the most sensitive part of the Scorpion; all other parts, as in the *Articulata* generally, being insensitive; — that these parts are used in the same manner as insects use their antennæ, for tactile purposes; and that they are necessary in this animal, that he may acquaint himself with the character of the body into which he is to inject his poison."

beneath stones in the vicinity of Bluefields. It appears inert when exposed, and perhaps hopes to escape observation by remaining motionless; on being touched, however, it shuffles away rather nimbly.

In the damp woods on the summit of Bluefields Ridge I find under stones the genera *Polydesmus* and *Julus*. The former are difficult to preserve, from their excessive fragility when dead. By the time I arrive at home, I commonly find the specimens, collected an hour or two before, completely disintegrated, the box presenting only a heap of fragments; and those segments that yet remain entire, separating at the slightest touch. They are inert animals, of slow motion. The living *Julus* when touched, gives out at the part, a brown fluid oozing from beneath the segments. This is of a rank pungent odour, and stains the fingers of a deep yellow hue.

SINGING-BIRDS, AND SWEET FLOWERS.

"In tropical countries, where brilliant and varied colours have been granted to the birds and flowers, song has been denied to the one, and fragrance to the other." This is one of those flippant generalisations which people are fond of repeating, originally made without investigation, and perpetuated without inquiry. In Jamaica it is certainly very far from truth; and I suspect would be found as groundless everywhere else. The groves and fields of this sunny isle ring with the melody of birds, to a degree fully equal, in my judgment, to that of Europe. In the lone forests of the mountain-heights the Glass-eye Merle

(*Merula Jamaicensis*) pours forth a rich and continued song; and that mysterious harmonist, the Solitaire (*Ptilogonys armillatus*) utters his sweet but solemn trills, long-drawn and slow, like broken notes of a psalm, so perfectly in keeping with the deep solitude. In the woods that cover, as with an ever-verdant crown, the lower hills, the Black Shrike (*Tityra leuconotus*) and the Cotton-tree Sparrow (*Pyrrhula violacea*) enunciate their clear musical calls, so much alike as scarcely to be distinguished; four or five notes running up the scale, so rapidly as to be fused as it were together, and suddenly falling at the end. Here, too, sits the Hopping Dick (*Merula leucogenys*), and whistles by the hour together a rich and mellow succession of wild notes, clear and flute-like, like his European cousin, the Blackbird. The constantly reiterated call of the Red-eyed Flycatcher (*Vireosylva olivacea*), "John to whip! John to whip!" heard at different distances from all parts of the woods, makes their green glades lively; and the loud varied voice of the White-eyed Flycatcher (*Vireo Noveboracensis*), sometimes soft and subdued, sometimes shrill and piercing, is always heard with pleasure.

But birds are particularly social animals; and it is chiefly in the neighbourhood of the presence of man, that their melodious voices are heard, as if to cheer him in his toil: the fields, and pastures, and meadows; the hedges, and hedge-row trees, that border and map out his domains; the orchards and groves that surround and embosom his dwellings, affording grateful fruit and shadow from the heat;—these are the situations in every inhabited country that most resound with the voices of feathered songsters. The beautiful park-like estates of the southern slopes of Jamaica present scenes peculiarly inviting and suitable for the winged orchestra to exercise its vocal talent; — and the notes of melodious joy are pouring forth in them from earliest dawn to sunset; — aye, long before dawn, and long after the veil of night has been outspread. The Swallows (*Hirundo pœciloma*) that shoot along in their arrowy traverses over the plains, now darting across the placid stream, now coursing far up in the thin air, almost lost in the glaring sun-beam, twitter sweetly as they pass, and now and then one and another sitting on the summit of a low tree commence a stammering song by no means deficient in music.* The Blue Martins

* "Ordinarily, in our livelong sunshine, our House Martin (*H. pœciloma*) is seen careering about his haunts, — in and out of caverns in the solitude of cliffs and rugged hills, and in and about sheds and galleries, in our social dwelling-places. When the *vernal* equinox has blown over, and has brought fitful showers of rain, and the House Martin has profited by the little puddles round about to collect mud to patch and extend the stucco-work of his grotto, he chatters to his mate between-whiles as he toils, a low muttering song, very guttural, and just barely musical; and he continues this twittering talk during the unwearied hours he spends by her side, and with her nurslings, all through the *summer*. When, however, the *autumnal* season gathers stormy, and the overcast sky prepares us for deluging rains, he changes his habit, and with his habit his very voice. He then quits the cavern or the shed, and making a party of four or five, is seen perched on the upper dry limb of some neighbouring tree, singing a loud-voiced song, so different in tone and character from anything you may hear from him at any other time, that you cannot recognise the same bird in the wild and deep-toned ecstasy of what is then his musical humour. My attention was first called to this peculiarity last year by some friends who noticed the unusual song, in a road that

(*Progne Dominicensis*), too, sit side by side in close rows on the dead frond of some tall palm, or on the

wound beneath some cliffy hills, and were surprised to discover that it came from the common red-gorged Martin. Since that,—in Spanish-town, — on the trees about the government buildings, I have observed him constantly, at this time of the year, after careering about, perching and pouring out a loud ecstatic song, quite different in tone and manner from his ordinary lowly twitter. His music arrests just as much attention as a solo from the Mocking-bird does in such a place. The singing is full of stammering cadences, continued and repeated with vehemence. When one bird has poured out his fit, another quits the wing, and perching near by, delivers himself of a similar strain of ecstasy. They sing but one at a time, the company on the tree-top being only listeners. These vehement bursts of song continue only during the tumultuous rains in the latter months of the year. When these are over the rhapsodial frenzy ceases, or only very casually occurs. After the weather is again tranquil, and the atmosphere assumes that unrivalled purity, which prevails during our *winter*, and the air is cool, and the sky and the earth fresh and beautiful, — the Martin is observed to resume his usual gentle habit; and to twitter again his lowly muttered song in his customary galleries and sheds." — *Letter from Mr. Hill*, 20th Nov. 1846.

"On occasional mornings, lately, I have heard the House Martin, but only a single bird at a time, singing that loud peculiar vehement song, with some fine clear musical tones in it, which I had described as the rhapsodial humour of Autumn. The seasonal outburst just now is, however, not so long sustained as that of Autumn. The ecstasy is less absorbing, and there are no congregated listeners on the tree-top, but the singer sits as he did when he sang to the October rains on the uppermost and outermost dry limb, and delivers himself of his vehement soul of melody; and then he takes to the wing, and making three or four circles, scuds under the eaves of the neighbour building, to join the lodgers and loiterers within doors." — *Ibid.* Feb. 20th, 1847.

In these communications, my friend has since informed me that he has confounded the Cave Swallow with the Blue Progne. Both species inhabit the public buildings in Spanish Town, the former exclusively tenanting the lower, the latter the upper story, neither species intruding on the domain of the other.

roof-ridge of the dwelling-house, and utter a shrill but not unmelodious chant. From the green tussocks of the Guinea-grass fields comes the singular hollow cry of the Tichicro (*Coturniculus tixicrus*), and now and again he runs to the summit of a stone, or jumps upon a wall, and warbles a sweet and low song. The clear whistle of the Banana-bird (*Icterus leucopteryx*), like the tones of a clarionet, resound from the fruit trees, among whose deep green foliage his gay hues, rich yellow, white, and black, glance fitfully as he shoots to and fro; and his companions, the little Blue Quits (*Euphonia Jamaica*), equally devoted admirers of a ripe sour-sop or custard-apple, accompany his loud notes with strains of their own, full of soft warbling music. And the most minute of birds, the tiny Vervain Humming-bird (*Mellisuga humilis*), not larger than a school-boy's thumb, utters a song so sweet, but of sounds so attenuated withal, that you wonder who the musician can be, and are ready to think it the voice of an invisible fairy; when presently you see the atom of a performer perched on the very topmost twig of a mango or orange-tree, his slender beak open and his spangled throat quivering, as if he would expire his little soul in the effort.

But there is one master-musician, whose varied notes leave the efforts of his rivals at an immeasurable distance behind him. It is he that makes our sunny glades and shady groves eminently melodious, by night and day, sustaining almost the whole burden himself.

Δενδρέων ἐν πετάλοισι καθεζομένη πυκινοῖσιν,
Η τε θαμὰ τρωπῶσα χέει πολυηχέα φωνήν.

Odyss. xix. 520.

He is the Nightingale of the Western World, the many-voiced Mocking-bird (*Mimus polyglottus*). Abundant in almost all situations, from mountain-peak to sea-shore, but especially common in the orchards and about the homesteads of the lowlands, the voice of the Mocking-bird is heard all through the year, even when other birds are silent; and all through the day; and that not by ones, or twos, but by dozens and scores, each straining his melodious throat to outsing his rivals, and pouring forth his full, expressive strains in all the rich variety for which this inimitable songster is so famous. Wilson has truly observed of this delightful bird, that "the ear can listen to his music alone, to which that of all the others seems a mere accompaniment." If all the birds of Jamaica were voiceless, except the Mocking-bird, the woods, and groves, and gardens would still be everywhere vocal with his profuse and rapturous songs.

In those brilliant nights, when the full-orbed moon shines from the depth of the clear sky with such intensity that the eye cannot gaze upon the dazzling brightness of her face, shedding down on plain and sea a flood of soft light sufficient to enable one to read an ordinary book with ease in the open air,—how sweet, how rich, how thrilling, are the bursts of melody that rise from the trees around, the serenades of wakeful Mocking-birds. Nothing to be compared to it have I ever heard in England; the night-song of a single bird, however fine may be its execution, is no more to be put in competition with such a chorus, than the performance of a single musician, though a master, with that of a band. Nights so lovely are seen only in the tropics, and the music is worthy of the night.

The Water-Thrushes (*Seiurus*), one of which at least is said by Wilson to be so exquisitely sweet a songster, that he was never tired of listening to it, though common in Jamaica, during a portion of the year, I have never heard sing, perhaps because the months that they spend in the island are those of autumn and winter. But the Wood-Thrush (*Turdus mustelinus*), or May-bird, as it is provincially called, is recognised as a songster rivalling even the Mocking-bird in the brilliant execution of its melody. This sober-coloured, but delightful bird does not extend, so far as I am aware, to the neighbourhood of the Bluefields, in its transient vernal sojourn; but confines itself to the sea-side groves and plains of the windward end of the island.*

* "On the morning of the 10th [of February] I heard a Wood Thrush singing his measured clear-voiced song from among a lofty cluster of trees that border a ravine near by [Spanish-Town]. The hills to which these Thrushes almost exclusively confine themselves in this and the neighbouring parishes, are the low range of mountains along the sea-shore. They come, however, occasionally, within the glens at the foot of the mountains inland, passing over the plains between, with only just resting to pour out a morning song, in some sequestered knots of trees that dot the intervening savannas. In these excursive rambles away from the customary sea-side haunts, it is our chance to hear them near about the town. The Wood Thrush sings always from some lofty exposed branch, having a thick shelter of foliage beneath his perch. His tone is clearer and louder than that of the Mocking-bird, and when he sings, the latter is a listener; not a single one attempting a strain of rivalry. The Wood Thrush's song, taken up immediately, is constantly trilled by the Mocking-bird for some days afterward, and is his peculiar mimic-melody for March and

Besides all these, which, in various measure, perform their parts in the music of our woods, and not to mention the multitudes of Warblers, and Fly-catchers, and Finches, whose notes, insignificant in themselves, help to swell and vary the general harmony, — there is another series of voices that must by no means be overlooked in an enumeration of our woodland music, — the plaintive cooings of our numerous wild Doves. In the recesses of the mountain-forests the silence is broken by the loud hollow calls of the Ring-tail and Blue Pigeon (*Columba Caribbea* and *rufina*), and by the mournful cadences of the lustrous Mountain Witch (*Geotrygon sylvatica*). The woods, that densely clothe the inferior summits, and sheet the sides of the sloping hills, resound with the energetic coo of the Baldpate (*Col. leucocephala*), the short reiterated moans of the Partridge Dove (*Geotrygon montana*), the querulous call of the Ground Dove (*Chamæpelia passerina*), and the tender, melancholy, sobbing fall of the gentle Whitebelly (*Peristera Jamaicensis*).

But, as it is in the lowland plains and cultivated

April. I had attempted to write down some of the very marked ecstatic cadences of this song, long before I met with Nuttall's description of it; and I had, like him, resolved the sounds into *a-vree-u*, and *a-vilhia*, frequently repeated. My spelling, however, differs from his, but we work out the same sounds. These are a part of his song only; the intermediate passages are surpassingly sweet, and all the tones, though clear, are mellow, and flute-like, and exceedingly harmonious; — and sustained with an agreeable flow of melody. The bird on the morning of the 10th, sang for a full half hour in the cluster of trees within the ravine." — *Letter from Mr. Hill*, Feb. 20th, 1847.

estates that we most abundantly hear the melody of singing-birds, so here do the plaintive voices of the Doves fall most frequently upon the ear. The Whitewing (*Turtur leucopterus*) and the Pea-dove (*Zenaida amabilis*) are essentially lowland birds; and these, with the exception of the Whitebelly, are the most incessant, and the most tender of all our cooers. Not, however, that we hear their voices immediately around the homestead; when they come into the open pastures to feed, they are usually wary and silent; but from the surrounding woods, the tall thick trees of the pens, and especially the impenetrable mangrove-morasses, their loud, but sweetly gentle, moanings fall with a mournfully pleasing cadence upon the ear. The Pea-dove's voice is the more tender, and is particularly prominent in the evening, while the blustering sea-breeze gradually lulls itself to repose; the longer, sharper and more impatient call of the Whitewing is most heard in the morning, though each season brings the notes of both birds, from all parts of the woods around. They are respectively characteristic of the quietude of the late and early hours.

> "Nec tamen interea raucæ . . . palumbes,
> Nec gemere aëria cessabit turtur ab ulmo."
>
> VIRG. *Ecl.* i. 58.

None but those who have listened to these gentle voices can tell what an effect they produce upon the mind. Their tender melancholy communicates itself to the hearer; and though reason tells him that they are the expressions of buoyant joy and health he can scarcely fail to feel a pang of sympathy for what seems to be the complaint of gentleness in distress.

Nor is it true that our groves and fields are destitute of fragrance. In spring the Oranges and Limes that are planted in such profusion upon every estate, both on mountain and in plain, and even border the public roads, are covered with their abundant blossom, and the air all around is loaded with the richest perfume. So it is in the upland districts, when the Coffee plantations are in bloom, the flower of which tree is as fragrant as it is delicately beautiful. In the edges and borders of woods there is a common shrub called Wild Coffee (*Tetramerium odoràtissimum*), nearly allied to the cultivated species botanically, as it is both in beauty and fragrance. Butterflies, Moths, Bees, and Flies throng around its lovely white blossoms, the delicious and powerful odour from which is diffused to a great distance.

I have observed that many flowers in Jamaica possess the aromatic odour so much admired in our pinks and carnations, that of the Clove. The beautiful plants called the red and the white Spanish Jasmine (*Plumieria rubra* and *P. alba*), common shrubs, whose thick stems, leathery leaves, and noble spikes of blossom form so striking objects in the smaller woods, have this odour. I found it in the blossom of a species of *Pancratium*, with small bulbs and large oval leaves, growing on the St. Elizabeth's mountains; the fragrance, which was very abundant, I should not have been able to distinguish from that of a carnation. That gorgeous flower, the Night-blowing Cereus (*Cereus grandiflora*), is another notable example of the same prevalent odour. The long trailing stems of this Cactus are very commonly seen in the lowlands, sprawling to a great length over the stone fences, hanging in irregular festoons from the forks and limbs of the trees, the great Cotton-tree in particular, and intertwining its tough and prickly vines among the shrubs, helping to give the woods that formidable, repellent, impenetrable character which a tropic "bush" is known to present. The magnificent flowers are, however, rarely seen; the plant seems to be a shy bloomer; and when the blossoms do meet the eye it is in nine cases out of ten either as unexpanded buds, or in that miserable drenched condition, which the flowers of a Cactus always assume when fading, looking exactly as if they had been dragged through boiling water. In order to see it in perfection, one must make it open in the house, or visit it at midnight, which is inconvenient. I have several times marked a maturing bud, and when it appeared nearly ready to burst, cut a few inches of the stem on each side, and brought it within doors. Soon after dark it begins to open, and towards midnight expands in its noble beauty, a disk six inches in diameter, very double, the exterior rows of petals of a yellowish-brown hue, gradually paling in tint to the centre, where the petals are of the purest white. Meanwhile the delicious clove-like perfume is diffused in such abundance, that a delicate person can scarcely sit in the room, and the very house is filled with it from one end to the other. In the morning beauty and fragrance are both gone, and the blossom, lately so gorgeous, possesses no more of either than may be pretended to by a boiled cabbage.

THE CHIGOE FLEA.

May 24th.—About this time the Chigoe or Jigger (*Pulex penetrans*) is numerous and very annoying. These parasitic Fleas may be seen hopping about among the dust of sheds and similar places, and the naked feet of the negroes suffer constantly from their attacks. But even the stockings and shoes of Europeans are not proof against the insidious assaults of this tiny flea. On several occasions I have found them ensconced in my feet; to-day I discovered that one had chosen the bend of my little toe as the scene of its domestic economy. The negroes, from mutual practice on each other, are quick at discovering, and skilful in extracting them; and, accordingly, to one of my servant lads I entrusted the operation. Taking my foot on his knee, he began with a sharp needle to open and widen the minute orifice in the epidermis, between which and the cutis the swollen body of the pregnant female had taken its place. Slowly and cautiously the lad exposed the depredator, giving no pain, and not drawing the least drop of blood, until at length he removed the insect uninjured. The great danger to be guarded against is the rupture of the delicate skin of the Jigger's abdomen, stretched and attenuated as it is by the great increase of its contents. If this should occur, the nits would escape into the wound, and produce a dreadful ulcer; such, however, is the skill of the sable practitioners, that it very rarely occurs. The negroes talk of two kinds, the White and the Poison Jigger. Mine was of the latter kind; and therefore a little grease was rubbed into the empty cavity, after the operation.

The insect, when removed, on being subjected to examination with a lens, greatly resembles a distended Tick (*Ixodes*); white and shining; there is this dissimilarity, however, that whereas in a Tick the little feet are pushed apart in the process of distension, and are scattered widely around the swollen abdomen, in the Jigger they remain close together in the centre; in other words, the skin of the abdomen between the bases of the feet is expansible in the former, and not in the latter. The feet sprawl helplessly after extraction.

The presence of a Jigger beneath the skin, during the process of its gradual increase, is commonly described as a titillation, rather pleasing than painful. This does not at all agree with my experience. I on no occasion felt any itching, but, as soon as I became conscious of any sensation at all, a dull pain with tension, somewhat like the rising of a small boil, which increased until the cause was removed by extraction.

THE SMOOTH SHEATH-CLAW.

June 3rd. — Coming down from Bluefields Peak about noon, my attention was attracted to two of these Geckos (*Thecadactylus lævis*), resting near each other on the trunk of a tree, beside a deep cleft. From a slight difference in their size, I supposed them male and female. I struck at one with my riding switch, but failed to reach it; and they both glided into the hollow tree. Soon, however, they began to peep out again; and I desired my servant to mount the tree in order to drive them out; which he did by thrusting in a stick at the opposite side. The larger darted out, and on being touched, precipitated himself to the ground, where he began to wriggle with the usual awkward agility of the species; I struck him lightly with a stick, but in so doing unfortunately severed the tail from the body. The amputated member, however, continued for some time to writhe in rapid contortions. I now put down a tin canister on the ground with the open mouth just before the animal's head, and on being touched behind, it darted in, according to its natural propensity to take refuge in dark holes and crevices; and I thus secured it. Its colour had been till now dark brown, handsomely mottled with black, but on my arrival at home, I found it of a dirty white or drab, with the mottlings few and almost obsolete. * The stomach was found on dissection to contain fragments of beetles, but in the intestine was a leguminous seed.

The appearance and physiognomy of this Lizard are unprepossessing; there is a savage sullenness in

* In my little volume on "REPTILES," (Soc. Prom. Chr. Kn. 1850), I have given a figure of this, as well as of some other Jamaican species, engraved from my own drawings.

its look very different from the meek countenance of the *Ameiva*. It is very common, particularly in out-buildings and offices, where it inhabits crevices in the roofs and rafters, a pair commonly living either in the same hole or near together. On the approach of night one hears on all sides the singular, cracked, cackling call of these animals, somewhat like the sound produced by drawing a stick across a comb. M. Duméril's suggestion, that this voice may be produced by the tongue smacked, as it were, in the concavity of the palate, is the less unlikely from the fact that this organ is large, flexible, and fleshy. The name of Croaking Lizard commonly applied to the species in Jamaica is derived from its peculiar voice. In the woods the voice is also heard at night proceeding from hollow trees, and continued through the whole of the hours of darkness. The large prominent eye without any eyelid, whose pupil contracts in strong light to a perpendicular line, indicates their nocturnal habits; yet they are frequently seen by day, as in the case just mentioned. In the old mill-house at Bluefields they are numerous; and two or three pairs may be seen day after day at the same spot, peeping out of their crevices, and remaining perfectly still for hours. Sometimes they venture forth, and may be observed crawling slowly along the beams and rafters; moving with excessive deliberation, and never going far from their holes, into which they dart on the least alarm with swift rapidity.

The curious structure of the feet in the *Geckotidæ*, by which they are enabled to walk on reversed surfaces, has been often described. The toes in this species radiate from the foot, are dilated into broad oval disks, and have the under surface covered with transverse laminæ, the edge of each overlapping its successor. Minute hooked claws, very acute, like those of a cat, doubtless assist the animal in its feats of this kind. I do not, however, remember to have ever seen this Gecko actually back downward, but often crawling on the vertical side of a beam.

The skin of this animal is very loosely attached to the muscles; and is so soft and fragile that it tears with a mere touch, like wetted brown paper. The head and back are covered with minute conical tubercles very closely set, which, on the sides and towards the tail, become more inclined, and flattened in a posterior direction, so as to form overlapping scales, which are most regular and largest on the belly and tail; each is in contact with six other surrounding ones. The under side of the tail is crossed by a series of broad plates, here and there varied, however, by two broad scales supplying the place of one plate. The tail has no transverse folds, but the scales are arranged with regularity there. I have not been able to detect any femoral pores.

The periodical casting of the skin takes place much as in other Lizards. The head and fore parts begin to assume a white appearance, and the next day the skin of these parts separates from the surface of the body, often irregularly, and lies in loose ragged folds around it. It now looks like muslin of the most delicate fineness: it is slit down the back and separated, but not yet thrown off. At the same time the hinder parts have the same whiteness as the head on the previous day; for there seems always to be the difference of a day in the sloughing of the fore and hind parts. When the cuticle, however, is manifestly detached, it is not thrown off at once, but hangs around the Lizard like a ragged garment, for several days; apparently to its no small annoyance.

The reproduction of the tail in Lizards, after it has been accidentally lost, is a very curious phenomenon, which seems not to have been observed with sufficient precision. In this species it takes place with great rapidity. The facility with which the tail separates has been already alluded to: it is said that the animal will frequently cast off this member spontaneously in its contortions on being put alive into spirits; and that the contraction of the tail into a globular form has given occasion to the supposition that a distinct species existed, called the Turnip-tailed Gecko (*Thecadactylus rapicauda*); this, however, I have not seen.

One day, at Grand Vale, I observed, on a gate, a Gecko with a new tail, not more than an inch and a half in length, abruptly tapered. The animal had a singular appearance, the tail being of a bluish grey hue, marked with longitudinal black stripes; it had a silky gloss, but was closely covered with minute transverse wrinkles. (The ordinary length of the tail when perfect is about five inches.)

About the middle of September, I caught in a noose one which I had deprived of its tail a few days before in attempting to secure it. The separation had taken place about half an inch behind the vent.

I put the reptile into a gauze-covered box for observation. In less than a week the new tail was manifest in the form of a bluish tubercle projecting from the centre of the wound, the edges of which had, in drying, shrunk up, and so lost their sharpness. The tail slowly increased; about the end of October it was an inch long, the base nearly commensurate in diameter with the wound. About this time I captured one with a renewed tail, which member *was covered with tuberculous scales* as the original had been, and the inferior surface of which displayed *the ordinary transverse plates*. In fact I should not have known that it had been severed, but for the dark grey colour, the peculiar character of the striping, the manifest suture at the point of junction, and the smaller size than normal of the scales and plates. On comparing the tail of my own living one with that of this specimen, I perceived that it differed *in the absence of scales*, the surface being silky, and covered with fine transverse wrinkles as the one observed at Grand Vale. On the 10th of November the new tail was about an inch and one-eighth long; when it threw off its skin, the sloughing being confined to the recent part; and I was surprised and pleased to observe that *the new surface displayed both scales and transverse plates*, but both small. The colour was still dark grey, with pale irregular longitudinal stripes or dashes. About the beginning of December the animal escaped, the cage having been incautiously left open: the tail was then fully an inch and a half long, *and the tip had become compressed.*

On the 21st of October I had found adhering to the inside of the door of the cage, an egg of a short-oval form, shelly in texture, and of a pure white hue. It adhered to the wood by a flattened base, as if it had been deposited in a soft state; when I saw it the top had been slightly crushed in, probably by accident; and this appeared to have been fatal to it, for its contents gradually dried up. I conjectured at the time that this egg had been deposited by the Gecko, but I could not be quite certain, because one or two Anoles were kept in the same cage. But on the 21st of February, as I was riding to Savanna-le-Mar, and passing a large Fig-tree that overhangs the sea-shore at Cave, I observed in a little crevice in the trunk, — or rather in one produced by the singular anastomosing of its supra-terrestrial roots,— several eggs. On dismounting I found they were about eight in number, and evidently of the same kind as the one above noticed. Their form was irregularly oval, round, or rondo-triangular, all flattened, very much resembling in size, shape, and colour, those comfits called *lemon-drops:* their greatest diameter was about half an inch. They were shelly, but the shell was thin and very brittle; yet it was evident that they had been soft when laid, for they adhered to each other, and the side of one was, as it were, *let in* to that of its neighbour; and at the base of some was a thick mass of shelly matter, as if a semifluid substance had run down, and then hardened. That all had not been deposited at once seemed apparent, from their diverse degrees of maturity; some containing only a white cream or soft curd, others the Lizard fully formed, but immature; while one, which I broke, displayed the Smooth Sheath-claw perfectly formed and coloured, which presently crawled out, being quite independent of the vitellus. The head of this new-born Gecko was large; the belly (of course) thin; the toes well formed; the tubercles perfect; the markings dark and beautifully distinct; the pupil linear and perpendicular. The length was two inches and a quarter; one-third of an inch of the extremity of the tail was pure white, abruptly defined from the darker colour. The surface of the eggs, under a lens, was covered with minute tubercles of lime, something like a whitewashed wall.

The cavity in which these eggs was found was so narrow, that I cannot understand how the Lizard had managed to deposit them in their situation, unless (which seems very improbable) they were first laid, and afterwards placed with the feet. After a few weeks, being again in that neighbourhood, I noticed that another egg had been laid since the removal of the former, and in the course of another month two more, occupying the place of the first deposited. We thus perceive that the Gecko lays her eggs at considerable intervals of time, but selects the same spot for their deposition.

The ill fame under which this genus labours in most parts of the world where it is known, attends it here: an indefinite dread of its being in some way hurtful generally prevails, perhaps mainly dependent on its repulsive aspect and stealthy motions. It is, however, perfectly harmless; feeding on insects and berries. I have already mentioned the contents of the stomach in one specimen that I examined; in another this viscus, a membranous sac, nearly as long as the trunk, contained nothing of an animal nature, but only some pulpy berries, and several stony seeds closely like those of a grape. One which I captured with a hair-noose, indeed, on being turned into the lizard-box, seemed to manifest hostility to the Anoles which were there; these fled from it, and the Gecko appeared to have a desire of pursuing, for it crept towards one and another with a stealthy step, the belly and head being on the ground, like a cat watching a mouse. Yet these actions might be nothing more than the manifestation of caution on being put into unwonted circumstances. While thus engaged, it ever and anon licked its lips with its protruded tongue; an action common to all our *Geckotidæ.* The ordinary length of the species is about nine inches.

THE POND TURTLE.

Between Shrewsbury and Content is a pond of about one fourth of an acre in extent, choked up with *Potamogeton,* and covered with Duckweed. It is embosomed in the high woods which rear themselves around it on every side; a giant Cotton-tree rises from the margin, and the long branches of a vigorous Bastard-cedar reach almost quite across it. A luxuriant gourd called the Duppy's melon *, has

* Duppy is a spiritual personage of whom the negroes are terribly afraid; it might be rendered "hobgoblin." The habit of naming

entwined in the limbs of the cedar, and sends down a hundred vines which dip their tangled mazes into the water. Cattle resort hither to drink, and hogs to roll in the cool mud; water-insects swarm in it, and pond-turtles are abundant; yet this is the only water drunk by several families in the dry season. The negro girls of the neighbourhood come hither, each with a large shallow basket or tray of wicker on her head, in which stand many calabashes. She wades into the pond, pushes aside the duckweed, and fills her vessels, which are then again transferred to her head, and carried home to settle. The water is turbid and tepid of course, yet not unwholesome; or else custom has habituated the people to its influence. But it is with the inhabitants of the pond that I have to do: on the logs and branches that from time to time have fallen from the overhanging trees, and now project here and there from the green surface, the Turtles (*Emys decussata*) may be seen sitting to enjoy the sunshine in the heat of the day. Some are as large as a dinner-plate; others no bigger than a crown-piece; but all are watchful, though still: the head is protruded and elevated to the utmost stretch of the neck, and the eye maintains a sharp look-out on the shore. Throw in but a stick, — and every one is gone; each has dropped from his seat into the water without a sound, and almost without a ripple on the surface. If we sit down for a few

things that are considered uncouth, or in some way unpleasing, as *Duppy's* this or that, is analogous to the practice which prevails in our own country, of appropriating things to the Devil; as "Devil's Squeaker" (the Swift), "Devil's Coach-horse" (the Rove-beetle), &c.

moments, remaining quiet and still, — that is, as still as the musquitoes will permit, for they are tolerably regnant in such a situation, — we soon see here and there the duckweed open, and bubbles of air ascend: presently a little black head projects, so little that you will hardly see it if you are not intently watching; but if you watch you will observe the little creatures crawl up on their logs again, and sit as before.

The flesh of these animals is esteemed superior even to that of the famed Green Turtle, by those who have abundant opportunities for judging of both; hence many are caught for the table. The pond is, however, deep and boggy; and recourse is had to the trap. As I entrapped two or three here myself, I will describe the mode of procedure. A long pole is prepared, such as the stem of a young tree in the contiguous forest, to one end of which is attached a flat piece of wood as a float. A common rat-gin is likewise lashed to the end of the pole, and baited with a piece of flesh. A small bird with the feathers singed in the fire is reckoned a very successful bait, its odour attracting the Turtles. The extremity of the pole thus armed is carefully pushed out into the deep water, and the other end is fastened with a line to some object on shore. The weight of the trap sinks the pole, but the float keeps it within a few inches of the surface. The attraction soon becomes apparent; a loud snap announces that the iron jaws have closed, and a struggle in the water tells us that a prisoner is taken. We haul upon the line, and perceive an unfortunate Turtle, caught either by the foot or by the neck. Sometimes a wicker pot or creel, made on the same principle as a wire mouse-trap, is set, which allows the ingress, but precludes the return of the animal.

A young one of this species, a pretty, brightly-marked little thing, much flatter than the adult, and about as large as the palm of one's hand, I brought alive to England in a chip-box. In this its prison, whose walls it could feel all around at the same moment, it performed its voyage of transportation, without food; and arrived in good health. Through the remainder of the summer and autumn it was lively, and was often indulged with a swim in a tub of water, which it evidently enjoyed. The instinctive stratagem by which the species takes its prey was well exemplified in the manners of this infant Turtle. On my dropping a small earthworm, or a little piece of meat into the water, the animal, on being aware of its presence, would creep with the utmost slowness and caution towards it, until within reach; then, quick as lightning, the head was darted out, and as suddenly retracted, bearing the morsel in the horny, beak-like jaws, to be swallowed at leisure.

This specimen, kept in a box in a warm corner of the kitchen, lived through the coldest part of the winter; but died at last about the end of February, after a sojourn here of nearly seven months.

Along with the above described, frequenting the same situations, having the same habits, and nearly equally common, another species is taken, *E. rugosa*, distinguished by having the pale-yellow plates of the *sternum* each edged with a broad border of dark brown. Cuba possesses both species in common with Jamaica.

CURIOUS FISHES.

July 2nd.—It is very pleasant to stand upon the rocks and admire the beautiful fishes of many kinds and gay hues that play about at one's feet, or explore the crevices and tortuous channels for food. My admiration was especially attracted this morning to one which I had not before seen, but which, as I had no means of capturing it, I had no opportunity of identifying. As well as I could judge it was of a percoid form, rather high, and about a foot in length. The back was marked with ring-spots of black on a pale ground, and the under parts were of a rich crimson. He played about in a narrow passage between the rocks for half an hour, occasionally nibbling at the projections of the rock, and now and then turning up his radiant belly to smell at some weed. He was evidently seeking his breakfast. This manner of feeding resembled the browsing of the *Sparidæ.*

A lovely little Chætodon (*C. striatus*) banded with black and rich yellow, plays about in the creek at low water; occasionally picking something from the surfaces of the stones, and sometimes butting at them repeatedly, so as to rebound a few inches. This same propensity has been noticed in another hemisphere; M. Freycinet, in his Voyage round the World, records, that when wading over the coral reef encircling the island of Guam, in the Indian Archipelago, in search of *mollusca,* he was assailed by a small Chætodon, not bigger than his hand; *it butted at his hand*, and pertinaciously refused to be driven away. In the case of our Jamaican species, it might, perhaps, be presumed that the fish was collecting some object or other, animal or vegetable, desirable to it, in these repeated strokes; but what could the naked hand of the worthy naturalist yield in the way of food? We must be content to reckon the action among the thousands which we observe in animals, to which our habits, instincts, and reason, afford us no clue whatever.

FIG-TREES.

One end of the old building formerly used as a boiling house at Bluefields is covered with the roots of a large Fig-tree. Its great limbs stretch out horizontally to an enormous length, and cover a vast space of ground with the deep shadow of their dense foliage. Its height is not at all proportioned to its expansion, yet the dark-green hue and shining surface of its large oval leaves, and its immense spreading boughs set with clumps of *Tillandsiæ*, give it a very noble appearance. The constant shadow cast beneath it imparts a deep gloom to the spot, and invests the vegetation there with a rank luxuriance of character that reminds one of the glades in the mountain forests rather than the sunny pastures of the lowlands. But it is of the roots that I would especially speak, for these form quite a curious spectacle. They spread over the wall in every direction from the roof downwards to the earth, all in the same plane, clinging to the wall; the chief roots are as thick as a man's leg, but subordinate roots proceed from one to another, anastomosing in all directions (if I may use such a term), so as to make a most elaborate network of a multitude of meshes of various angular forms and sizes. These cross-roots are *at each extremity* united with the larger roots, and look as if the whole network had been skilfully carved out of one solid plank of wood, by cutting out the areas or meshes, and rounding the component bars; the very bark that covers the whole is continuous, where the roots unite, as if they had been always integrally one.

The only mode in which I can account for this singular phenomenon is the following hypothesis. The seed of the tree was originally deposited on the summit of the wall, beneath the eaves. As it germinated, the roots ran down towards the earth, some perpendicularly, some diagonally; but all creeping along the surface of the wall, no roots having shot out from its perpendicular. As these roots increased, they sent out side rootlets, which, still running on the face of the wall, by and by came in contact with another of the primary roots. Then, instead of creeping *over* it, as the roots of other trees would have done, the soft tip of the rootlet actually united with the substance of the root at the point of contact, the fibres of the two becoming interlaced, and their united surfaces gradually becoming covered with a common bark. The repetition of this process has produced the very curious wooden net which I have attempted to describe.

The tendency of the roots of the Fig to envelope closely any surfaces that they meet with, produces other singular results. One of these is what the natives delight to show to a stranger as a great curiosity, facetiously calling it "The Creole in the embrace of the Scotchman."

By the side of the road that leads in a zigzag line up from Black River to the summit of the Luana Mountains, about midway between Shrewsbury and Content, is a Fig-tree perhaps still more remarkable than the one above described. About thirty feet above the ground is the base of the trunk, which thence rears itself up pillar-like towards the heavens, and spreads abroad its vast horizontal array of branches across the road. From the same point there descends to the earth a hollow cone of roots, interwoven and anastomosed, especially at the upper parts, in the same manner as those of the boiling-house wall, but forming towards the bottom only three or four flattened irregular columns. Into the area inclosed by this network of roots a person may enter, for it is about six feet wide, and, looking up, behold the base of the trunk eight or ten yards above his head.

The explanation of this curious phenomenon depends upon the tendency just mentioned. On this site once stood a large tree of some other species, probably a Cotton-tree (*Eriodendron*), or some other soft-timbered kind. The little scarlet berry of a Fig-tree was carried by some vagrant Banana-bird or Pigeon to its boughs, and there devoured. After the little truant had finished his morsel, he perhaps wiped his beak against the rough bark of the trunk, beside the branch on which he was seated. Some of the minute seeds, enveloped in mucilage, were thus left on the tree, which the rain presently washed down into the broad concavity of the forks, where among moss and rotten leaves it soon germinated and grew. The roots gradually crept down the trunk of the supporting tree, closely clinging to its bark, and by their interlacement at length formed a living case, enveloping it on every side, and penetrating the earth around its base. The growth of these, and also of the enclosed tree, daily induced a tighter and tighter pressure, which at length arrived at such a degree as to stop the circulation of the sap between the bark and wood. Death of course was the result, and speedy decay reduced the supporting tree to a heap of mouldering dust; while the parasite, now able to maintain its own position by its hollow cone of roots, increased in size and strength, and over-topped its fellows of the forest; — *a tree standing upon stilts.*

THE ANT-LION.

July 16th. — One of the old buildings, now fast going to decay on Bluefields estate, was, in the time of sugar cultivation, the mill-house. The wheel was turned by water power, a stream from the upper part of the rivulet having been led through a long aqueduct into the mill, and passing off through a deep and narrow trench to the lower course of the river. Through this winding trench, cut to a depth of fifteen or twenty feet, but not more than a yard wide, and now so entirely choked up and overgrown with rank vegetation as to be quite dark, access is with some difficulty obtained to the basal floor of the mill, which is covered with a layer of impalpable sand, the residuum, no doubt, of the water that shot upon it when the wheel was in operation. The flooring planks of the upper level have been removed, leaving only the rafters; and the walls of the mill consist now of scarcely more than the posts and beams, so that sufficient light descends to the lower level notwithstanding its depth. Here I found many little conical pits in the fine sand, which upon examination proved to be the traps formed by the grubs of a species of Ant-lion (*Myrmeleon*), and inhabited by them. The appearance of the crafty insect, its motionless vigilance at the bottom of its den, the curved tubular jaws expanded to their utmost stretch, and the broad body concealed in the sand; the alertness displayed when an unfortunate ant slipped over the edge, the struggles of the prey to escape, the reiterated showers of sand vigorously cast up from the head of the expectant Giant Grim, and falling on the miserable victim, and the slow but sure sliding down of the latter, until the formidable jaws closed upon it; — I observed with intense pleasure, not only for the interest attached to so curious an example of insect cunning, but also for that of repeating observations long ago made in a distant part of the world, and, no doubt, on another species. The manners of these Jamaican Ant-lions agreed minutely with those of the *Myrmeleon formicarius* of the South of France, as recorded by the accurate Réaumur.

The singularity of the spot chosen in this case for

the exhibition of the stratagem, strikes one at first sight. But on reflection, we perceive that this very circumstance is but a further display of unerring instinct; for the frail pits on which the insect's success depends, would be filled up and effaced by a breath of wind, spoiled by a shower of rain, and destroyed, with their ingenious architects, by a passing footstep of man or beast. The depth of this locality was a protection against the first contingency, its inaccessibility precluded the last, while rain was kept off by the remaining roof of the building!

How inexhaustible are the resources of Divine wisdom, when the outgoings of it in the meanest insects are so wonderful!

I took two or three of the grubs into the house, and put them into a small box partially filled with sand, hoping to witness the construction of the pitfall. They soon began to work, proceeding backward, and shovelling the sand exactly as described, but only in irregular lines, leaving one after it had proceeded for some distance, and beginning another; so that they did not make even one complete circle. I was called to a distance, however, and the insects were thrown away. The species was probably *M. Leachii*, of which I have taken a single specimen near Bluefields, the only one I ever met with in a perfect state.*

* This rarity of the imago, contrasting with the abundance of the larva, of this insect, has been noticed by Guilding in St. Vincents. He observes that not a single perfect insect had been found by him in a state of liberty, though the larvæ swarm under every rock or shed calculated to protect their pitfalls from the rain and wind. (*Linn. Trans.* xvi. 47.)

The late Lansdowne Guilding has described an insect closely allied to the Ant-lions (*Ascalaphus Macleayanus*), but readily distinguished from them by its long thread-like antennæ, terminating in an abrupt knob. That zoologist found it at St. Vincents; it is just now rather common around Content, in St. Elizabeths. It rests like a Noctua, with deflexed wings, and carries the abdomen elevated and pointing upwards at a high angle. The larva of this genus, though in form much resembling the Ant-lion, differs from it in habit, never constructing a pitfall.

THE COTTA-WOOD.

The narrow bridle-path which descends steeply into the dense bush from the high road just below Content leads us down a precipitous and dangerous track to the foot of the hill. Here we emerge into another track crossing the former at right angles, scarcely more passable. It leads, if we turn to the left, through the most gloomy and savage scenes, —a valley dark at noon from the loftiness of the pillar-like trees, whose foliage intertwines far above; the bed of a turbid and impetuous torrent in the rainy season, half filled with enormous masses of limestone embraced and overspread by the fantastic roots of the huge Mahoganies, Broadleaves, and Santa Marias, perhaps a thousand years old. Little underwood grows here, from the lack of light; so that there is no interruption to vision, save that arising from the gloom and from the columnar trunks of the trees. Some of these are indeed gigantic, especially such as send out immense spurs and buttresses, like diverging walls. Animal life is almost unseen: the solitude is scarcely broken by the voices of birds, except that now and then the Rainbird and the Hunter (large cat-tailed cuckoos that love the shade) sound their startling rattle, or the Mountain Partridge utters those mournful cooings which are like the moans of a dying man. Such a scene however is the favourite haunt of *Erebus odora*, a gigantic Moth, which often, as we pass, darts out, like a great bat, from its concealment in the dark corner of two tree-spurs, or a hollow in the rock, and dances with a perplexing irregularity hither and thither, till it suddenly settles again, or is lost in the maze of stems.

Returning to the foot of the hill, if we follow the right-hand track, the scene is widely different. Here were formerly the provision grounds of the negroes belonging to Grand Vale; but these having been thrown up, a dense growth of small wood has accumulated, consisting chiefly of Bastard Cedar, Sweetwood, Prickly-yellow, with a scattering of Mahogany and Logwood. The path is narrow; and though the bushes are low, they meet overhead, so that we ride in shade; but it is the "greenwood shade," that the old ballads delight to describe; for a soft, green light penetrates through the pellucid foliage, and the rays of the sun, reflected from the glossy surface of the dancing leaves, flash and sparkle like a thousand diamonds, throwing an indescribable air of cheerfulness over the scene. A little way from the path there is a natural well of considerable depth, at the bottom of which is an unfailing spring of clear water, always cool; the negro children scramble down its rugged sides and fill their gourds and calabashes; and birds of many kinds resort hither, particularly in the droughts, to quench their thirst. For a long time a frog of large size had made his dwelling in some hole in the rocky side of the Hallow-well, as it is named, and used often to appear, to the no small terror of the water-children, who would scamper back with precipitancy. In these cool glades many birds build their nests; the neat little cotton cup of the Humming-bird, and the purse of the John-to-whip, and the hollow globe of the Banana Quit may be found around the spring: and the air resounds with song; multitudes of voices, some, it is true, inharmonious in themselves, combine to cheer the traveller as he threads the rocky, winding path. Among them the cooing of various Doves is prominent; and ever and anon, he perceives the gentle White-belly walking on the road before him, or detects the wary Bald-pate watching his motions from the summit of a tree, or is startled by the ruddy Partridge whirring across, just before his horse's face. At length the woods cease, and we suddenly emerge into the sunny canefields of Grand Vale estate.

THE BAMBOO.

Humboldt, if I mistake not, has mentioned the Bamboo as standing pre-eminent among the features which distinguish tropical from European scenery. It is an object which once seen can never be forgotten, especially when growing in those isolated clumps that look like tufts of ostrich plumes magnified to colossal dimensions. A thousand of these noble reeds, standing in close array, each four or five inches in diameter, and rising in erect dignity to the height of forty feet, and all waving their tufted summits in diverging curves, moved by every breeze, — form indeed a magnificent spectacle. Growing in the most rocky situations, the Bamboo is frequently planted in Jamaica on the very apex of those conical hills which form so remarkable a feature in the landscape of the interior, and to which its noble tufts form a most becoming crown. But it is scattered over all kinds of situations, from these elevated summits to the green plains that border the sea. On the steep sides of the mountains it is applied to a singular use, the preservation of the roads, which are cut in zigzag lines upon the rocky face of the mountains. The gradual disintegration of the exterior edge of such a road by the influence of the weather, and the wear caused by travelling, would soon destroy its level, and necessitate the cutting of it afresh. To prevent this, it is found sufficient to lay down lengths of green Bamboo just below the edge of the road, along the mountain side, and cover them with earth. These germinate at every joint, roots strike into the earth, binding it firmly, and a rampart of young shoots springs up, which, increasing every year in number and size, effectually prevent the crumbling away of the edge, and by throwing their feathery arches over the road, form beautiful green avenues, under whose grateful shadow the traveller may journey for miles, and scarcely feel the toil of the steep ascent. These avenues are called Bamboo walks, and their appearance is so peculiar, that they can be recognised at a great distance, almost as far as the surface of the mountain-side can itself be seen. There is one on the side of the Luana mountains above the picturesque estate of Grand Vale, a small one about half way up the acclivity of Bluefields mountain, and a much more extensive one near Haddo, on the road from Savanna-le-Mar to Montego Bay, on the other side of the Island. Some small but interesting clumps are growing in Bluefields pasture, which wave their plumes over the romantic little river that meanders through it. A beautiful contrivance connected with the growth of the Bamboo has been frequently noticed, but I may be excused for mentioning it again, for its interest; especially as, in my own independent observations, it had excited my admiration. Any one looking at a dense Bamboo clump, the polished rigid stems standing but a foot, or even less, apart, and each bristling with stiff branches shooting out horizontally in every direction, — would ask, "How is it possible for fresh stems to rear themselves through such a labyrinth of crossed and re-crossed branches? Surely their side shoots would catch some of these horizontal rods before the stem was well out of the ground, and either be broken off while young and tender, or be irreparably distorted!" Yet we never see such distortion; each stem bears its whorls of horizontal branches, and each branch finds its place among its fellows, adding to the maze, and apparently to the impenetrability.

The contrivance, however, which obviates all difficulty is most simple. The new stem shoots up from the root-stock and attains its lofty stature, before a single lateral branch has budded. In this simple form its sharp top and polished surface find no difficulty in threading the crossed branches; and when once its elevation is gained, the lateral branches find their horizontal course no less unresisted. I have seen in the dense clump at Bluefields, in the middle of July, vigorous shoots of the present year, fully three inches in diameter, and from twenty to thirty feet high, without the least trace of a lateral shoot.

The appearance of such a growing stem is singularly beautiful. The delicate green hue, set off by the smoothness and even polish of the surface, the swelling joints at regular intervals, the cylindrical internodes, and the uniform thickness of the whole, strike the mind with an emotion of pleasure, the slenderness conveying the idea of elegance without weakness. The basal part of every joint is partly inclosed in dark brown triangular sheaths, having a hairy or downy surface; and these increase in number, and are closer together towards the summit of the shoot, where the joints are still extending: the contrast of colour and of surface between the woolly brown sheath and the polished vivid green of the reed adds to the beauty of the whole.

The various uses to which the Bamboo is applied, it would be quite foreign to my purpose to notice; but I will mention one to which I myself applied it. I found the slender stems, well-ripened, excellent for the handles of insect nets, combining lightness with firmness and elasticity, and affording lengths from twenty feet downwards.

PERIODICAL RAIN.

July 26th. Content. — The regularity of the daily rain during the season is a remarkable circumstance. The mornings are almost invariably fine, or if there be a few clouds at sunrise, they are soon dissipated, and the sun burns with unmitigated rigour through the forenoon. About midday fleecy clouds begin to appear in the north; — these increase and coalesce, and begin to blacken: — thunder mutters, and waxes louder: — by and by one or two terrific claps are heard, and we look out, but there is a good deal of blue sky yet. Soon, however, the mountain top becomes enwrapped in thick misty cloud, which rolls rapidly down; more thunder; vivid flashes of lightning gleam through the daylight. Now we hear the rain clattering upon the leaves of the forest above us, like shot; but not a drop falls on us yet. Now it comes, however, not so much in drops, as in a sheet of water; and in an instant the gullies and channels of the rocky hill-side are filled with discoloured and frothing rivulets.* We trace the progress of the

* The fine description of Virgil will hardly fail to recur to the classic reader: —

"Sæpe etiam immensum cœlo venit agmen aquarum,
Et fœdam glomerant tempestatem imbribus atris
Collectæ ex alto nubes; ruit arduus æther,
Et pluviâ ingenti sata læta boûmque labores
Diluit; implentur fossæ, et cava flumina crescunt
Cum sonitu; fervetque fretis spirantibus æquor.
Ipse Pater, mediâ nimborum in nocte, coruscâ
Fulmina molitur dextrâ; quo maxima motu
Terra tremit; fugêre feræ; et mortalia corda
Per gentes humilis stravit pavor."

Georg. i. 322.

rain cloud onward, enveloping the immense forest between us and the sea in one misty mass, bringing out much more distinctly the varieties of distance of the hill-tops beneath us, which before were undistinguishable in the clear and almost equal visibility of all. During the storm the lightning is vivid, zigzagged, and frequent; the thunder instant, crashing, and terrifically loud. In an hour or two the rain has ceased, the sky presently resumes its clearness and transparent azure, and the coolness of the evening air is delightful, loaded as it often is with the perfume of various flowers.

After the rain has ceased the fog begins to rise, and this is sometimes the grandest part of the whole scene. I have said that the little cottage where I sojourn looks from its eminence, like an eagle from its aërie, over seven miles of almost unbroken forest: it is irregular ground; but hill and dale looked down upon from a mountain elevation cannot be distinguished from a plain. But on such occasions as this we discern the form of the ground; for from every dell and bottom arises a dense mist of snowy whiteness, throwing the summit of each little intervening hummock into strong outline. As these local mists arise, they unite; and often the whole range of country becomes overlaid with a mass of white cottony cloud, through the openings in which, here and there, we catch a glimpse of the dark forest below. This has a novel and charming effect. Sometimes the setting sun throws his horizontal rays upon the mass, and the whole becomes suffused with a soft, transparent, orange-coloured light, which gives the finishing touch to the magnificent picture.

Nearly the same uniformity prevails in the lightning which almost every night at this season is wont to play in the western sky. A dark gloomy-looking cloud towers up from the horizon, which every two or three seconds becomes a flood of soft light like the concentrated glory which we sometimes see in paintings representing heaven. Sometimes the light gleams fitfully from behind the cloud, revealing its outline in stern detail, and gilding the edge; at others a faint glimmer peeps as it were round one corner, and tremulously quivers. Then a full blaze appears again, and a dazzling zigzag cleft in the midst of it, darting upwards. This zigzag track is in almost every broad flash, as if the sky, like a solid wall of light, had split and closed again, revealing the most intense lustre behind it. All the time, perhaps some hours, not a sound of thunder is audible.

FISHES AND FISHING.

I accompanied an old negro one morning, when he paddled out in his canoe to examine his fish-pots. The canoe was, as usual, a single log of the Silk-cotton tree, shaped and hollowed by the hands of the fisherman himself, partly by the aid of the axe, partly by fire. It was long and narrow, and brought to a rounded point at each end. The owner squatted down in the stern, and, with a single paddle held in both hands, gave two or three short strokes on one quarter, then on the other, alternately; by the force of which the slender skiff shot rapidly out from the shadow of the shore into deep water beyond the reef. The clear water rippled off behind with a whispering sound, varied by the short quick plashing of the paddle; but every thing else was noiseless and motionless. The sky scarcely showed any glow in the horizon, the land-breeze had gone to sleep, and the sea-breeze had not yet thought of awaking; the woods on the slopes of Bluefields and Mount Edgecumbe looked black and indistinct, and were scarcely relieved from the huge mountain behind them; while the whole was reflected, an unbroken mass of deep shadow, in the silver sea.

Two or three specks began to appear on the sleeping surface, towards one of which the fisherman directed his course, and presently we were close to it. It was a billet of soft light wood, with a rope attached, which went down into the deep water; it marked the position of the sunken pot. The old negro began to haul upon the rope, and presently drew to the surface his device for entrapping unwary fishes.

The fish-pot was a four-sided box, about four feet square and one foot deep, divided by a partition into two compartments. It was made of strips of very tough wood, torn off, about an inch wide, and no thicker than card-board, interwoven much like the cane-bottom of a chair, through the interspaces of which the contents of the interior could be readily discerned. In each compartment an orifice of eight or nine inches wide was left, where the strips were ingeniously worked inwards to form a narrow tubular entrance, extending a little way into the interior; and so managed that the elastic ends should meet each other in a point, offering no obstruction to a fish pushing his way in, but resuming their position the moment his body had passed, so as effectually to preclude his return. A lid, moving on a hinge of the same material, and tied down, afforded the fisherman an opportunity of introducing his hand to take out the prey captured, or of putting in the offal, which constituted the temptation to enter.

We examined three or four in succession, and all had taken something; perhaps half-a-dozen fishes in one, three or four in another, and so on. Some of them were transferred to me, and were saved from the ignoble destiny to which the cook would have consigned them, to be (literally) embalmed in the honourable service of science. One of these was sufficiently hideous, and was no favourite with the old negro, who gave it the name of Poison-grouper. It was a species of *Apistes*, the more interesting because belonging to a rather numerous genus, hitherto recognized only in the Oriental seas; a circumstance which I would commemorate by calling the present species *Apistes exul.** Another fish of

* It may be thus described. *Apistes exul*, MIHI. Body well covered with scales. Head very spinous; sub-orbital spines short, scarcely more than serratures: profile gently descending; mouth opening diagonally upwards. Dorsal commencing above the base of pectoral; ending considerably short of the base of caudal; moderately hollowed in the middle; pectoral with no free ray. Fin-ray formula: B. 7; D. 12—10; A. 3—5; C. 18; P. 19; V. 1—5.

rather unpleasing aspect he named Sucking-fish, though without reason, as far as I could find. I rather think he confounded it with the Remora, but it was a species of *Elacate;* and, as I believe, like the former, undescribed. I shall name it, from the form of the caudal and pectoral fins, *E. falcipinnis.** Besides these, there was a pair of King-mullets (*Upeneus maculatus*, Cuv.), called also Queen-mullet, and, from its depending beards (which are, however, in some specimens entirely wanting), Goat-fish. This is a beautiful fish; its general hue is pink, fading to white below, with three large livid spots on each side; the central portion of each scale on the upper parts is of a pale pearly azure tint. A specimen of the beautiful Angel Chætodon (*Holocanthus ciliaris*), and several Snappers and Grunts, made up the booty; together with a large crustacean

Length 8 to 10 inches. Body and fins marbled with red, brown, black, and grey; the marblings on the huge pectorals resemble the pencilling of some moths; the basal part of their under surface is deep-black, marked with round white spots. The spots and marks on the body are very irregular in form, sinuous, and confluent; with darker edges. The caudal is pale reddish, with three transverse bands of dark brown. The irides are yellowish.

* *Elacate falcipinnis*, MIHI. Height of body in proportion to the total length, as 1 to $9\frac{1}{4}$: length of head to total length as 1 to $5\frac{2}{3}$. Pectorals falcate, about as long as the head; ventrals straight, pointed, about two-thirds as long as the pectorals: præ-dorsal spines ten, small: dorsal and anal low; caudal large, somewhat forked; the lobes falcate; the upper lobe considerably exceeding the lower. Length eleven inches. Irides pearly white. Body blackish above, grey beneath; a broad band of deep black runs along each side from muzzle to tail, bounded both above and below by a whitish band of similar breadth. Fins black: the caudal tipped obliquely with white on both lobes.

)*Scyllarus æquinoctialis*), that had found his way into the prison among the fishes. A few shrimps, and other creatures of smaller size, were seen in some of the pots as they were drawn up; but these waited not to be lifted out of the water, darting through the interstices of the wicker-work as it approached the surface. The others were left disconsolate, to mourn their imprudence, in getting in, without first inquiring how they were to get out. Poor things! as they roamed round their narrow walls, and looked through the grating at their fellows enjoying freedom, perhaps they philosophically repeated, in piscine phrase, the poet's reflection: —

> ——— " Facilis descensus . . . . ;
> Noctes atque dies patet atri janua *corbis*;
> Sed revocare gradum, superasque evadere ad *undas*,
> Hoc opus, hic labor est.

The most esteemed fishes for the table, and the most common, are the Snappers, Yellow-tails, Silks, and Hinds (various species of *Serranina*, or Marine Perches), and the Grunts and Squirrels (species of *Sciænadæ*). Many of these are fishes of brilliant colours: the Yellow-tail (*Mesoprion chrysurus*), for example, is pale azure on the back, and pearly white below, with a broad band along each side, of the richest yellow, which is the hue also of the dorsal and caudal fins. The Dog-tooth Snapper (*Mesoprion cynodon?*) is white, with six transverse dark bands; each scale on the belly has a border of golden lustre; and all the fins are rich yellow. The Spotted Snapper (*M. uninotatus*) is white, traversed by longitudinal lines of yellow; the dorsal and caudal fins have borders of rose-pink, and there is a large oval black spot on each flank. The Hind (a species of *Serranus*) is a handsome fish; it is studded with scarlet spots on a greyish-white ground: the fins are yellow, especially the caudal, with black borders having a narrow white edge; sometimes the pectorals are brilliantly scarlet.

But all of these yield to the different species of *Hæmulon*, which, under the name of Grunts, are well known and highly esteemed throughout the Caribbean Sea. The names scientifically bestowed on many West Indian species, such as *elegans*, *formosum*, *xanthopteron*, &c., indicate their pretensions to beauty. Their characteristic markings and hues are oblique parallel lines of gold, on a silver or metallic azure ground, with delicately tinted fins, and sometimes spots of peculiarly intense lustre: the whole interior of their mouth is generally of the finest scarlet.

All of these are taken with the line, and with the seine, as well as in pots. The Snappers are perhaps more highly esteemed than the Grunts, but both are excellent. They chiefly affect what is called "broken ground," where patches of white sand alternate with masses of rough rock, and fields of grass-like weeds. They range from deep water to the rocky shore; are taken abundantly with the seine, and bite freely at a bait of Sprat (*Harengula clupeola*); but only fish of small and middling size are commonly caught in pots. The Snappers occasionally attain a length of two feet and a half; but fish of such dimensions will rarely bite at a hook worked in the usual manner. For them the negro fisherman takes a wire hook (No. 1. or 2.) as large as a goose-quill, which he throws overboard, baited with a sprat, but in a peculiar fashion. One side of the sprat is split nearly off, remaining attached only by the tail; this is allowed to hang free; and a slice from the back and one from the belly are allowed to hang in the same way. The hook is then passed in at the mouth, out at the gills, and again through the middle, and the head is tied to the top of the hook: another slice is then put upon the hook, and made to hang down. This is designated a "full bait." No sinker is attached, but its own weight is sufficient to carry it nearly to the bottom. The line being passed with two turns *round the fisherman's great toe*, he lies comfortably down in his canoe (Blackie will always lie down wherever he can), and awaits the bite of the first large fish that may choose to essay the baited hook; which it usually does by taking in the whole at a gulp.

The seine is here, as elsewhere, the chief resource of the fisherman; and many kinds of fishes are taken by this means, that rarely enter a pot or seize a bait; together with many species that he calls *rubbish*, as being of no esteem in the market, though often interesting to the naturalist. There are several places in the vicinity of Bluefields, where the peculiarities of the beach and of the shoaling water are favourable to the hauling of the seine; one of these is Belmont beach, already spoken of; another, still more used, is the beach at Cave, a few miles to leeward, where a lofty spur from Bluefields ridge juts out in a bold promontory to the very edge of the sea. The wooded mountain side descends abruptly, almost precipitously, leaving only a narrow beach of white coral sand curving round its foot, along which the high road winds from the windward parts to Savanna-le-Mar, and onward.

I have often admired the loveliness of the scene presented by the termination of this promontory, frowning down upon the beautiful bay beneath; especially in the early morning, while the sun, if risen above the actual horizon, has yet far to climb before a single ray can shoot over the shaggy summit of the ridge; and the whole mountain-side, covered with a dense forest in every part, except when a little white cottage is perched at mid-height, casts a deep, black shadow, reflected as in a mirror from the calm water below. Beyond the promontory, the low mangrove-shore trends away to the westward, and from the level country behind rises in majestic elevation the steep mountain-mass known as the Dolphin's Head, clear and distinct, but empurpled by distance.

The foreground of the landscape, of which I have made a sketch, is the summit of a shelving cliff at Lindo, overgrown with bushes and herbaceous vegetation in rude luxuriant wildness; from the midst of this tangled bed of weeds and shrubs the singular forms of the Papaw (*Carica papaya*), and the Trumpet-tree (*Cecropia peltata*), erect themselves; and a huge Silk-cotton tree (*Eriodendron anfractuosum*), hoary with age, towers and spreads overhead, with

P. H. GOSSE., DEL.T M & N. HANHART IMP.T

CAVE

LONDON LONGMAN & C.o 1851.

many slender lianes depending like long strings from its branches.

In the bottom of the bight, upon the narrow beach that looks like a thread of silver between the black mountain and the equally black reflection, we see several moving atoms; and a little speck slowly glides out into the still calm bay. Those are the fishermen, and this is their canoe, in which they are carrying out their ground seine. One end of a long rope is made fast on shore, and the seine is attached to the other extremity; when the canoe has got as far as the rope will allow, the seine, loaded and corked, is gradually dropped parallel to the shore; and a rope of similar length to the former, but attached to the other side of the net, is brought to land.

If we leave our post of observation, and walk leisurely down towards the beach, we shall arrive by the time they begin to haul. It is a pleasant road, and at this hour, beneath the cool shadow of the mountain, walking is a very different thing from what it would be in the after part of the day,

> ——— "sol ubi montium
> Mutaret umbras,"

when the fierce beams will glare upon us with twofold rigour, one sun in the sky, and another in the sea. Tall trees overhang the road on both sides, for part of the way, many of them covered with beautiful blossom; fine flowering shrubs display their charms amidst the bush, and some magnificent butterflies hover about and flap their heavy wings over the summits of the trees, now and then alighting on the lofty leaves. Here and there tall cliffs of rugged rock rise perpendicularly from the road-side, their roughness half-concealed by the multitude of shrubs and slender trees that jut out from the crevices, and by the climbing and trailing plants that throw wild graceful festoons over their sides. Among them grows in profusion the *Portlandia grandiflora*, having much the aspect of a climber, from its height and slenderness, and from its growing close to the face of the rock; conspicuous above all for its magnificent trumpet-shaped flowers of purest white, eight inches in length, and its large glossy oval leaves of deepest green.

We cross a streamlet, which, from some machinery formerly erected here, passes by the name of Water-wheel, and where a rude aqueduct carried out a few hundred yards into the sea enables ships' boats to fill their water-casks without the danger of beaching. Here a deep morass borders the road, inhabited by myriads of Land-crabs (*Gecarcinus ruricola*), whose burrows riddle the ground so completely that, even by the road-side, it is dangerous for a horse to pass. The morass is covered with trees, among which the Cork-wood or Alligator-apple (*Anona palustris*), is abundant, displaying its beautiful and fragrant, but noxious, fruit; which nevertheless affords food to these Crabs, to Morass-Galliwasps (*Celestus occiduus*), and, as is believed, to the formidable Crocodile.

The sound of human voices in melody falls now upon the ear, the song of the negroes who have begun to haul in the seine. Rude their music is, and artless their tune; yet, mellowed and softened by distance, now swelling in chorus, now feeble and faint, it has considerable sweetness, as the human voice always has under such circumstances. Yonder we see them, forming two lines in the water, ten or a dozen men in each row, hauling upon the two ropes; the outmost up to the neck in the sea, and the inmost on the beach; all naked, regardless of the burning sun that now pours down his beams upon their woolly heads and glossy backs. It is a slow operation; and as they all throw their weight upon the line together, they sway backward and forward in time with the wild air whose notes they are singing.

In an hour or two the fishes that the seine has inclosed are dragged on shore, and lie gasping and fluttering on the wet sand. Let us see what they have taken. Here is the usual predominance of Grunts and Snappers, Hamlets and Hinds; — two pretty Chætodons, *C. capistratus,* with its eye-like spot on the tail, and *C. striatus,* with its black bands; two kinds of Doctor-fish, so called from the curious glassy lancets that they carry in a sheath on each side of the tail, *Acanthurus chirurgus,* and *A. cæruleus;* and a Parrot-fish (*Scarus cæruleus*), remarkable for its abrupt, almost vertical, profile, white eye, and brilliant azure hue; I observe that the two divisions of the upper jaw, in this fish, are capable, during life, of separate motion, up and down; a circumstance, I think, not before noticed. Here is a *Muræna,* looking as if it had been varnished; another lengthened fish, of curious form, and remarkable style of colouring, rust-red with longitudinal white lines, and numerous black dots, a species of *Aulostoma**, to which the negroes give the name of Soap-fish. Here, too, is what they call a Flounder, but truly a kind of Turbot (*Rhombus argus*), a handsomely marked fish, being studded all over the upper side with large blue rings, inclosing pale yellow areas on a dusky brown ground-colour. There are many other things, young Sharks, Hedgehog-fishes, Trunk-fishes, *et hoc genus omne*; — but these we must leave, and make the best of our way back to Bluefields, or we shall not be in time for "second breakfast."

ANOLES.

The faculty possessed by certain small species of the Iguaniform tribe of Lizards, of effecting rapid and strongly marked changes in the colours of their bodies, is exceedingly interesting. The extent of these changes is scarcely inferior to that to which the same singular power is exercised by the Chameleons of the Old World; and if the latter display peculiarities of structure more curious to the naturalist, the Anoles have the advantage of a form and motions as graceful and elegant as the *coup d'œil* of the Chameleons is hideous. I have already alluded to this *metachrosis* in the *Dactyloa Edwardsii*, and it no less remarkably characterizes the smaller and more agreeable species of the restricted genus *Anolis*. Two

* This is, doubtless, the *A. coloratum* of Müller and Troschel (Ann. and Mag. of N. H., July, 1848); but the character by which they distinguish it from *A. Chinensis*, — the want of brown spots on the head, — is valueless; as I have had specimens in which the spots were abundant in that part. The fact is, that individuals vary much in the amount of the maculation.

of these, to which I have given the names of the Purple-tailed, and the Pearly-bellied Anoles*, are the commonest Reptiles in Jamaica, at least in St. Elizabeth's and Westmoreland, the districts with which I am familiar. About the walls and rafters of out-buildings, the sides and summits of fence-walls, and similar places, they are continually running, reminding one, as they dart about in their changeable beauty, of Moore's description, —

> "Gay Lizards glittering on the walls
> Of ruin'd shrines, busy and bright,
> As they were all alive with light."

They are particularly numerous in the *lieux*, which

* I have described these, with some other species of Jamaican *Sauria*, in the Annals and Mag. of N. H. for Nov. 1850, under the names of *Anolis iodurus* and *A. opalinus*. Though I feel assured that these are distinct species, it is difficult to establish any unvarying mark of distinction. Colour, in creatures that are so variable, seems an unsatisfactory foundation for comparison or contrast; yet, except the pale band that runs down the side of the latter, I can discover, even by minute examination under a lens, little of difference. In *opalinus*, the general form is more slender, the belly rather flatter, the head narrower and longer, and the muzzle more pointed. The tubercular plates on the head are less definitely shaped; the large central superciliary plates are divided from the coronal ones by not more than one perfect row of small ones, where in *iodurus* two or three rows intervene. The scaling on the goitre is coarser in *opalinus*; that is, the scales are larger and more pointed, and the naked interspaces wider. Most of these diversities are, it is true, minute, and even microscopic; yet it is not difficult to distinguish the species when alive, at sight, even at a considerable distance. The scales of the inferior surface are iridescent in both, but chiefly in *opalinus*, the light from the belly of which, when the angle formed by the incident ray and the reflected ray is very wide, glows with a ruddy golden hue, exceedingly beautiful and opaline.

in Jamaica are usually at some distance from the dwelling house. A very handsome fly with lustrous green thorax and purple abdomen (*Syrphus obesus*), breeds in such offices very abundantly; and it is possible that the Anoles and Geckos may feed on it, though I have never seen them taking such prey. Indeed very rarely have I seen them take food at all. Some kept in captivity in a box with a gauze front, would occasionally seize a small butterfly (*Terias* or *Hesperia*) that was now and then put in; and I once lost a butterfly by the means of one of these little Lizards rather singularly. I was pursuing a specimen of *Calisto Zangis*, one day at the summit of the Bluefields Ridge, and had watched it to its place of alighting: I was about to throw the net over it, when, on a slight rustle among the leaves, I observed that it was fluttering as if unable to get away; my impression was that an invisible spider's web was holding it, but looking closer I found that a little green Anolis had the butterfly in its mouth. Its colour was so exactly that of the verdant leaves of the bush, that I had not perceived it before, although my eyes were fixed on the spot. On my approach it darted away and was seen no more. I once saw one rob a spider's hole-web; and I have taken from the stomachs of individuals, fragments of ants (both the large rufous kind, and that much dreaded species called the Corromantee ant), caterpillars, maggots, flies, ichneumons, and other insects.

At Spanish-town, where, as well as at Kingston, the Purple-tail attains a size and vividness of colour much superior to any that I have seen to the leeward,

I observed this species feeding on ants. On a gate-post a number of scattered ants of a small kind were running to and fro, as they very frequently are seen to do. A beautiful male Purple-tail had stationed himself on the post, perpendicularly, with the head downward; and as the ants one by one came near him, he snapped them up. He did not run after them, but waited till each one came within reach: in a minute or two I saw him thus take a dozen or more. Each capture was the work of an instant; he touched the post with his muzzle, and the ant was gone: they were evidently seized with the lips, not with the tongue. I afterwards observed the *A. opalinus* employed in the very same way.

These little creatures are as playful as they are pretty. As they creep about, they often catch sight of another of the same species; immediately one raises and depresses suddenly the head and fore parts, flirts the tail from side to side, and extends the goitre by means of the elastic arched bone in front, till its tip reaches nearly as far as the muzzle. The brilliant goitre is thus extended and relaxed alternately several times; an action, I incline to think, intended as a provocative, or else a manifestation of sexual desire. After having thus "signalised" for a few seconds, one darts towards the other, who usually runs away, apparently as if wishing to be caught.

> "——— latebras fugitiva petebat,
> Non tamen effugiens, tota latere volens;
> Sed magis ex aliquâ cupiebat parte videri;
> Lætior hoc multo, quod male tecta foret."
>
> MAXIM. Eleg. i. 66.

Frequently one having the advantage of a more elevated position, will jump on the other's back, but the latter manages to scuffle away. Their agility is very great; they take long and high leaps in the most graceful manner, always alighting with precision: as they run on trees and bushes a good deal, and jump from twig to twig, this accuracy of eye in measuring the distance, and of muscle in clearing it, is important. It is quite common to see both the species in dwelling-houses, and even leaping on the persons of the inmates; their well known inoffensiveness, combined with their sprightly motions, and gentle aspect, causing them to be viewed without any of that horror or disgust, which so often accompanies the sight of a reptile. When taken into the hand, however, they bite most courageously, though with impotent violence; for the teeth are not long enough even to penetrate the epidermis of the fingers, nor are the muscles of the jaws sufficiently powerful to enable them to pinch with any inconvenience. Their hold, however, is tenacious.

We still find in works of scientific authority much misunderstanding on the nature and colours of the goitre. It is continually spoken of as being *inflated* at the will of the animal: but this is quite erroneous. I have occasionally seen the skin of the *body* puffed out, and also the skin around the neck, but the *goitre* never. It is a portion of the common integument running down the mesial line of the throat, which is capable of being stretched to a remarkable degree. Down the centre pass the *rami* of the *os hyoides*, lying side by side close under the skin, and at length becoming cartilaginous, the extremities reaching to the abdomen. In excitement the front portion of the *os hyoides* is drawn backwards towards the abdomen; by which action, the cartilaginous extremities remaining fixed, the slender elastic horns are forced out in a bowed form, carrying with them the extensile skin with its two internal surfaces in contact, forming a thin lamina, frequently so far projected as to constitute half of a long ellipse with an oblique base. When the bone is relaxed it resumes its straight direction by its own elasticity, and the goitre shrinks up again under the throat so as to be scarcely visible.

The changes of colour for which these little reptiles are remarkable, are often stated to take place with special intensity on the *goitre,* which is said to present an endless succession of ever varying hues. This also is entirely a mistake. The extensible skin, which forms the goitre, is always of one permanent hue, generally some rich tint of red or yellow. In *A. iodurus* and *opalinus,* the two species before us, the colour is a brilliant orange or red-lead, becoming pale towards the margin, so as not at all to be perceived when the skin is relaxed, the edge itself being whitish, like the rest of the breast and belly. This alternate appearance and disappearance of the bright colour has no doubt been mistaken by careless observers for changes of hue. The scales which cover the goitre are whitish like those on the neighbouring parts; they lie close together in the relaxed state of the skin, but when the latter is stretched, the little white pointed scales are widely separated, forming longitudinal rows, with broad interspaces of the rich ground colour, greatly increasing the effect.

What the object or peculiar function of this protrusile goitre is, remains still obscure. It has been suggested that as the Anoles prey upon butterflies and other insects, the brilliant disk-like surface may present an attraction to vagrant insects by its resemblance either to a richly coloured flower or an equally gay butterfly. It may be so; though in the hundreds that I have observed, I never saw an insect approach it; nor do I think, if such were the object, the animal would keep projecting and withdrawing it, as it does, with considerable motion of the body, but would rather let it lie extended as motionless as possible, which it does not do. I certainly think it is connected with the intercourse of the sexes; and am pretty sure that the exposure of the goitre is meant as an expression of anger and defiance to one of the same sex. As far as I have observed, the female Anolis is almost, if not quite, destitute of the goitre. I have had a Purple-tail, which when held in my hand protruded its goitre, while the energy with which it would now and then fiercely seize upon my fingers, and endeavour to gnaw them, manifested its anger and impatience at being confined.

After having been actively engaged in running and leaping, whether in play or alarm, we occasionally see them remain still, loll out the tongue, and pant with open mouth very much in the manner of a dog.

Our little Anoles are not aquatic, nor have they, as has been stated, any predilection for marshy situations. I have never seen them take the water spontaneously, but in order to try their natatory powers, I caught a male Purple-tail, and placed it in the midst of a deep pool of clear water. It floated like a cork, as might have been expected from the volume of the lung, and from its capability of inflation. It was not more than half immersed, and as the whole skin repelled water, the fluid stood around it, in the form of a little heap or bank. The animal struck out vigorously for a few seconds, moving the feet alternately, the right fore with the left hind foot, and *vice versâ*, as in running; and this motion threw the body into lateral serpentine undulations. It made way thus, but soon relaxed, and then intermitted its exertions, and floated motionless. On my taking it up, I found it much exhausted; and though, on being again put in, it struck out anew, yet it was each time more languidly, until I finally rescued it, and restored it to dry land, where, though it lay helpless and almost motionless awhile, it soon recovered its agility in the warm sunshine. From this experiment I judge that this species could not swim the distance of a yard before it would become exhausted, and drown.

The Anoles, and I believe all the quadruped Saurians, progress only by the alternate use of the right and left feet; never galloping, how swiftly soever they may move. In rapid running, they commonly elevate the tail above the general line of the body. Perhaps this is what Lacépède means when he says of *Anolis bullaris*, that it carries the tail habitually raised above the back.

The green hue of the Purple-tail when enjoying

itself in the intense beams of the vertical sun is a beautiful tint; and the surface being covered with minute tuberculous scales, which reflect the light, have a frosted appearance, which is exceedingly rich. It is fond of leaping about the serrate leaves of the Pinguin, and of lying in their hollows; and in March and April, when the interior leaves of this formidable plant are spreading out, all glossily scarlet as they are, the bright green form of the little Anolis, couchant or passant, is displayed to much advantage.

The female of this species differs so much from the male in size and colours, that if I had not had indubitable evidence of her identity, I should have supposed her quite distinct. She is much less commonly seen than the other sex.

Before I bid adieu to these little engaging reptiles, I will transcribe a note of Mr. Hill's on the interesting subject of their changeability of colour. "It is now pretty satisfactorily determined, that the direct or more immediate cause of this peculiarity is physiological, and dependent on the action of the lungs upon the circulatory system. Their lungs are large, dilatable, and prolonged; and the phenomenon itself is always most remarkable among Lizards, whose general cutaneous covering does not adhere closely or uniformly to the muscular layer beneath. A large portion of air enters below the skin; and as it is variously distributed, according to the state of the reptile, in respect to tranquillity or disquietude, it gives the many-coloured hues we see so instantaneously occurring in them. Cuvier says, that 'in effect, their lungs, rendering them more or less transparent, urge the blood to rush more upon the skin, and, according as the fluid fills itself or empties itself of air, its colour becomes more or less lively.' By the investigations of Mr. Houston (Trans. of Roy. Irish Acad. xv. 177.) the proximate cause of changeability is connected with the circulatory system. The skin of the changeable Saurians is not only very thin, but highly vascular; and he thinks that the colour of the blood appearing through the semi-transparent covering, and being variously modified by its more permanent hues, is of itself sufficient to account for every diversity of tint which the *Chameleon* can assume. He maintains the opinion that these effects are produced by vascular turgescence, 'just as the increased redness in blushing is caused by a rush of blood to the cheeks.' I would seek in addition an illustration from the changeable hues in the *caruncles* of the Turkey. There, too, aeration produces a diversity of influences on the circulatory system. *The red blood distributed through these parts* increases and diminishes its intensity of tint by the different aerations at the caprice of the bird, and, being sometimes wholly deprived of its red particles, flows colourless, as the fluids circulate in the white of the eye."

I have seen the Pearly-bellied Anolis only in Westmoreland. In those parts of the coast of St. Elizabeth which lie to the eastward of Black River, and in the district around Spanish-town and Kingston, it appears to be replaced by the Zebra Anolis, (*A. maculatus*), a very handsome species, of rather larger size, with fine contrasts of pale yellow and deep brown disposed in irregular transverse bands, and a very large goitre of pale orange. This is the most common Lizard around Kingston, but is unknown at the leeward end of the island. The Purple-tail, on the other hand, attains a superior development both of form and colour in the Kingston district. The vegetation in these parts and in the eastern parts of St. Elizabeth (*ex. gr.* Starvegut Bay and Pedro Bluff) have a common character: species of *Inga*, *Acacia*, and *Prosopis* abound, intermingled with the beautiful Lignum vitæ (*Guaiacum officinale*), and enormous Cacti, (*C. repandus*, and *C. Peruvianus*) known as Dildoes. This is a botanical character widely different from that presented around Bluefields.

The other Anoles that I have met with in Jamaica are rare. The one, a new species of *Draconura* (the Chain-marked Anolis), I know only by a single specimen. The other, of which three individuals have come under my observation, is the Plate-headed Anolis, which exhibits characters that I have thought to demand the constitution of a new genus for its reception.* It is a handsome Lizard, of a delicate greenish-white, crossed by irregular bands of black, particularly conspicuous on the tail; the goitre is large, and of a dark lake-crimson. The male is bold and fierce in self-defence, biting with such energy as to pierce the skin of the hand: the female, which is less distinctly marked, is timid, making no effort to bite when held in the hand; is more slender, and has

* These are described in the Annals and Mag. of Nat. Hist. for November, 1850, as *Draconura catenata* and *Placopsis ocellata*.

the crimson goitre very slightly developed. The first specimen obtained, a female, was crawling on the stem of a herbaceous plant, by the road-side at Cave, and was easily captured, displaying little agility. After I had possessed it a little while, it gradually changed the ground colour to a dusky umber, the marks still being visible. Of the markings, which vary much in different individuals, a conspicuous black spot behind the ear seems to be constant. This female was accidentally killed a day or two after I had captured it, and I took the opportunity to dissect it. I found on the left side one large egg nearly ready for exclusion. Its form was oblong, the diameters about $\frac{5}{10}$ by $\frac{2}{10}$ inch ; its hue a cream-white ; its integument parchment-like. Two or three other ova were present, but minute. The animal had begun to slough its cuticle before it died. I think it likely that any unwonted excitement, such as fear or inordinate anger, may bring on the periodical sloughing in reptiles : in nearly every case the capture of a Lizard has been presently followed by the commencement of this process.

FRUIT ON A MALE PAPAW.

Aug. 6th. — On the summit of Bluefields Ridge I took from a male Papaw-tree (*Carica papaya*) in the woods, two perfect fruits, an instance of accidental hermaphrodism in a plant ordinarily diœcious. The fruits were more elongated than usual. I determined the sex of the tree, not only by its form and general appearance, which cannot be mistaken, but also by an examination of the flowers, which were staminiferous only.

BEAUTIFUL COCOON.

Aug. 11*th.*—A friend sent me an insect production of great beauty. It consists of an immense number of cocoons of a small *hymenopterous* fly, set close to each other in exact order, the whole arranged around an axis, which in this case was the stalk of a Coco-leaf (*Colocasia*). The material is a downy silk of the purest white, and the congeries appears like an egg-shaped mass of the finest cotton-wool.

I afterwards obtained another specimen, precisely similar, from the twig of a Cotton-tree, at the Vineyard, near Black River.

THE BLACK SNAKE.

The most common Ophidian reptile in Jamaica is the Black Snake.* It is frequently met with in

* This has sometimes been confounded with the Black Snake (*Coluber constrictor*, Linn.) of the United States, but it is manifestly a very distinct species. It may be thus described:—

Natrix atra, mihi. Scales hexagonal, or sub-rhomboidal on the body, becoming broader on the tail, imbricate, smooth, convex. Tail one-third of the total length. The gape reaches to the middle of the occipital plates, but the rictal furrow extends beyond their tips. Gape nearly straight, slightly arching downwards, and rising behind. A row of fine teeth in each jaw, and one in each palatal, pointing backwards. Labial plates eight, of which the third, fourth, and fifth form the lower part of the orbit: they increase in size from the front to the sixth, and then diminish. Vertical plate long-pentagonal, straight in front, nearly rectangular behind. Superior orbitals large, projecting. Occipitals large, sub-pentagonal, their interior-front sides very short, posterior ends rounded. Rostral nearly erect, semilunar, with the all localities that I am familiar with, lying coiled up among the dead leaves that the wind has accumulated in the hollows of tree roots, or gliding swiftly through the herbaceous weeds at the borders of woods. It may often be seen hanging half out of the loose walls so much used as fences, and thus lying in the sunshine, watching motionless for the lizards that likewise frequent such places. I could never see a snake thus circumstanced, familiar as the sight was, without being reminded of the simile of the Prophet Amos, — "As if a man . . . leaned his hand on the wall, and a serpent bit him." (Amos v. 19.) It climbs with facility, mounting perpendicularly the smooth trunk of a tree, and gliding along the branches, on which it loves to lie in the sun. If alarmed it will sometimes move along the branch, but generally drops to the ground, lowering its foreparts gradually, but very quickly,

concavity downward. There are 17 rows of scales, arranged in oblique lines. Abdominal shields 180; caudal 137 pairs. The form is slender, very gradually tapering; without any ridge on the back. Colour mentioned in the text.

The scales are sub-rhomboidal, with rounded tips; they are attached to the skin by the whole under surface, so that they cannot be separated without laceration; they are thin and flexible. Each scale is marked at its posterior extremity with a row of minute depressed points, ordinarily two in number, but on the thick part of the tail, where the scales are broader, four is the common number, though some scales have five, and, more rarely, even six: towards the end of the tail, the number diminishes to two again. No trace of these is found on the shields of either the belly or tail; and on the lateral ranges of scales they become evanescent. The last scale of the tail is a minute cup or thimble.

The plates of the head in this species are figured in my "Reptiles," page 191.

and letting go with the tail last of all. The mode in which Colubrine Snakes (and perhaps others) mount trees is, I think, misunderstood. We see them represented in engravings, as encircling the trunk or branches in spiral coils, but this, though it may do very well for stuffed specimens in a museum, is not the way in which a living Snake mounts a tree. It simply glides up with the whole body extended in a straight line, doubtless clinging by means of the tips of the expanded ribs, as we can see that the body is perceptibly dilated and flattened. In fact a Snake finds no more difficulty in passing swiftly up the vertical trunk of a tree, than in gliding over the ground. I have been astonished to remark how slight a contact is sufficient for it to maintain its hold. The Black Snake will allow the greatest part of its body to hang down in the air, and thus remain still, while little more than the tail maintains its position by clinging (straight, not spirally, and not half round it, but longitudinally along it) to the upper surface of a branch: and it will often pass freely and gracefully from one branch to another at a considerable interval, projecting its head and body with the utmost ease across the interval. The motions of a Snake in a tree are beautifully easy and free, and convey the impression that the reptile feels quite at home among the branches.

This is a bold and fierce Snake, often turning when struck and approaching its assailant, with the head erected in a most menacing attitude; the mouth opened to its widest extent. I have seen one, thus endeavouring to attack, when foiled by being struck, and thrown off by a stick, at length become quite enraged, the neck being dilated to nearly an inch in width, and perfectly flattened, so that the white skin could be seen between the scales.

> "Tollentemque minas et sibila colla tumentem."
> VIRG. *Georg.* iii. 421.

It is this dilatation of the neck, but in a much higher degree, which gives so remarkable an appearance to the deadly Najas or Cobras of Africa and India. A Black Snake which I had tied by the neck with a string while I made a sketch of it, struck fiercely at me with gaping jaws as far as its cord would allow, every time I looked up or down. The creoles say that if a dog attacks it, it always strikes at his eyes, and not infrequently produces blindness.

Though not venomous, the bite of this Coluber is rather an unpleasant affair. In Mr. Purdie's botanical tour in the eastern and central parts of the island, he has recorded a case of some interest. "When walking with Dr. Bromfield," says this gentleman, "in St. Ann's, I pointed out a fine Black Snake lying under a stone wall, which he insisted on capturing alive with his unprotected hand, in the belief that the reptile was innocuous, like the common ringed serpent of England. But it proved otherwise; the seizure was strictly mutual; the Black Snake fastened on his hand as he laid hold of it, and bit him severely. The wound swelled for some days, though with little pain, and no dangerous consequences ensued."*

* Comp. to Bot. Mag. ii. 40.

The Black Snake is subject to considerable variation in colour. The most common variety (α) is black, highly polished; with the ventral plates bluish grey, and iridescent. Var. β is black with a dead lustre, like the bloom on a plum; and has a row of large square palish spots on each side of the back, but not on the tail. It appears as if obscurely banded. The chin is white, mottled with brick-red. Var. γ is polished black, like α, but has a single scale here and there pale yellow. Var. δ has the upper parts dark brownish grey; the abdominal shields lead grey; the chin and sides of the face whitish, mottled with brick-red and dark grey. This was a remarkable deviation from the normal colouring. It was killed in January, at Mount Edgecumbe grass-piece, by the seaside. I took from its stomach a common *Anolis*, and a young Woodslave (*Mabouya*) both considerably digested. Lizards constitute the principal food of our Snakes. The greatest length attained by any Black Snake that I have measured was thirty-nine inches, of which the tail occupied exactly one-third. The irides are dark hazel, or golden brown during life, and the pupil is circular.

NEGRO PROPER NAMES.

I learned a very curious fact from an old coloured lady, which may probably be as interesting to others as it was to me. The names which in anecdotes and tales we often see applied to negroes, as Quashy, Cudjo, &c. are not promiscuously appropriated, nor are they meaningless. *They indicate the day of the week* on which the individual was born, and being, as they doubtless are, heathen and African in their origin, they afford an interesting illustration of a weekly division of time among Pagan nations.

An infant born on a Sunday would be named if a male, Quashé, if a female, Quasheba, and so on, each sex receiving a name proper and peculiar to each day of the week according to the following table.

	Male.		*Female.*	
Sunday - -	Quáshe	(Cooa-she) - -	Quásheba	(Cooa-she-ba)
Monday - -	Cúdjo	(Coo-jo) - -	Jóba	([Coo]-jo-ba)
Tuesday - -	Cúbena	(Coo-bena) -	Bénaba	([Coo]-bena-ba)
Wednesday -	Quácco	(Cooa-co) - -	Coóba	([Cooa]-co-ba)
Thursday -	Quão	(Cooa-o) - -	A'bba	([Coo]-a-ba)
Friday - -	Cúffee	(Coo-fee) - -	Feéba	([Coo]-fee-ba)
Saturday - -	Quámin	(Cooa-min) -	Mim'ba	([Coo]-min-ba).

It thus appears that the affix "ba" is a mark of the feminine gender, while the prefix "Coo" or "Qua" (Cooa) is, less exclusively, a masculine distinction. These grammatical niceties indicate a language of considerable regularity.

In the Mandingo language, so extensively spoken among the negroes of West Africa, we are informed by Mr. Macbrair that "no distinction of gender, in so far as regards the termination of words, is known. Only one distinguishing word is used, viz. *muso*, a woman, or female; thus from *dingo*, a child, comes *dinke* (*ding-keo*), a male child or a son, and *ding-muso*, a female child or a daughter." So *jatto*, a lion, makes *jattemuso*, a lioness. (MACBRAIR's *Grammar of the Mandingo Language.*)

EREBUS ODORA.

That fine insect, *Erebus odora*, one of the largest of Moths, frequently attaining an expanse of wing of seven inches, is not uncommon in certain localities. It occasionally flies into the house at night, and I have sometimes found it resting with horizontally expanded wings against a wall under a piazza, or in the angle formed by a rafter, during the day. But it is in the deep and sombre woods that we chiefly see this fine Moth. I know several gloomy glens, where the tall trees nearly shut out the rays of the sun, in which I could be pretty sure to rouse one or more, on any day in July or August. A large log or fallen trunk lying on the ground in the woods, will often harbour one on its under side; in the angle formed by the buttress-like roots of a Cotton-tree, we may see one resting; or even against the dark trunk of some rough-barked tree, without any shelter or concealment, except the resemblance of its own colours, variegations of brown and grey, to the lichened and weather-stained surface on which it rests motionless. It is easily alarmed; and then usually dances to and fro, on rapid wing, without leaving the sight; and suddenly alights on a similar spot to that from which it rose; often the very same corner. These habits of frequenting the dark woods, of suddenly alighting on a tree without hovering, and of resting on a dark surface with horizontal wings, so as to be with difficulty found even though watched to the resting-place, are almost exactly those of the *Catocalæ*, near which I presume these *Erebi* stand, connecting the *Noctuæ* with the *Geometræ.*

Sometimes one of these large Moths is known to reside in a certain hole in a rock, or a hollow tree, to which it resorts with such uniformity, during its hours of repose, that it may almost with certainty be dislodged on any afternoon, by giving a smart rap on the outside, when out it rushes with such a startling suddenness, and with so irregular and zigzag a motion, as often to defy capture, even though we are on the watch for it.

A NATURALIST'S WORK-ROOM.

Let me describe a working naturalist's laboratory. Suppose the time to be 2 P.M., after a morning's excursion to the mountain. In the room are three large tables, one of them against the window, at which a negro youth is sitting. Before him lie half a dozen *birds,* one of which he is skinning; beside him lie scissors, knives, nippers, forceps, a pepper-box of pounded chalk, a jar of arsenical soap, needles and thread, cotton-wool, and other apparatus, with several cones of paper ready to drop each skin into, when finished. Across the room are strung lines in various directions, from which are suspended some hundreds of similar paper cones, each tenanted by a bird-skin; they are thus placed in order that they may dry out of the reach of rats, which nevertheless sometimes manage ingeniously to scramble along the slender lines and gnaw the feet and wing-tips of the specimens. On another table is a large bowl half full of water, in which are thrown the *land shells* collected in the morning. The animals in the dry season withdraw themselves into the recesses of their shells, covered either with their operculum, or, — if this be wanting, as is the case with the numerous tribe of Snails, — with an epiphragma of hardened mucus that answers the same purpose. They are thus difficult to extract; but a few hours' immersion in cold water stimulates them to action, and they crawl about freely. A boy sits at this table with another vessel containing *boiling* water; into this have been put the shells collected on the previous day, after having been all night in the cold water. The hot water kills them immediately, and the lad takes out each in succession, extracts the animal with a needle of suitable size, and separates the *operculum* if there be one. He then takes a soft tooth-brush, and well cleanses the shell from earth, mucus, and all other defilements, rinses it in a basin of clear water, and lays it on a cloth to drain, with its *operculum* in it. When thoroughly dry, each shell is wrapped in soft paper separately, (the operculum having been confined in its cavity by a bit of cotton wool,) and carefully packed with others in boxes.

In one corner of the floor a third youth is engaged in laying out the *botanical specimens* brought in, or in shifting the papers of those already drying. A board three feet square, made double to prevent warping, lies on the floor; and sheets of coarse brown paper of the largest size are spread *open*, on it, one over the other, each with a plant (or more than one if small) between it and the next. The whole series is then covered with a board similar to the bottom one, and loaded with large stones for weights. Every day these are shifted; the upper board becoming in turn the foundation, fresh paper being supplied, on which the plants are laid one by one, as before, and the damp paper taken away. When the shifting is performed, this paper is spread out in the sun to dry, and laid in a heap to be used in turn to-morrow. The new plants are taken from the large portfolio in which they were placed when gathered, and added to those in the press; while such specimens as are sufficiently dried are successively removed to the store-box.

Perchance the curious visitor might see the naturalist himself busy with his *insect-spoils;* immersing the *beetles* in boiling water, subjecting the *Lepidoptera* to the vapour of prussic acid, pinning them in the setting-boxes, and fastening down the wings of the butterflies with little braces of card-paper. Or he might be recording the facts observed in the morning's tour, before their freshness had faded from the memory; or taking sketches of forms and colours that death would destroy; or occasionally glancing a master's eye over the operations of the subordinates.

But other than human tenants occupy this room. The visitor would see hanging against the wall a long low cage containing a dozen or so of the native *Columbadæ*, among which the noble Baldpate and gentle Peadove are conspicuous. Another large cage is inhabited by some of the more gaily coloured fruit-eating birds, as the Cashew-bird, the Blue Quit, &c., and in a gauze-fronted box on one of the tables are half a score Lizards of different species, crawling along the sides or roof, as it happens. Three or four of the gorgeous Long-tailed Humming-birds are flitting here and there snapping up the invisible insects that dance like motes in the sunbeam, or flinging back the light with the lustre of an emerald from their lovely bosoms as they sit contentedly upon the lines, or hover in front of the cup of syrup placed for them on the table's edge. Sometimes several of the minuter Vervain Humming-birds are buzzing like bees in the corners of the ceiling, hanging on wings that are visible only like an undefined cloud on each side. Beautiful Orchideous plants growing on clumps of wood are hanging from pegs in the wall, some throwing out their fantastic spikes of blossoms, others in a state of rest, displaying nothing but a maze of intertwined roots and the shrivelled pseudo-bulbs. Large sacks containing *Orchideæ* newly brought in lie on the floor, and many specimens of the same curious tribe of plants are heaped up under the tables, with *Cactoideæ,* awaiting the time when they may be shipped for England. On the trunks of the trees around the dwelling house, and more especially on the top of a broad buttress of one of the outbuildings, are placed specimens of some twenty or thirty species of epiphyte *Orchideæ,* fastened in various ways, partly that I may enjoy the beauty of a race which has always been a favourite of mine, and partly for a more practical purpose, that of identifying them. In exploring the woods, at all seasons, of course one often sees clumps of *Orchideæ* attached to trees, but not in flower; at these times so great is the resemblance of the bulbs of one kind to those of another, that a close personal acquaintance would be hardly sufficient to determine the species; in such cases my custom has been to bring home a specimen in a growing state, and suspend it either within doors or without until the appearance of its blossom should enable me to identify it. The watching of the daily development of the plants, and the pleasant suspense and expectation of what the flower may turn out to be, are enjoyments that will readily be appreciated by every one who has ever cultivated a flower-garden.

THE BRUSH-FOOTED SPIDER.

A little way beyond Sabito Bottom, on the road to Savanna le Mar, the ground slopes upward from the sea-level to an elevation of about forty feet, and is covered with a dense growth of timber trees. A group of Ebby Palms (*Acrocomia sclerocarpa*), formidable from the long pointed spines with which their trunks and the mid ribs of their spreading fronds are armed, — grew along the sides of the highway, and mark the approximation to a spot known as the resting and roosting place of Boobies, Pelicans, and Frigate birds. The deadly Manchioneel (*Hippomane mancinilla*) grows abundantly in this belt of woods, mingled with Button-wood (*Conocarpus erectus*), Mahogany, Sweetwood (*Laurus leucoxylon*), the singular Prickly-yellow, or Club of Hercules (*Xanthoxylon clava-Herculis*), horrid with bristling spines, and many other trees. The wood bears the common character of inpenetrability, being choked up with lianes, spinous Sarsaparillas, tough Supplejacks, and whip-like Cacti, in great numbers. In addition to these obstructions this wood is particularly infested with the great long-bodied Spider with brush-tufted feet (*Nephila clavipes*), that I have already mentioned as common at Alligator Pond. If we succeed in pushing our way with much difficulty through the briers, our faces are pretty sure to come into contact with the strong threads of these Spiders, and we see them spread over the bushes and between the trees along the road side. The appearance of this beautiful Spider I have already described: I will merely add that its web is perpendicular; the part on which the Spider sits, head downward, is geometric, but this is surrounded on all sides by a vast array of irregular lines, the frame of which consists of compound threads, stretching from the surrounding trees and shrubs. Some of these threads are twelve feet long, of a yellow colour, and nearly as thick as sewing silk; I have found them able to resist a great pressure without breaking. But I think it utterly improbable that the rapid and powerful flight of even our most minute Humming-bird could be for a moment arrested by the web of this or any other Spider.

One of my servants informed me in September that he had been bitten by one of these large and handsome Spiders. Coming through the woods at early dawn, his face came into collision with one of the strong webs. He stopped to brush it off, and immediately felt some large insect run down his body, which presently bit him on his great toe. The pain was less severe than that following the sting of a wasp, or even than the puncture of a *Tabanus;* but he described it as having three distinct paroxysms (if I may use such a term for so small a matter). The pain was not of long duration.

THE NURSE SHARK.

About the end of September my lad Sam informed me that the time was now arriving for striking "Nurse," which he so described as to make me interested in the observation of one. A day or two afterwards he told me that he had just seen one brought in at Crab-pond, about two miles from Bluefields, and that he had left it alive, rolling and writhing in agony on the beach; the captors having cruelly cut out the liver, which was all they cared for, without taking the trouble to kill the wretched animal. I immediately rode off to the place, and found the creature just dead; it was a species of *Scyllium,* seven feet six inches long; and its fleshy cirri, about two inches long, depending from the extremity of the muzzle, indicated it to be the *Sc. cirratum* of Cuvier.

Having made a sketch of it as it lay, I returned, determining to secure it on the morrow. Early in the morning, I proceeded thither with my two servants, furnished with knives, a hammer, a bill, and other implements for dissecting and skinning it. The operation was laborious and unpleasant, as decomposition had already made the odour offensive, though twelve hours had scarcely elapsed since death;

and the flies were innumerable, a handsome *Musca*, of a metallic green hue, with brassy reflections. No shelter was near from the scorching sun, which was unmitigated by a cloud, and reflected up from the coral sand of dazzling whiteness. We removed the skin, however, and transferred it to Bluefields across the back of a horse, by the afternoon; but the sun was dipping in the western sea by the time the cutaneous muscles were removed. I strewed the interior with pounded alum, and then with wood-ashes, and the next day stuffed it, and suspended it from a tree to dry.

This fish is sought only for the liver, which is not a very important item, as that of such a fish as I have described will not yield more than a gallon of oil: it is of a fishy smell, and is used only for burning. The animal, however, sometimes occurs of much larger size, attaining the length of twelve feet or more. It is seen, though rarely, at all times of the year, in deep water; but in the month of October, it is said to resort to the shallows; that is to say, the outer side of a shelving bar, which girds the coast in some situations, at a short distance from the shore. The fishermen say the Nurse comes to sleep on the sandy bottom, and that when one has once been seen in a particular locality, they calculate with certainty on finding it there again. The fish is seen lying motionless, or as they say, *asleep*, on the bottom, for many hours at a time.

It is when lying thus, heedless of what is going on above them, that these Sharks are struck with the harpoon, dragged to the surface, and then, the bight of a rope having been passed around the body, behind the pectorals, towed to the beach. Since, however, the one above described, and another lying on the shore, not far off, were both females, full of egg-cases far advanced to maturity, I think it not improbable that the periodical resort to the warm sunny shallows is for the purpose of depositing the ripe capsules; and that the animal may be intent on this process, when supposed to be sleeping. Many animals, when once actually engaged in the operation of ovipositing, become insensible, or at least indifferent, to all that takes place around them, and will allow themselves to be taken or killed without making an attempt to escape.

The egg-capsules were about five inches in length, somewhat pear-shaped, but flattened, destitute of tendrils, of a clear brown hue, and horny texture, but not of great strength: in some of them the young Shark was fully formed, and about three inches long. Many of the capsules were lying on the sand, around the second specimen, which was more advanced in decomposition than mine; and some were floating on the surface of the shallow sea, among the arching roots of the Mangrove trees which were staining the water of the colour of brandy.

The Nurse is of a dull brown hue on the upper parts, without spots; the inferior parts are whitish. The singularly small white eye, with the horizontally linear pupil, imparts a remarkable, and certainly not a pleasing expression to the countenance; but as the eyebrow is not projecting, there is nothing of the sinister and malignant aspect common to the true Sharks. The skin is covered with minute round stony scales, of extreme hardness, and somewhat pellucid, individually. The hide is said to be hence used for polishing cabinet work, but it seems to me hardly rough enough for such a purpose; I have been informed, also, that it makes a durable leather, but I am not aware how it is rendered pliable. The teeth, which are small, have a single high central point, and three small points on each side.

THE HEDGEHOG-FISH.

Specimens of the Hedgehog-fish, or Sea Porcupine (*Diodon*), are frequently carried home by mariners, and it is not uncommon to meet with one hung up as a curiosity in houses in England; where it is sure to attract attention from its globose figure, bristling on every side with stiff sharp spines. In September I obtained one alive, the *Diodon atinga.* I observed that the spines, which vary in length from one-fourth of an inch to an inch, the longest being behind the pectorals, are transparent in their own substance, but are inclosed in a sheath of the common integument of the body; this sheath of skin is either normally pervious at the tip, or made so by the points of the spines, which for the most part protrude a little. This specimen was caught in a fish-pot, sunk to the bottom of the sea. It lived some hours after it had been taken out of the water; but did not inflate its oblong body. Its form was somewhat that of a low cone, the apex being in the middle of the back; the line of the belly nearly straight; the forehead high, and the profile as vertical as that of a Surmullet, with protuberant lips. The air-bladder of this fish is very beautiful; it is shaped somewhat like a heart with the point split instead of the summit; it is of the consistence of kid leather, and of the purest white, with a brilliant satin lustre, but in drying it becomes stiff, semi-transparent, and bladder-like. A playful imagination might trace in it when fresh, a resemblance to a sporting gentleman's wash-leather breeches, tied at the waistband.

THE MANGROVE.

Eminently characteristic of a tropical shore is the dense belt of Mangrove bushes with which it is in many places lined. To an European it is a strange sight to see a grove of trees growing actually out of the sea, and his admiration is not diminished when he examines more closely the structure of these singular plants. The extensive morass at Crab-pond Point, a flat of fetid mud over which the tide flows daily, is closely covered with Mangroves. The trunk of every tree springs from the union of a number of slender arches, each forming the quadrant of a circle, whose extremities penetrate into the mud. These are the roots of the tree, which always shoot out in this arched form, often taking a regular curve of six feet in length before they dip into the mud. The larger ones send out side shoots which take the same curved form at right angles; and thus by the crossing of the roots of neighbouring trees, and of the subordinate roots of each, a complex array of arches is produced, on which one may securely walk for hundreds of yards, probably in some places for miles, about eighteen inches above the mud or above the surface of the water when the tide is in. The average thickness of these natural bows is about an inch, and if stretched straight, they would hardly support the weight of a man; but their vaulted form greatly increases their strength, and though they frequently swerve a little under the foot, I never knew one to break.

On the branches overhead, depending from the tips of the twigs, we see the no less curious seeds. Each is a long club-shaped body with a bulbous base and a slender point more or less drawn out. *They germinate and grow while attached to the parental twig;* those which hang near the water gradually lengthen until the tip reaches the mud, which it penetrates, and thus it roots itself; those which depend from the higher branches, after growing for a while, drop, and then sticking in the mud throw out rootlets from one end and leaves from the other. In the process of growth, the roots gradually assume the arched form, and raise the common centre or base of the trunk considerably above the soil.

The foliage of the Mangrove is dense and leathery, and the aspect of the swamps in which it grows sombre and dismal in the extreme. The sea-water which flows among the roots, though clear, is frequently of a dark brown colour, like strong beer; putrid exhalations continually arise from the daily exposed mud, which, being prevented from dispersion by the density of the foliage, load the air in these situations with poisonous miasmata that frequently form the prolific source of disease and death.

There seems to be a continual encroachment of the land upon the sea in certain parts of the coast by the agency of this tree. The Mangrove growing irregularly projects its sombre shrubberies into the sea in capes and points, inclosing little bays; which by the gradual growth of the encircling points, by and by become lagoons, or shallow salt-lakes. On the sheltered expanse of these beautiful but treacherous lakes, the seeds of the surrounding groves begin to root, and presently we see rising here and there rounded clumps of Mangroves, like little wooded islets spotting its broad bosom. These continually increase in extent, approach each other, and in the course of years unite into a continuous grove.

The unbroken silence and sheltered retirement of these lagoons offer temptations to wading birds, of which they are not slow to avail themselves. The shallowness of the water, which often does not exceed eight or ten inches in depth over a surface of many acres, the abundance of marine animals that inhabit the mud, and the facilities for roosting and incubation tion presented by the arching roots and spreading branches everywhere around, enable these semi-aquatic fowl to pass their lives here in security and content. The timid Water-rail, which from its size, its form and colour, and its habits, so much resembling those of a pullet, has received the appellation of the Mangrove Hen (*Rallus longirostris*), may frequently be seen slowly running over the vaulted roots, or hurrying through the shallow water from the shelter of one clump to another, or wading about, picking up the small crabs on which it habitually feeds. Many of the little Gambets and Sand-pipers (*Tringæ* and *Totani*) also run about here, and occasionally that very curious bird the Rosy Stilt (*Himantopus nigricollis*), whose enormous length of leg so eminently fits it for such situations. The Scarlet Ibis and Roseate Spoonbill are said sometimes to exhibit their brilliant plumage in these dark recesses, though I never met with either: and many of the *Ardeadæ*, from the little Yellow Bittern, and the elegant Egrets, both blue and white, to the Great White Heron with its sweeping plumes of snowy whiteness, and the still larger Herodias, rivalling the stature of man, are commonly to be found here. One of these, the Crab-catcher, Night-raven, or Qua-bird (*Nycticorax Americanus*), is peculiarly characteristic of these sombre solitudes; for though its jealous wariness precludes it from being often seen, its hoarse voice, loud, sudden and startling, not infrequently surprises the benighted traveller as he passes near their obscure depths.

THE VIOLET FLAT-CRAB.

There is another inhabitant of these localities, which I would describe a little more in detail. It is the Violet Flat-crab (*Goniopsis ruricola*, De Geer), perhaps the most beautiful of all our Crustacea, its colours being bright without being gaudy, and its markings striking and elegant. It is abundant in those dismal morasses which border the shore, and which communicate with the sea by means of creeks, the tide running in and flooding the morass, or running out with as rapid a stream and draining it. On the roots and up the trunks of the Mangroves, the Flat-crabs may be seen crawling by thousands, moving with exceeding deliberation if unexcited, but if need be, darting up, or down, or round the tree, as swiftly as a lizard. This at once attracts notice; for we are apt to think a crab's feet not the most adapted for climbing, and perhaps have wondered at the accounts given of the Tree-lobsters (*Birgus*), of the South Sea Islands, which are said to rob the cocoa-nut trees. With our Flat-crab, however, it is a matter of constant occurrence; he runs up any perpendicular surface, *not by grasping*, but on tiptoe, as he runs on a level, not making much use of the claws, though these occasionally aid the progression. Sometimes when watching for birds, seated on one arching root, with my feet on another, remaining quite still, with my fowling-piece across my knees, I have been entertained by observing the manners of these pretty Crabs. When the tide is out, the water collects in little pools, particularly near the margins of the morass; and in and about these pools the Crabs are crawling, seeking for food. The manner in which they pursue their search is curious, and seems to indicate a sensibility of touch in the hard and stony points of the claws, which we should not expect to exist there.

The Crab crawls slowly along, in no defined direction, while the two claws are held, points downward, in front of the face, and lightly feel the surface of the mud, as we should by using the finger and thumb. At very short intervals, one or the other claw picks up some little morsel, — often so small that the spectator can only guess its presence by the action,—and carries it to the mouth with so easy, *so human-like* a motion, that I have been greatly pleased with it; exactly like a person feeding himself with his fingers. That the eyes are not the guides to the situation of the morsels, I feel assured, for they are placed high up on the forehead, and point upwards; and moreover, I have repeatedly seen the claws feel, and even pick up, from *under the body.* I have watched the progress of the Crab, too, to some morsel that I had thrown in; no notice was taken of it, until the claw touched it, as it were, accidentally, in feeling round and round; but the instant it was touched it was conveyed to the mouth.

That the eyes do occasionally aid in the search is apparent; but then the proceeding is different. The Crab leaps suddenly upon the object, and huddles, so to speak, over it, as if afraid of its getting away. In this, and in other actions, there is much resemblance to Spiders. I have sometimes thrown down to the Crabs crumbs of bread, or little bits of meat, to observe their actions: if the piece be too large to be at once transferred to the mouth, it is held with one claw, while the other detaches morsels from it, and conveys them to the mouth; just as I have seen a little pelagic Swimming-crab (*Lupa*) dismember a shrimp that he had caught, holding it in one claw, and picking it to pieces and feeding himself with the other. The Flat-crabs eat slowly: a fragment as large as a pea will occupy a minute and a half or more before it has all disappeared. But a morsel of such a size is not held with one claw, but with the jaws, and *nibbled away.* I think it likely that they catch and devour small animals, as fishes or reptiles. I disabled a small Anolis to prevent its escape, and threw it near to a Flat-crab: the latter leaped fiercely upon it, and seized it with both claws, one holding it by the head, and the other by the loins. It then began to munch the fore-foot with a perceptible crackling of the bones, eating up towards the shoulder. As if fearful of losing its prey, however, it presently retired with it under the root of a tree, and I unfortunately lost sight of it. When two Flat-crabs approach each other, they usually manifest distrust and timidity, retreating or turning aside: but sometimes they cautiously feel each other by stretching out the feet, *not the claws.* In this action, again, they strongly remind one of Spiders.

A SWARM OF DRAGON-FLIES.

October 8th. — What appeared to me an unusual trait in the economy of Dragon-flies occurred to my notice. In the afternoon, I observed, over the stream that runs through Bluefields, and near the point where it plunges over the limestone rock in a little cascade, a swarm of these insects in the air, about twenty feet from the level of the ground. They floated and danced much in the manner of gnats, which they resembled also in the immense numbers which were associated; a most unusual circumstance, for I conjectured that there were not fewer than from three to five hundred, assembled within the space of ten yards square. They seemed to be all of one kind, true *Libelluladæ*, not *Agrions*. I had never before seen any association, except the pairing of the sexes, in this predaceous family of insects: conformably to the usual habits of animals that live by violence, each Dragon-fly generally quarters his own hunting-ground, hawking to and fro in his rapid traverses in grim indifference to his fellows around, whose courses cross and re-cross his own: but these were acting in concert, as was evident not only because they herded close together, but by the unity of their motions. The lower stratum was uniformly moving in one direction, while the upper moved in the opposite; yet the integrity of the mass was, somehow or other, still maintained. They were close to me; but my surprise and interest were so much excited by the unwonted spectacle, that I did not think of disturbing, or attempting to capture any of them.

A DRIVE TO MONTEGO BAY.

October 22nd. — A friend drove me across the island to Montego Bay. The road, which is excellent, leads through a cultivated country almost the whole distance, and offers comparatively little of interest to the naturalist. A magnificent Bamboo walk at Haddo, the most extensive I have seen, is worthy of mention. On the broad undulations of the interior little forest is left standing, and the country has a very open appearance. I saw a single specimen of that curious slender Palm called the Prickly-pole (*Cocos Guineensis*), and a pond was covered with the leaves and flowers of the beautiful Water-lily (*Nymphæa ampla*). My companion pointed out the estate called Shuttlewood, where the few bird-seeds were thrown out, the fruit of which has proved an inestimable benefit to the island. The owner of this estate had received a cage of Finches from Africa, and a bag of grass-seed had been put on board for their food. Soon after their arrival, however, they died, and the few seeds that remained in the bag were thrown away. After a while it was noticed that the horses and cattle eagerly seized every opportunity of devouring the verdure that covered the bank where the bag had been shaken out. The grass was protected, and allowed to go to seed; when it proved to be what is now known as Guinea-grass (*Panicum jumentorum*); it is now cultivated all over the island, especially on the lowland plains of the south side, where it covers even the most rocky soil with its dense tussocks of juicy, nutritive, and ever-verdant pasturage.

Near Mount Carey, one of the stations of the Baptist Missionary Society, there is a singular little grotto close by the road side, half concealed by the climbing plants and slender creepers that hang and trail over it, like that of the Mantuan swain, —

> "——— antrum
> Silvestris raris sparsit labrusca racemis."
>
> VIRG. *Ecl.* v. 6.

The contrast between its refreshing coolness and the burning heat of the dazzling white road I found most delightful, as I stood for a few minutes within its fern-embowered shelter, and drank of the rivulet of clear, cold water, that runs along its bottom.

The approach to Montego Bay is striking and beautiful: the high road is in some parts cut through the marly soil to a considerable depth, leaving high perpendicular banks on each side, crowned with forest trees, and fringed and festooned with ferns and flowering plants differing much from those of the north side. Through the narrow avenue before us, noble views of the coast, of the bay and town, and of the sea beyond, are obtained. I even persuaded myself, more than once, that I saw high land in the horizon, which if real, could be none other than the mountains of Cuba, though about ninety miles distant. But the power of observation was unhappily almost quenched by a violent headache, brought on doubtless by eight hours' exposure to a nearly vertical sun, combined with the glare that was reflected from the white chalky road.

THE RINGTAIL PALLETTE-TIP.

This brief visit to the north side made me acquainted with a little Saurian of great beauty, the Ringtail Pallette-tip (*Sphæriodactylus Richardsonii*). Its ground colour is pale red; the head is marked with irregular bands and stripes of brilliant yellow, and the body and tail are crossed by transverse broad bands, those on the body reddish-lilac, becoming brown on the tail; and at length towards the extremity of that organ, when the ground colour pales to pure white, the transverse bands or rings become deep black. The yellow stripes and the lilac bands are all margined with a narrow line of black, which greatly increases the effect of the rich colours. The scales are large, and form a netted pattern over the body.

This most lovely little reptile was taken in the parlour of my hospitable friend, J. Lewin, Esq., at Montego Bay. On disturbing some papers on the table, it darted out and I secured it. As I kept it in captivity, I noticed the fleshy tongue used to lick the lips, just in the manner of the little *Sphærio-dactylus argus*. It could not walk on the under surface of glass, and slipped slowly down a perpendicular pane. On a painted board it could maintain its hold back-downwards, if the board were turned slowly, to allow it time for preparation; but if it were turned suddenly, or if the reptile attempted to walk, it fell. On rough surfaces it succeeded better. The skin repelled water; and it swam, rather *on the surface*, than immersed. I put flies into its box, but it disregarded them, manifesting impatience only, when they crawled over its face. No change of hue was exhibited; I believe this genus of Geckos have all their colours permanent. The pupil was ordinarily almost circular, but contracted to a perpendicular ellipse in the sun. The claw is sheathed, or protruded at will, on the *inner* side of the pallette, wherewith each toe is tipped, supposing the foot to be directed forwards.

I never met with another example of this pretty

little creature in the island; but Mr. Lewin informed me some months afterwards, that a second individual had been found in his house. There is a specimen in the British Museum, marked as Brazilian; it is curious that so feeble a reptile should be so widely spread. It may, however, have been accidentally introduced into one of these localities, with goods or passengers' baggage; the more readily, as like its congeners, it is an inhabitant of houses.

BEAUTY OF NEGRO VILLAGES.

One cannot look on a little negro hamlet without being struck with its extreme picturesqueness. The peasants who commonly labour on the same estate usually have their huts congregated together, not by the side of a high road, but retired into some secluded nook, approachable through a narrow winding path. You might pass within a stone's throw of the village, and hardly be aware of its existence, except by the hogs which scamper away on the sight of a stranger into the bush, or the poultry that strut and pick about the vicinity. This love of seclusion is almost invariable, and is no doubt a habit inherited from "slavery-time," when it was an object to keep the domestic economy as much out of the way of Buckra as possible. If you purposely seek the collection of cabins, you will probably have some difficulty in threading the maze of Pinguins into which the original fence has spread. This plant (*Bromelia pinguin*) is very commonly cultivated as a fence, being absolutely impenetrable; when not in flower or fruit it can hardly be distinguished from the Pine-apple, but is more vigorous and formidable, the recurved spines with which the edges of the long leaves are set being exceedingly sharp, and inflicting terrible scratches. When flowering in March it is a beautiful object; the central leaves being of the most brilliant glossy vermilion, and the thick spike of blossom of a delicate pink-white. This is replaced by a dense head of hard woody capsules, not united into a compound succulent fruit as in the Pine-apple, but separate, though closely packed. They contain an acid juice, which is pleasant to moisten the lips or tongue, but is found to be acrid and caustic if used in any quantity.

The picturesque beauty of which I have spoken as characterizing the peasants' hamlets does not depend on the habitations themselves; these are small huts, generally made of wattle, or hurdle-work, and thatched with the fronds of some of the Palms. But it is in the variety and grandeur of the various trees in which they are embowered. It so happens that the tropical trees most valued for their fruit are also eminently conspicuous for beauty. The Papaw, whose large fruit has the singular property of rendering tender the toughest meat with a few drops of its juice, and the Cocoa-nut which supplies meat and drink, are fine examples of tall and slender grace. The glossy evergreen of all the Citron tribe, from the great Shaddock to the little Lime, — how beautifully it throws out into relief the noble golden fruit, or serves as a ground for the delicately white blossoms, studding the dark trees like stars on a winter night's sky, as fragrant too as lovely! The Star-apple with its parti-coloured leaves, shining green on one surface, and on the other a bright golden bay, has an indescribable effect, as its mass of foliage, all quivering and dancing in the breeze, changes momentarily in a thousand points from the one hue to the other. But there are two other trees which help more than all the rest to produce the admired result. Both are of stately form and noble dimensions. The one is the Mango, which, though introduced at no very distant period, now grows almost everywhere, at least around every homestead, gentle or simple. It forms a towering compact conical head of foliage peculiarly dense and dark, through which no ray of the sun penetrates. He who has once seen the Mango growing in its own ample dimensions, will never mistake it for another tree, nor ever forget the impression produced by its magnificent form and massive proportions. The other is the Bread fruit; like the Mango, a foreigner made to feel himself at home. The negroes cultivate it more than the higher classes: I was myself disappointed in the fruit; it has a sort of woolliness not agreeable; but I bear willing testimony to the fine appearance presented by it when hanging by scores from the thick many-jointed twigs. The enormous leaves, eighteen inches in length and breadth, elegantly cut into fingers, and of a beautiful green, well set off the large depending fruit, and seem to suit its colossal dimensions.

These are the grander features of the scene, which, mingled with other trees, form groves of many tinted foliage, and much variety of light and shadow. The under growth, however, is no less pleasing. The lively tender green of the Plantains and Bananas planted in regular avenues, the light tracery of the Yams, the Cho-chos, the Melons and Gourds, the numerous sorts of Peas, and other climbers, among which several species of Passion-flower throw their elegant foliage, magnificent blossoms, and grateful acid fruits over the branches of the trees,—the delicate forms of the Castor-oil tree and the Cassavas; the noble flower of the esculent Hibiscus or Okra—these are the ordinary, almost I might say universal, features of a Jamaican negro-garden; and when I add to these fine Convolvuli and Ipomeæ of rainbow hues, the pride of our conservatories, and large white and yellow species of Echites, that, altogether unsought, trail in wild luxuriance about the fences,—I shall be justified in pronouncing the scene one of more than common loveliness, even in the grandeur and beauty of a tropical land.

THE COCOA-NUT PALM.

A grove of Cocoa-nut Palms is a very interesting scene to an European. The radiating tuft of fronds which surmounts the tall stem like a crown, is so unlike any other object, that even a single Cocoa-nut tree stands out conspicuously from the surrounding vegetation, (on a hill side for example, where it is backed by the common forest,) so as to catch the eye at a great distance. There is such a grove behind Bluefields, halfway up the dark mountain. It is always a pleasant sight, for it tells of cultivation and human habitation in the midst of the wilderness.

In walking through an extensive plantation of Cocoa-palms, such as that one which borders the beach to the west of Savanna le Mar, we are strongly reminded that we are in a land remote from home. It is strange to see on every side the tall straight stems, like slender columns, unvaried by any other vegetation, and to be canopied by the rigid pinnated fronds rustling and rattling against each other as they are swayed by the breeze. In common engravings of tropical scenery, the Cocoa-palm has frequently a flexuose character given to the stem which is not natural. A double sigmoid curve is figured, very graceful and pretty, to be sure, but not consistent with truth; while at other times a hanging curve is given to it more like that of a fern springing out of a wall than the habit of a palm. One would suppose from these representations that the character of the stem was extreme flexibility, bowed and curved with every breath of wind.* Such, however, is far from the case. The Cocoa-nut frequently springs from the earth in a very oblique direction, but as it grows it soon bends into a perpendicular, and then shoots up like an arrow towards the sky. Its character is rather a rigid straightness, very little subject to be bent even before the strongest breeze. I have seen the huge fronds toss and flutter like flags, in a

* Thus too Southey makes Madoc describe, more elegantly than correctly, the Cocoa-nut Palms of Mexico:—

"Their tresses nodding like a crested helm."

furious gale, so that it seemed as if they would be torn from their strong bases and carried away; yet even in such circumstances, the bending and rocking of the stem was much less than that of an ordinary tree of correspondent dimensions. The iron-like firmness of the outer wood of a tall palm, the necessary result of its endogenous growth, doubtless is the cause of this rigidity.

The Palms have no proper bark, nor is their surface scored with longitudinal furrows; but they are frequently roughened with transverse projections, the bases of the fronds that have been successively thrown off. In the Cocoa-nut these are strongly marked, and form a sort of rude steps which afford great facility for climbing the tree. Around these bases a curious substance lies in large sheathing irregular pieces. At first sight one would pronounce it a coarse loose cloth, so like a textile fabric are the strong but slender fibres which are interwoven at right angles to each other. It is however a natural tissue, which spontaneously separates from the base of the huge leafstalk.

Another thing that strikes one forcibly in a grove of Palms is to see in the trunk the same thickness associated with all gradations of stature. Nor is the novelty of the appearance much diminished by the knowledge that such a phenomenon is also an invariable accompaniment of the endogenous structure already alluded to. We are so accustomed to see the size of a tree-head, the height, and the thickness of the trunk, always bearing (at least approximately) the same proportions, that when we see the diverging crown of fronds here set on the summit of a stem sixty feet high, and there one of the same extent three feet from the ground, and close by another set of fronds equally expansive, springing from a central spot of earth, and radiating, *without any visible stem at all,* we cannot help an emotion somewhat akin to surprise. The fact is, however, that the stem of a Palm has acquired its full diameter, before it begins to rise from the earth, and its subsequent increase is merely in height, by the progressive development of one great terminal bud. It is this bud, which in the Cabbage Palm is eaten as a delicacy, either boiled, or raw as a salad: the young unexpanded leaves are wrapped over each other so closely, as to acquire a crispness and a tenderness, which with the delicate whiteness produced by the exclusion of light, somewhat resemble those of the heart of a large cabbage.

It may be worth remarking that the frond of the Cocoa-nut is *entire* when first expanded, though plaited. It is the growth of the mid rib which splits each side of the leaf into sword-shaped pinnæ, that by and by are widely separated from each other. The pinnæ and fronds of all our Palms are used by the negroes for thatching, and, being split, for the making of hats and baskets, resembling those of straw.

THE TWO-HEADED SNAKE.

In digging the ground or removing stones that have lain long half-imbedded in the soil, a little Serpent is often found by the negroes, which from the uniform thickness of both extremities, and its power of moving backward or forward with equal facility, they designate as the Two-headed Snake. Naturalists distinguish it as the *Typhlops lumbricalis*, the former term alluding to its apparent want of eyes, the latter intended to mark its resemblance, a rather slight one, to an earthworm.

It is a pretty little animal, and perfectly harmless, though, with the common prejudice against serpentiform reptiles, viewed with dread by the uneducated. It reaches to about thirteen inches in length, with an average thickness of one-fourth of an inch, the fore parts, however, being rather more slender than the middle of the body. The whole form is slightly depressed, the head especially; the head is lengthened and covered with plates; the tail is one-third of an inch long, terminating in a very minute horny nipple, on a shining round plate. When we hold the living animal in the hand, this terminal point of the tail is pressed with some force against the fingers, as if it were a weapon of offence; a slight pricking is produced, but it cannot pierce the human skin. The colour of the upper parts is a chaste bluish grey, that of the belly yellowish white; the two colours abruptly divided, not, however, by a straight line, but by one of that form which in heraldry is technically called embattled, but somewhat irregular. The whole surface is beautifully even and polished while alive, but after having been kept awhile in spirits, the edges of the minute scales become raised, and this smoothness is quite lost, the surface being rough both to the eye and the touch. The colours also become dingy, and scarcely a vestige of the elegance and beauty of the animal is left. Unhappily this is the case with a great proportion of the reptiles of foreign countries; they are known to European naturalists only by specimens sent home in spirits, or by still more hideous skins: they are described in this condition, which admits, it is true, of a recognition of their technical characters; but the beauty of their varied hues, and often the grace of their living forms remain as unknown as ever. Hence reptiles, more than any other animals, are indebted to descriptive observations and carefully coloured figures made from them, while in the enjoyment of life and health.

Several specimens of this little animal, one of the forms that connect the true Serpents with the Saurian tribes, have occurred to me. The first I took myself, when some men were digging out and removing some stones from Bluefields pasture. It was not without difficulty that I secured it, as its agility and power of burrowing were very great. Another was brought me alive, inclosed in a long bottle, otherwise empty; whence on my removing the cork, it poked up its head, and glided out repeatedly, with perfect ease, and was near escaping several times, it was so slippery and so agile. While crawling, it frequently protruded and retracted quickly the little white forked tongue. I perceived no vestige of eyes, nor did it appear to have the power of vision; for on any object being presented to the head, no notice was taken until it came into actual contact, when the creature shrank back in sudden alarm.

As nothing has as yet been recorded, that I am aware of, concerning the early stages of this reptile's existence, the discovery of its egg and of the production of the young possesses some interest. On the 3rd of September, I found in the secluded woods of Auld Ayr, behind Bluefields, an egg to which I was a stranger. I was out shooting with my negro servants, when we heard in these lonely woods what we supposed to be the voice of the Ringtail Pigeon. As this fine and rare bird is said to resort to the smoke that ascends from any fire that may be kindled within its haunts, for the relief which is thus afforded to it from the incessant torment of the musquitoes, we determined to make a fire, in order to get a shot. The lads had collected some deserted nests of Termites for fuel, and on breaking them up, I discovered in one of the cavities an egg of a long-oval form, and of a clear buff hue, with a stiffly membranous integument. The breaking of the surrounding mass had ruptured also the egg, and disclosed a young *Typhlops*, which writhed nimbly about, and soon crawled from its prison, to which it remained attached, however, by the vitellus. It was very active, fully formed, similar in colour and appearance to the adult, except that the inferior surface was tinged with a delicate rosy hue. The eyes were very plainly discernible, though in the adult I had searched for these organs in vain. The frequent protrusion of its forked tongue gave it a snake-like character, which its general aspect did not possess. It was four and a half inches in length, and one eighth of an inch in diameter; depressed in form as the adult. The tail was one sixth of an inch in length; and the *umbilicus* was exactly one inch distant from the caudal point. The egg measured an inch and one eighth in length, and five twelfths in diameter: and this size struck me as surprisingly great, seeing that the greatest thickness of the body in the adult animal is considerably less than that of this egg!* The appearance of a female *Typhlops* with such an egg contained in the abdomen, must be singular, even if but one is developed at a time; but if many are synchronously matured, her dimensions must be immensely enlarged during pregnancy.

In another, of about the same size as this prematurely born young one, or rather less, which was taken on the 1st of November, wriggling quickly along on the ground near Bluefields house, the *umbilicus* was not perceptible, except by an exceedingly slight depression.

I am not aware that this reptile, or any of the allied species, is aquatic in the slightest degree; but its natatory powers are considerable. One, which I put into a vessel of water for observation, swam rapidly and gracefully, throwing the body into elegant vertical undulations like a Leech. Snakes, I think, swim in this way. The fæcal discharges I found to have the white creamy appearance common to Serpents.

Mr. Hill informs me that, in course of the cuttings for the laying down of the railway between Spanish-town and Kingston, the labourers laid open the sub-

* This young specimen, still attached to the egg, is now in the magnificent collection of the British Museum.

terranean habitations of several of these reptiles; a circumstance which afforded subject of wondering comment to the vulgar, who suppose that they are truly monsters, with a head at each extremity. The depth of the burrow, which would be interesting to know, my friend has not mentioned.

THE NASEBERRY BAT.

It was at the Vineyard, near Black River, the residence of Sandford Forrest, Esq., that I first saw this little quaint Bat. I was spending a few days with him in the latter part of December, when it was struck down by one of the children in the house after sunset. The sultriness of a tropical climate obliges the inhabitants to dispense with the closeness of windows, at least to a great degree, and to substitute for them the large Venetian blinds called jalousies. One inconvenience arising from this arrangement is, that no sooner are candles brought, than insects of all orders, particularly minute moths, *Pyralidæ*, *Tineæ*, &c., swarm around the lights, and spread themselves over the table. The entomologist regrets the annoyance less than others, as he thereby occasionally obtains specimens of great beauty and rarity. Those industrious insect-collectors, the Bats, resort thither also on such occasions, to pursue the same game, and perhaps, too, impelled by the propensity which they have to enter narrow orifices. Hence we frequently see these sombre visitants, flitting on swift but silent wing around the room, usually retiring after taking two or three turns.

With an insect-net, however, I have repeatedly captured specimens, having first closed the doors and jalousies. On the evening referred to, more Bats than usual resorted to the house, one and another flying in until the family retired to bed; yet from the openness of the rooms, I obtained only the present specimen, a little species of *Artibeus.*

On my return to Content, however, my attention was directed to some Bats, which every evening swarmed around a large and fruitful Naseberry tree, that overshadowed one corner of the yard. The Naseberry, or neesberry, the *nispero* of the Spanish colonists, and, I believe, of the native Indians (*Achras sapota*), is one of the richest and most agreeable of West Indian fruits. In size and appearance it resembles a very rough russet apple, firm and fleshy, of a rich sugary sweetness; when young, the fruit contains eight or nine cells, diverging from the axis, most of which become abortive, from one to three being usually found when ripe, each containing a large flat oval black seed. When green the fruit yields by incision, as does also the bark of the tree, a viscid milk, which soon acquires, by exposure to the air, a strong tenacity, and makes an effective birdlime. It is much used for the capture of the frugivorous birds, such as the Blue Quit, the Cashew-bird, the Banana-bird, and others, by the negro youths, who call the substance *naseberry gum.* The tree is large, spreading, and handsome, with glossy green leaves, having a tendency to crowd together in rosettes; the flowers form bunches, each being a deep narrow cup, with white fleshy petals, nearly hidden by the calyx.

Around this tree I watched night after night in the fading twilight, in the desire of obtaining one of the Bats that I could see around it; but owing to the rapidity of their movements, and the imperfect light, it was not until I had fruitlessly expended much powder and shot, that I succeeded in procuring a specimen, which I found to be a little Vampyre, and specifically identical with the individual captured at the Vineyard. I thus obtained, however, some interesting acquaintance with the manners of the frugivorous Bats.

About a quarter of an hour after the sun has disappeared, and while the western horizon is yet glowing with those effulgent peak-like clouds, which only a tropical sunset displays — we discover by attentively watching the tree the Bats begin to visit it. First one comes, takes a rapid flight around the tree, darts once or twice through the dense foliage, and winging away is lost in the light of the sky. Another and another comes immediately, and performs the same evolutions; and as the glory of the west fades away to a warm ruddy brown, like the blush of a mulatto girl, many dusky forms are discerned flitting round and round. By carefully following the flight of an individual with the eye, we perceive that now and then he alights for a moment on some object at the extremity of a bunch of leaves; but no sooner has the eye rested on the spot than the sooty wings are again spread, and he is pursuing his giddy course with his fellows. The object of his attention is a ripe naseberry, nestled in the midst of that rosette of leaves. Occasionally the weight of the suspended Bat dislodges the ripe fruit, and it falls to the ground, splitting with the shock. On picking it up we see that it has been just bitten, not gnawed as by the rodent incisors of a mouse, but nibbled in a ragged manner. Though the Vampyres often eat the fruit on the tree in this manner, detaching minute morsels, and again and again returning for more, it appears that not seldom they succeed in tearing out a large piece, which they carry away: for fragments of naseberry of considerable size, partly eaten by a Bat, are frequently found at the distance of half a mile from the nearest naseberry tree, dropped on the high road. The delicate scented Rose-apple (*Eugenia jambos*) is also a favourite fruit with these winged quadrupeds; and fragments of this are dropped about in the same way.*

* As well as I can determine by comparison of specimens with the published description of Dr. Horsfield (Ann. & Mag. N. H. iii. 238.) this seems to be *Artibeus Jamaicensis* of Leach. The description of the latter zoologist is far too vague for the discrimination of species. I will however subjoin admeasurements and descriptions of this, and of three other species of the same genus, all of which have similar habits.

The Small Naseberry Bat. *Artibeus Jamaicensis*, Leach. Central lobe of nose-leaf (see Pl. VI. fig. 3.) lanceolate, broadest near the bottom, thence tapering to a point; furnished at the base with an indistinct footstalk: lobe surrounding the nostril slightly sigmoid in outline: external lobe free, scolloped with about 5 regular crenations. Edge of upper lip nearly smooth; that of lower lip minutely notched. Tragus three-toothed. Forehead round, prominent.

Upper parts ashy-brown, slightly inclined to purplish; head paler, having a hoary appearance; a short dash of white on each shoulder. Wings black; fingers pale. Under parts hoary grey, darkening a little posteriorly. Nose-leaf dusky or reddish brown. Ears yellowish olive. Irides pale brown. Inner edge of interfemoral membrane fringed with free hair. A few fine bristles among the downy fur on the cheeks. Toes of hind feet about equal in length.

THE SILK-COTTON TREE.

Of all the trees that adorn this region of enduring summer, many of which are giants of vegetation, none is more imposing in stature and magnitude combined than the Silk-Cotton tree (*Eriodendron anfrac-*

The Dusky Naseberry Bat. *Artibeus achradophilus*, MIHI. ('Αχρὰς, the wild-pear, the botanical name of the naseberry, and φιλέω, to love.) Central lobe of nose-leaf (Pl. VI. fig. 4.) lanceolate, but less tapering; furnished at the base with a distinct projecting ridge, like a footstalk: lobe of the nostril quite round; external lobe hardly free, with a thickened, slightly-waved edge. Edges of both lips minutely notched. Forehead less prominent; muzzle broad: warts of the mouth more obvious.

General hue a dark dusky brown, nearly uniform, but a little paler on the under parts. Wings black; fingers pale. Toes about equal.

The Great Naseberry Bat. *Artibeus carpolegus*, MIHI. (Καρπὸς, fruit, and λέγω, to pluck.) Central lobe of nose-leaf (Pl. VI. fig. 5.) oblong, with parallel sides, abruptly pointed; a rounded lobe in place of the footstalk: lobe of the nostril much sinuated, with a deep indent near its summit: external lobe obsolescent, reduced to four low but broad warts, with depressed centres. Edge of upper lip notched all round; that of lower lip notched between the incisors, distinctly; elsewhere indistinctly. Head more brutal in form; muzzle large and prominent; forehead low and receding. Toes regularly diminishing in length from the hallux.

Colour a dark bistre-brown, above and below: wings black. Both of these species were taken at Content; they have the same predilection for the luscious naseberry as the first named, a predilection which I have recorded in the trivial names assigned to them.

The Brimstone Naseberry Bat. *Artibeus sulphureus*, MIHI. General form that of *Jamaicensis;* the forehead equally round and high. Upper parts a dusky yellow; head, shoulders, sides, and belly gamboge yellow, dingy; the face and breast tinged with blue: wings black; fingers pale yellow. The nose-leaf, the tragus, and the volar membrane of the only specimen that I met with, had been so much damaged by ants, before it was examined, that no characters could be *tuosum*). When young, it is covered with a green bark; the trunk commonly swells out in the middle, and is studded as well as the branches with great triangular spines, some of them an inch in diameter. But by the time the tree has attained adult age its appearance is very different: the bark is of a hoary grey hue, sometimes almost white; the triangular spines disappear from the bole, and are found only on the upper surfaces of the limbs; the ventricose form of the trunk is generally lost, and, what gives this tree a very remarkable aspect, the basal part of the stem sends out vast spurs radiating in all directions and extending to a great distance. Their outline is very irregular, but their usual form is that of perpendicular walls of timber, often not more than six inches thick, and not commonly exceeding a foot, retaining an even thickness in all parts, but frequently

deduced from those organs: the colour, however, is very marked and peculiar. This specimen also was obtained at Content, in January.

ADMEASUREMENTS OF THE ABOVE FOUR SPECIES.

	Jam.	Achr.	Carp.	Sulp.
Expanse of wings - - - inches	$12\frac{1}{2}$	$13\frac{3}{4}$	$16\frac{1}{2}$	?
Length from crown to fork - - -	$2\frac{1}{2}$	$2\frac{1}{2}$	4	$2\frac{1}{2}$
„ muzzle to tip of calcareum -	?	?	$5\frac{1}{4}$	?
Ear from hinder base of tragus to tip -	$\frac{13}{20}$	$\frac{14}{20}$	$\frac{17}{20}$	$\frac{14}{20}$
„ front base " " -	$\frac{11}{20}$	$\frac{12}{20}$	$\frac{11}{20}$	$\frac{12}{20}$
Tragus, longest side - - - -	$\frac{3}{10}$	$\frac{3}{10}$	-	?
„ shortest side - - - -	$\frac{2}{10}$	$\frac{2}{10}$	-	?
Nose-leaf, from base to tip - - - -	$\frac{9}{20}$	$\frac{10}{20}$	$\frac{10}{20}$	?
Length of fore-arm - - - -	$1\frac{1}{2}$	$1\frac{6}{10}$	$2\frac{1}{8}$	-
„ leg from knee to calcareum -	$\frac{12}{20}$	$\frac{14}{20}$	1	$\frac{13}{20}$
„ foot from calcareum to tip of claws	$\frac{10}{20}$	$\frac{12}{20}$	$\frac{12}{20}$	$\frac{12}{20}$
„ calcareum - - - -	$\frac{3}{20}$	$\frac{3}{20}$	-	$\frac{4}{20}$

twelve or fifteen feet high at their origin in the trunk. I have often thought that very commodious temporary dwellings might be made of the broad areas inclosed by two spurs, with a very little trimming with the axe, covering the top with a sloping roof of palm-leaves on two or three cross-poles. The projection of a dozen or more of these wide, and high, but thin, expansions gives a monstrous bulk to a tree which without them would be of vast dimensions. I have seen the stump of one recently cut down, the solid timber of which, exclusive of the root-spurs, I judged to be not less than forty feet in circumference; though I did not measure it. Its altitude and expanse are equally gigantic; its enormous crown is frequently elevated far above the general level of the forest, and hence particular trees are often specified in nautical guidebooks as land-marks, being conspicuous objects at sea. It is by no means uncommon for the colossal trunk to reach to eighty or a hundred feet in naked majesty, before a single branch is sent forth; and I should think its total height not infrequently reaches a hundred and fifty feet. The limbs are of the bulk of ordinary forest-trees; they commonly break out from the bole three or four upon the same plane, and radiate nearly horizontally to a vast distance. They are often much contorted, and full of sudden angular inflections. Long ragged-looking Cacti (*Cereus triangularis* and other more whip-like species) creep and hang loosely from these limbs; immense numbers of Wild Pines, from the rough hairy tufts of "Old Man's beard" (*Tillandsia usneoides*) to the noble *Æchmeæ* and *Vriesiæ*, are never wanting, and slender flexible Lianes of great length dangle in the air from the lofty branches.

In general a tropical forest knows no such phenomenon as the fall of the leaf; that is, the decay of some and the unfolding of other leaves proceed simultaneously and constantly, so that the foliage is ever full, and ever verdant. There are however a few exceptions of trees, which are periodically denuded, and stretch out their naked arms, as if they had experienced the chill blasts of a northern November. The Tropic Birch (*Bursera*) is one of these, and the Cotton-tree is another. The latter I have seen almost wholly stripped of its full foliage in the course of an hour, during a tempestuous gust of wind in July. It remains bare for several months, the young leaves appearing about the end of May. The green pods in the meantime have formed, ripened, and burst, liberating a quantity of fine silky filamentous down of a pale brown colour. At first sight one is ready to say, What a pity that a material so cotton-like and so abundant should be suffered to fly about upon the breeze, instead of being collected for manufacturing purposes! But in reality it is incapable of being so used, the fibres being found to possess no power of mutual cohesion. Under the microscope they are seen to be simple cylindrical transparent tubes, sometimes flattened, but having neither the twisted form of true cotton, nor the jointed appearance of linen, nor the imbricated surface of wool. They are sometimes used to stuff pillows and mattrasses; and the Palm Swift and the different species of Humming-birds find in them a material sufficiently soft and warm for the reception of their tender young. The latter may often be seen pursuing on the wing the tufts of down as they float along on the wind, and collecting the filaments in their beaks, for employment in their domestic economy.

The negroes regard this magnificent tree with superstitious reverence. They believe that if a person throw a stone at the trunk, he will surely be visited by sickness or some other misfortune. When they intend to cut one down, they first pour rum at the root as a propitiatory offering. In Demerara, I have been told, the African negroes will not lift the axe against a Cotton-tree on any consideration, but in Jamaica the suitability of the wood for canoes overcomes their scruples. The immense trunk is shaped and hollowed, and thus even canoes of large size are made out of a single piece. The softness of the timber facilitates the operation.

These superstitions are doubtless of African origin; their genuine object we may suppose to be the great *Bombax pentandrum*, which is common to that continent and equinoctial Asia; an immense tree so closely allied, as well in appearance as in botanical characters, to our *Eriodendron*, as to be readily mistaken for it.

Since the above was prepared for the press I have been favoured with some details of great interest respecting this noble tree, from Mr. Hill; elicited by an allusion in one of my letters to what appeared to me so singular a circumstance as a tropical tree remaining defoliated ten months in the year. I shall quote my friend's remarks at length.

"I am glad you have asked me to explain the occurrence of the unleafed *Eriodendron* in the month of July, because it enables me to detail some peculiarities in the economy of that magnificent tree, not usually noticed by naturalists.

"The *Eriodendron*, or Silk-Cotton tree, is characterised by the remarkable property of *producing leaves* and *flowers* in *alternate years*. When seen with its seed pods at the terminal twigs, dotting its immense mass of stems and branches all over, it has not yet expanded into leaf; the foliage is still enclosed in the leaf-bud. At this time it is much more an emblem of Hope, than Moore's Almond-tree in Lalla Rookh; it not alone blossoms, but matures fruit upon a leafless stem.

"The *Eriodendron*, or Ceiba, as the Indians called the tree, exhibits a growth by concentric layers of wood, of more or less regulated thickness, *only in the early period of its life*. It is observed soon to become *ventricose* at a short height in the trunk, being thicker about the middle than lower down towards the root. Up to this time the bark is armed with strong spines, which are obliterated when the ventricose character disappears. After this state of progression, it commences throwing out buttresses from the trunk to the large radiated roots, which now show themselves on the surface of the soil. When the growth has advanced to this condition, the wood is no longer deposited in the lower part of the tree in concentric lines of regulated thickness, for the sap, both in ascending and descending, instead of being equally distributed under the bark, is now running in streams from the main branches, and forming those projecting spurs that obliterate the angles from the main roots; and the result is a ligneous deposition, of greatest density where the streams strike off from the trunk. With this unequal deposit of wood, is accomplished those flattened buttresses which become so remarkable a feature in this gigantic tree.

"The *Eriodendron* is one of our few deciduous trees. By deciduous trees I mean trees exhibiting that suspension of vegetative energy for a determinate period of time, which not alone is an exhibition of hybernation, but of a state in which the leaves are *entirely* shed, and the tree stands bare and verdureless, — an economy very unusual in tropical trees. Now this hybernation is annual; but the succession of leaves, though annual too, is biennial as *foliage only*, or as *flowers* and *seed-pods*, and eventually as leaves.

"In every third year, the foliage comes out early after the latter rains, and continues thickening and darkening, and finally becomes sere before the autumn. It is, perhaps, as early as July in these years, when no further sap flows from the roots to the leaf-buds, that it 'declines into the sere and yellow leaf.' The leaves having by that time ceased to perform their office of vegetative lungs, and to give out oxygen to the atmosphere under the action of the sun's rays, the sap is descending as pulp to the roots. The twigs are beginning to harden and shrink, and if not emptied, are now so drained by the terminal buds, which are elaborating flowers for the year to come, that the leaves, of what I will call the *foliage year*, as distinguished from the *alternate flower and seed year*, become soon detached and fall, particularly if the season has been such as to thicken the juices by rapid exhaustion. The frail-bound vegetation withering or not adhering firmly in such a season, would be shaken off in a shower of leaves under any one of those fitful tornadoes, that sweep by so often and so gustily after the sun has a second time reached the zenith of our island, and is hastening with its train of storm-clouds to recross the equator and to enter the southern hemisphere.

"It frequently happens that one half of a Silk-Cotton tree, or some particular cluster of stems and branches, has an alternation of leaves and flowers in a different sequence of years from other parts of the tree. This deviation from what we have laid down as the economy of the *whole* tree is very intelligible as a new condition of *parts of the tree.* It must have been seen that in the long run an Eriodendron or Ceiba, in distributing its sap in streams and lines from the main roots to the main stems, must change from an united to a divided economy of vegetation: — that instead of regulating its functions as one tree, it would set up an order as a bundle of trees clustered together in one column. Now it happens from some factitious circumstance, that one side of the tree, or one set of branches, have suffered some interruption, or have been forced into some acceleration of function as great evaporating organs. This may have been a diminished growing property in the terminal twigs, or an increased nutritive power in a part of the roots; and nourishment may have been retarded, or oversupplied, in the portions of the tree which have gone out of equilibrium. Be it whichever it may of these causes, the alternation of foliage and flowers has been changed; and a half, or less or more than a half, of a Silk-Cotton tree may be seen in luxuriant summer foliage, and the other parts bearing flowers and fruit on a leafless stem, or suffering hybernation, and shedding its leaves partially, while the other portions of the vegetable giant stand up a marvellous pile of verdure.

"I do not wonder that you were not acquainted with this characteristic economy of the Eriodendron, for it requires a residence of five years in the colony, or observations during two successive alternations of flowers and foliage, and the commencement of a third, to gain a knowledge of it."

THE RED HAIRY-TAILED BAT.

Immediately behind Bluefields, the road leading to the mountain is crossed by the little river, which spreads itself out in a brawling, rippling sheet of considerable width, but scarcely more than mid-leg high at the deepest. Here the black women and girls assemble to wash their linen, choosing to come several together, that gossip and jocularity may lighten their labour. The berries of the Soap-tree (*Sapindus saponaria*) supply the place of soap, and make with water a lather in no respect inferior to that (to us) indispensable article of household economy. The little *piccaninnies* play about on the banks, exhibiting their plump oily faces to the sun, or lie on the grass beneath the fruit-trees that are so numerous, sucking the oranges and mangoes that grow in profuse abundance.

One evening in the beginning of December my lad Sam, having gone out to chat with the young women before they left for home, and to assist to place the immense baskets of wet clothes upon their heads, where the negro carries everything, heavy or light—had his notice attracted by a great number of Bats flying a little before sunset under the Avocada Pear-trees. With an insect-net, which he hastened to procure, he captured one, and on the next evening another, both of which were of one species, the Red Hairy-tail of the United States (*Lasiurus rufus*).

One of them had its wing broken by the net, and soon died; the other was uninjured. Its bright rufous colour, round head, *short ears*, prominent nose, and small eyes, gave it a remarkable aspect, especially since it usually kept these last closed; when their position could with difficulty be recognised. When held in the hand, the long tail and ample interfemoral membrane, (which is clothed on its upper surface with rufous hair like the body,) were continually bent up towards the belly. In its impatience of restraint it was perpetually *clawing* with the hind feet at anything within reach, by which it tore several holes in the membrane of its own wings; for the same reason it strove to bite, seizing its hind feet or tail with its jaws, or snapping at my fingers; but I found that though the little teeth were very sharp, they were not moved with sufficient muscular power to penetrate the cuticle of my hand. The ear altered very little; sometimes it was slightly wrinkled posteriorly.

In order to discover whether it possessed cheek-pouches, I caught a small flesh-fly, smaller than our house-fly, and presented it to the Bat. He bit at it instantly, and thus maimed it, but it appeared much too large for his mouth. The fly being viviparous, the little maggots were escaping from the body, and these the Bat picked up one by one from my hand, and ate. At several attempts he at last got the fly itself dismembered, and thus devoured it piecemeal. He drank eagerly, on my holding him by the wings, and putting his nose to the water; his mode of drinking was curious, he just touched the surface with his muzzle, and then brought up his head with a jerk. A drop of water would adhere to it, sometimes projecting in a little globule as large as a pea; this he sucked in by a motion of the mouth as if masticating very rapidly; he repeated this process half a dozen times before he was satisfied. We may infer from hence that in a state of freedom, this, and perhaps other Bats, drink on the wing, like swallows, sweeping down, and just touching the surface with the mouth.

There was no appearance of cheek-pouches either in eating or drinking. While held, it frequently emitted a harsh hissing sound, with the mouth open, and occasionally a little peculiar "click." Both specimens were infested with numbers of a parasite (*Nycteribius*) rather large for the size of the Bat, which ran in and out among the fur with great agility; it appeared as if the hair actually separated and opened spontaneously for their admission.

MUSQUITOES.

These troublesome insects seem nearly equally annoying throughout the New World. I do not think them at all worse in Jamaica, than in Canada or Newfoundland, perhaps not so bad. In marshy places, even in England, the punctures of these minute tormentors (for Musquitoes are merely Gnats)* are as painful, and perhaps as numerous, as in many parts of Jamaica. Some situations are of course more subject to their presence than others. Bluefields, situated on a rising ground, open and exposed to the invigorating sea-breeze, enjoys a remarkable immunity from them. The humid forest harbours them, especially in the mountains; and in many cases the roads are almost quite free from them, where if you step into the wood on either side though only a few paces, you would presently be surrounded by their shrill trumpets, and covered with their bites. There is a good deal of difference in the character of the wounds inflicted by different species: those that frequent the lowlands (*Culex pungens* for example) are of larger size, sing with a graver sound, and insert the proboscis often without any present pain, but a

* Humboldt's remark that the term Musquito is not given to the *Culex*, but to the *Simulium*, though it may be true of South America, certainly does not apply to Jamaica, or to the Northern Continent, where both these genera are but too well known.

hard white tumour presently rises on the spot as large as a silver threepence, which itches intolerably, and remains attended with dull pain and tension, for many hours. The mountain Musquitoes are generally very much smaller, *C. fasciatus*, for instance, a minute species; they are more pertinacious, associate in more numerous swarms, emit a sharp shrill hum, and produce a sudden twinge as they pierce the flesh, as if a spark of fire had fallen on it. A violent itching is the immediate result, but it soon goes off, leaves scarcely any perceptible tumour, and is soon forgotten. These, however, are more intolerable than the former, the recurrence of the spark-like prick at every moment, or rather all over the exposed parts of the person at the same moment, is too maddening to be endured; and it is almost impossible to face a phalanx of these tiny adversaries, where they are numerous, without some device for keeping them off. A fragment of the earthy nest of the Duck-ants (*Termites*) answers well for this purpose; being ignited, it continues to smoulder a considerable time, sending forth a large volume of smoke; this carried in the hand, and waved to and fro, is the most effective weapon against these winged warriors.

A young friend, on whose powers of observation and veracity I have full reliance, has assured me that he once saw a swarm of Musquitoes clinging to each other, in the manner of Bees when secreting wax, till the mass was about as big as a man's fist. He was not at first aware of the nature of the assemblage, till on touching it the whole multitude took to flight in an instant, singing around him, leaving nothing but the point of the twig to which the first Musquito had been adhering.

I found on one or two occasions a little fly, much resembling a Musquito, but belonging to a widely removed genus, of rather singular manners. It is a little Crane-fly (*Limnobia*) * with the mouth produced in an extraordinary degree, so as to form a tube half as long as the insect. It hovers in the air, with the legs (which are of great length and slenderness) hanging down, a few inches from a wall, without altering its position, and almost without perceptible motion, except a slight vibration. It continues to do this for a long time, now and then passing a little to the right or left, quickly, but not so rapidly as a *Syrphus*. It is always solitary; the specimen captured was a male. On putting my hand towards it, it slowly moved away, but I easily took it by surrounding it with both hands. All the individuals that I have seen occurred in the same place, an outhouse at Bluefields; and had exactly the same manners.

THE EYED PALLETTE-TIP.

A very little species of Gecko (*Sphæriodactylus argus*), the smallest Lizard that I am acquainted with, is not uncommonly seen in the dwelling-houses and out-buildings of Jamaica. We see it running along with a wriggling agility in the corners of the floor, creeping out from beneath the doors, hiding in crevices and joints of boards, &c., and often, on opening a window, we perceive lying in the bed of the

* Mr. F. Walker has named the species *L. intermedia*.

frame, one or two of these little harmless creatures, beneath the sash we have lifted; their minute dimensions enabling them to insinuate themselves into situations where we should suppose they would inevitably be crushed to death. The frequency of its occurrence has made the species familiar to the sight of the inhabitants, and it is viewed with no more dread or dislike than we feel for the flies that alight on our window-panes and buzz in the corners. Though it bustles about with much contortion, its speed is small, so that it is very readily caught. Yet on being held in the hand, it writhes so pertinaciously as frequently to slip through the fingers: and so violent are its actions, that though held by the middle or fore parts of the body, I have known it twist its tail absolutely off, spontaneously, in these contortions. Such a circumstance strikingly proves how slight is the tenacity both of bone and muscle in the tail of Saurian reptiles, and in that of the *Geckotidæ* in particular. This member, so readily separated, turns and twists about, of its own accord, and continues to roll over and over, for some minutes after its dissolution of partnership with its body.

The tongue in the Geckos is large and fleshy; and all our species have the habit of protruding it to lick the lips and face, just as a cow does: in the present little species this is an action very frequently performed, and the tongue is thrust out to such a degree, that even the eyes are wiped by it.

Another of its actions reminds one of a cat; an animal to which the Geckos have been compared, on account of their retractile claws, and the pupil of their eyes contracting to a line in daylight. As this little creature runs about, it frequently stops, and creeps stealthily, with its chin and belly brought into contact with the ground, while it *whisks its tail from side to side* very briskly and repeatedly, exactly as a cat will often do.

Between the middle and end of February a good many little eggs were brought to me, covered with a white calcareous shell, of regularly oval form, and exactly resembling a bird's eggs in miniature, measuring $\frac{3}{10}$ inch by $\frac{2}{10}$. Some of these were found in old thatch, and other rubbish, others in crevices of boards, slightly concealed. After a week or two they began to become discoloured, and at length blackish. About the middle of March I opened one, and found the young of this Pallette-tip alive, and perfectly ready for exclusion, so as to run actively about when freed. In form, colour, and markings it precisely agreed with the adult, but was particularly bright. On the 2nd of April I found in the box another, very active, and the egg-shell empty with one extremity thrown off, but not otherwise shattered. Between that time and the middle of April, several more were born. I have also found at the end of May, several eggs with the young perfectly formed, in the cavities of dried Termites' nest.

In the gizzard of a Whitebelly Dove (*Peristera Jamaicensis*), which my servant was skinning, he discovered a small egg, which he brought to me. On opening it I found a perfectly formed Gecko of this same species; yet, strange to say, the integument was not shelly but tough and membranous, of a dirty yellowish hue. Could the action of the gizzard have dissolved the lime of the shell, so as to alter its texture thus? The time when this occurred was about the middle of summer.

THE SAND-GOOTOO.

In walking along the smooth sandy beach between Bluefields and the Creek, our attention is attracted by some little Fishes, the largest not exceeding four inches in length, which shoot hither and thither in the shallows close to the water's edge. They are somewhat fearless, and if alarmed do not flee far, and never into the deep water, but frequently seek to hide among the sea-grass, or attempt to bury themselves in the sand. The negroes call them Sand-Gootoo, to distinguish them from a kind of *Scarus,* which they designate as Eatable Gootoo; the former belong to the genus *Tetraodon.* They have little beauty of form or colour, the upper parts of the body being marked with large sinuous spots of olive on a white ground: the eyes however are brilliant, the irides being golden, tinged with orange, the pupils lustrous green or blue.*

* *Tetraodon ammocryptus,* MIHI. (*Ἄμμος,* sand, *κρύπτω,* to hide.) The upper jaw most advanced; the body not compressed; pectorals rounded; caudal truncate. There are minute points scattered over the belly, scarcely projecting beyond the surface of the skin; and during life so concealed by the loose, mucous integument as to present not the slightest roughness to the feel: they become obvious, however, after a long immersion in spirit. Lateral line invisible. A minute, erect, horn-like wart in front of each eye. The colours are described in the text.

In endeavouring to capture some of these little fishes, a curious habit came to my knowledge. Having in my hand a gauze insect-net, I clapped it over a Gootoo beginning to hide itself in the sand. I felt sure that I had it, but my servant could not feel it with his hand, through the gauze, as I held the ring tightly down upon the bottom of the shallow water. Presently I saw, emerging from under the edge of the ring, an object, that, in size, form, and colour, looked exactly like a hen's egg. The lad instantly seized it, telling me that it was the fish; and as he held it up, I saw with surprise the abdomen tightly inflated to the dimensions described, and the fish still inspiring more air with a sucking noise, and motion of the mouth. To the touch it was as tense as a blown bladder, and it was with difficulty that I could force it into a wide-mouthed pickle-bottle of sea-water, for it filled the neck like a cork. The instant, however, it touched the water in the bottle, it resumed its ordinary appearance, and the change of form was like the effect of magic.

THE PIPER.

A little further off from the beach may be seen that species of *Belone**, called indiscriminately Piper

* Perhaps *B. truncata* of Lesueur, or *B. gerania* of Cuvier and Valenciennes; but the colours do not agree with those of any one of the twenty-five species described by the latter zoologists. I regret that the specimen preserved for comparison is lost, but the following note of the colouring was made from the recent fish. Irides golden (sometimes silvery); the iris depends in a short pointed curtain over the top of the pupil. Back dark green, mottled; edges of the jaws,

or Long-jaw. They are fond of congregating in little parties, containing from two or three to a dozen, skimming leisurely along near the surface, or darting so swiftly as to shoot out of water. The hinder part of the body and the tail maintain a constant wriggling, or uniform lateral serpentine motion. I have procured specimens of these fishes by the aid of my fowling-piece, killing two or three at one discharge of small shot.

THE SPRAT.

Another fish common in shallow water is the beautiful little Sprat, a small species of *Clupea;* it congregates in large shoals over the sandy bottom, or, as caprice may dictate, resorts in smaller numbers to the fissures and caverns of rocks. A great many are taken with the casting-net to be used as bait, and the Pelican makes these shoals his standard food. But their most terrific enemies are the different kinds of Jack (*Caranx*), which attack and "beat" them with such ferocity, that, in their frantic endeavours to escape, the whole shoal of Sprats will often rush high and dry upon the sand. The species appears to be the *Harengula clupeola* of M. Valenciennes.

BEAUTY OF FISHES.

The beauty and gorgeousness of Fishes in the tropical seas have often been admired. These

crown, and lateral line, brilliant emerald green, iridescent. Lower parts silvery, with a most brilliant opalescence on the sides. Fins hyaline, bluish when in the water. Length eighteen inches.

charms are not confined to mere richness and variety of colour. In this little *Clupea* a singular effect is produced by the sides being marked with narrow longitudinal bands, not of colour, but produced by a slight difference in the mode in which the pearly light is reflected. In the "Herring" a species of the same Family, but belonging to a genus (*Chatoessus*) marked by the last ray of the dorsal being prolonged into a whip-like filament, the back is steel-blue, the sides and belly silvery, *with longitudinal lines highly polished;* the whole fish appearing as if made in metal.*

The Jacks, too, just named, are remarkable for the brilliant opaline reflections, which their bodies present, quite distinct from their permanent colours. The reflected hues are principally flushes of purple on the back, and of flame-colour on the sides. These bold and voracious fishes are members of the great Mackarel family: they are distinguished as Buntung Jack, Round-headed, or Ground Jack, and Cavally

* This seems to be new. *Chatoessus eumorphus*, MIHI. The Elegant Bristle-herring. Form very gracefully spindle-shaped; outlines of back and belly uniformly arched from muzzle to caudal; belly-line more curved than that of the back. Height barely one-fourth of total length (reckoned to the fork of caudal): head (from muzzle to gill aperture), to total length, as 1 to $5\frac{1}{4}$. Mouth very small, sub-vertical. Dorsal triangular, pointed, hollowed; filament reaching to base of caudal: pectorals small, pointed: ventrals minute; anal very low, of equal height all along: caudal deeply forked, with the lobes acutely pointed. Belly scarcely serrate. Length seven and a half inches. The colour is described in the text. Irides silvery.

This species differs considerably in form from *Ch. cepedianus*, the only occidental one mentioned by Cuv. and Val.; it approaches *C. nasus* of the Indian Ocean, but is distinguished from it by the longitudinal lines, and also by its proportions.

Jack. This last term is probably a corruption of *cavalho*, and may find an analogy in our "Horse Mackarel," an epithet by which a British species of the same genus is familiarly distinguished.

Some very little fishes are common in the same situations as those above described, which go by the name of Fry, but consisting of two species, so much alike, on a cursory examination, as to be readily confounded. They are all of about the same size, two inches and three-fourths in length, of a greyish, pellucid appearance, with the cheeks and gill-covers, and a broad band running down each side, silvery. Some, which have the mouth and gill-opening enormously wide, are a small kind of Anchovy (*Engraulis Brownii*). The others belong to another genus, and constitute, I believe, a new species of *Clupea*, remarkable for the great length of the head in proportion to the body.* It is a curious circumstance that fishes of different genera should associate together in shoals; is it on account of their close resemblance in form, size, general appearance, and colour?

* The Silver-banded Herring. *Clupea lamprotænia*, MIHI. (Λαμπρὸς, resplendent, ταινία, a ribbon or band.) Head one-fourth of total length, and nearly twice the vertical diameter of the body; back nearly straight: belly very slightly arched: body but slightly compressed: belly not serrated. Fin-rays, D. 13; A. 13; C. 24; P. 13; V. 8. Length two inches and three fourths. Irides silvery. Body pellucid, greyish; a broad band of rich silver runs along each side from the operculum to the base of caudal: cheeks and gill covers silvery: fins grey, transparent. (Plate I. fig. 2.)

I find teeth certainly on the intermaxillaries, and, I think, in a band on the vomer; hence I consider it a *Clupea*, as restricted by Cuv. and Val.; but from the minuteness of the specimens it is difficult to ascertain this point with accuracy.

In the shallows along the beach, and in the interstices of the rugged coral rocks, these tiny fishes swim in immense shoals, huddling so close together, that to a person looking down upon a shoal at his feet, not a speck, literally, of the white sandy bottom is visible beneath them; the group looking like an uniform brownish patch. These, like the Sprats, form a considerable portion of the regular supply of the Rufous-necked Pelican, as well as of the Jack; to avoid whose impetuous attacks it is, no doubt, that they swim so near the shore. When Jack are "beating" (as the phrase is) a shoal of Fry, the latter seem quite stupefied, and pressing together, may be taken with a bucket, or even with the hand: whereas, at other times, if ever so crowded, the pointing of a stick towards the water, in an instant clears a space beneath, so that they cannot be captured with a dip-net, with ever so much caution. They are too small to be put on a hook, but they are much used as ground-bait, and for this purpose are caught by the negro fishermen, who use a cast-net of very fine mesh.

THE MONK BAT.

At first sight I was disposed to think this identical with the Chestnut Mastiff Bat of Phœnix Park, notwithstanding its inferiority of size. Careful admeasurement and examination, however, soon proved it to be distinct, revealing essential differences in the external anatomy: especially in the form and proportions of the ear, its tragus, and the frontal membrane. The lips, in this species, are less pendent, and consequently the mastiff-like physiognomy is less conspicuous: the orifice of the gular gland is larger. The species is the *Molossus fumarius* of Spix, but this designation is not very felicitous, as the general hue is not at all smoky-brown, but a deep umber, rather lighter on the under side. The ears and all the membranes, as well as the face, are black, the last-named tinged with purple, as are the muscles of the limbs. It is remarkable that the volar membranes on both their surfaces have their basal part covered with the common brown fur as far as a line reaching from the elbow to the middle of the thigh; rather narrower, however, above than below.

This little species differs from the Chestnut Mastiff Bat, formerly described, in its manners too. The specimen which first came into my possession was more active on the ground than any that I had seen. On the least touch it ran, or rather crawled, about with such agility, that it was not easy to seize it; and that forward, by means of the wing-thumbs, or backward by means of the hinder feet. It would also strive to jump out of the hand, or would leap off the table; but in falling, it never opened the wings. I did not see it attempt to fly.

It differed from other Bats, also, in being silent; all the time I had it, though it was much handled and incommoded in various ways, it never uttered the slightest click, nor any sound at all. It was fierce, attempting to bite, but could not pierce the callous skin of the end of my thumb, though it drew blood from one of my fingers.

It was at Mount Edgecumbe that this Bat was

discovered. Some labourers, felling a decayed Thatch-Palm (*Thrinax*), found the hollow trunk to be tenanted by Bats, in such aggregated numbers, that my informant, who brought me the specimen, declared that he could have filled a large basket with them. He took five, but all escaped, except the present individual, which he brought to me. Many of them, he said, were larger than this, but he described them all as being "like" it. This was about the end of January.

A month afterwards, I had proof that these statements were worthy of more credit than I had at first assigned to them. On knocking down another Thatch-palm, there was found in the hollow of the trunk another immense assemblage of Bats. A large *cuttacoo*, or negro's basket, was presented to me, which, on being uncovered, displayed a pretty scene of dusky life. The "pie" of our infant days, that contained "four and twenty blackbirds" all ready to sing, was nothing to it. Fifty bats, all alive and kicking, were huddled into the narrow space; an arrangement which, considering their natural propensities, was probably not very disagreeable to them. I examined forty-three, a few escaping from the crowd; and if I was surprised before at the extent of their gregarious habits, I was still more astonished to find that of this number, every one was of the male sex, as had been the one formerly examined. The habit so strongly reminded me of a herd of monks, shutting themselves up in a convent, that I thought the species might well be designated as the Monk Bat.

All the host were of the same species, and there was no perceptible variation in size, but some in colour. Many were of a very intense brown, others paler and brighter, while one might almost be described as fulvous. One of them had irregular patches of white on the breast. As they huddled and crawled over each other, they emitted quivering squeaks. They all displayed the extraordinary activity mentioned above, preferring to run rather than fly, though a few took to wing. In climbing to suspend themselves they used the thumbs or the hind feet indiscriminately. In running along the floor, an action which they performed very swiftly, they rested on the wrists, elevating the fore parts of the body considerably.

The tongue is large and thick, with the posterior half elevated; the papillæ on this portion are large and mammillary; those on the anterior part are small.

In May 1846, my servant caught, in the evening, at Belmont House, two of this species, both of which were females, and one was pregnant. This was the only occasion on which the female came under my notice; its size, form, and colour do not differ from those of the male.

I found this little *Molossus* infested with a curious parasitic insect, a species of *Trichodectes*.

THE GREAT-EARED LEAF-BAT.

One of the most common of the Jamaican *Cheiroptera* is the Great-eared Leaf-bat (*Macrotus Water-*

housei). All the specimens of the species that have come into my possession were taken by their flying into lighted rooms at night, a practice to which it seems more addicted than any other. It is remarkable for the large volume, and for the erect position of its ears, which, however, are occasionally thrown into curves posteriorly, with fine *plicæ;* and also for the great *breadth* of its wings, in contradistinction to their *length* or expanse, which gives it in flight an appearance of much greater magnitude than it possesses. (Plate VI. fig. 2.)

The Great-eared Bat is active in confinement, but in a very different manner from the little Monk Bat (*Molossus fumarius*), for on the slightest cause it leaps up into flight, and that even during the day, when it flies about the room with ease and celerity, now and then hanging head downward against any little projection, or even from the bare plastered wall. One, which I had captured, I put within a glass candle-shade to secure it; and though the area inclosed by the bottom of this was little more than sufficient to allow it to turn its body, it kept continually leaping up perpendicularly to the height of six inches, expanding the wings at the same time as well as the confined space would permit.

Its flight has not the noiseless character common to the Order; the impulse of its great volar membranes producing an audible rushing sound. When placed on the floor it makes no awkward scrambling attempts to crawl, but leaps up at once into the air, and flies as readily as any bird. My specimens uttered no clicking, or other oral sound, when held;

Plate. VI

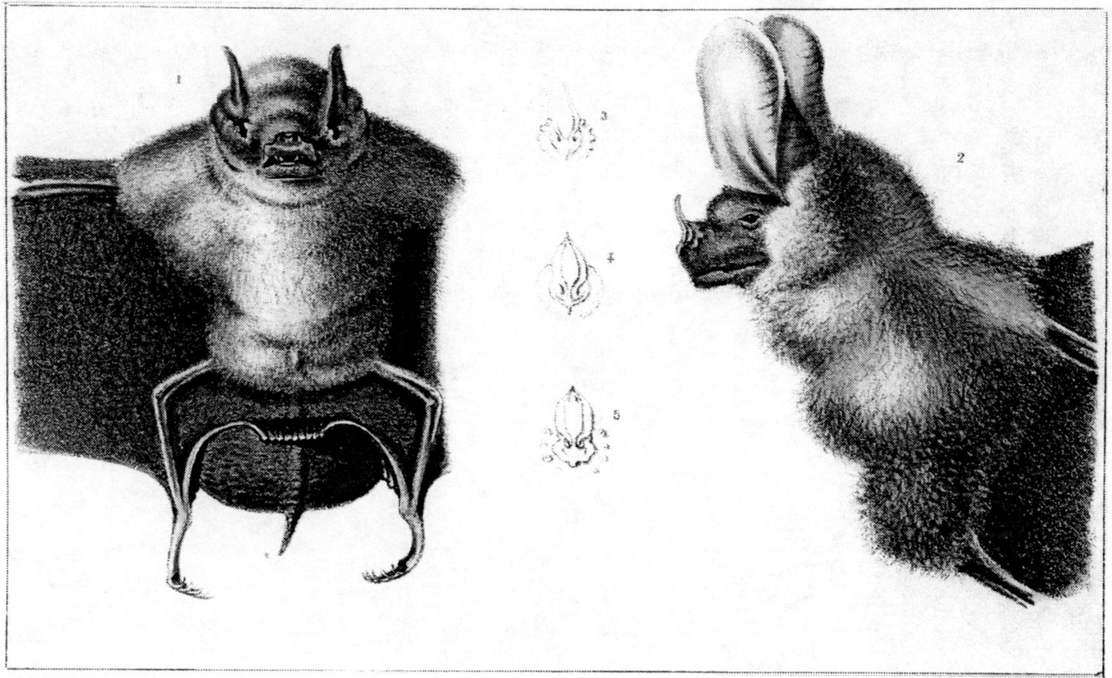

P H GOSSE DEL^T

M & N HANHART IMP^T

1, CHILONYCTERIS GRISEA, N.S. 3.4.5. NASAL LEAVES OF ARTIBEI. 2, MACROTUS WATERHOUSEI N.S.

LONDON. LONGMAN & C° 1851.

but yet were fierce, turning on the hand and drawing blood with their acute little canines. The eyes were rather larger and fuller than usual, and at evening shone brightly; but by day they were, for the most part, nearly or wholly covered by the eyelids. I found this Bat infested with a few winged insect-parasites of small size and buff colour (*Strebla vespertilionis*) which flew away from among its fur even while it was alive.

MOONLIGHT.

There is something exceedingly romantic in the nights of the tropics. It is pleasant to sit on the landing-place at the top of the flight of steps in front of Bluefields House, after night has spread her "purple wings" over the sky, or even to lie at full length on the smooth stones; it is a hard bed, but not a cold one, for the thick flags, exposed to the burning sun through the day, become thoroughly heated, and retain a considerable degree of warmth till morning nearly comes again. The warmth of the flat stones is particularly pleasant, as the cool night breezes play over the face. The scene is favourable for meditation; the moon, "walking in brightness," gradually climbing up to the very centre of the deep blue sky, sheds on the grassy sward, the beasts, lying down here and there, the fruit-trees, the surrounding forest, and the glistening sea spread out in front, a soft but brilliant radiance unknown to the duller regions of the north. The babbling of the little rivulet, winning its seaward way over the rocks and pebbles, comes like distant music upon the ear, of which the bass is supplied by the roll of the surf falling on the sea-beach at measured intervals,—a low hollow roar, protracted until it dies away along the sinuous shore, the memorial of a fierce but transitory sea-breeze. But there are sweeter sounds than these: the Mocking Bird takes his seat on the highest twig of the orange tree at my feet, and pours forth his rich and solemn gushes of melody, with such an earnestness as if his soul were in his song. A rival from a neighbouring tree commences a similar strain, and now the two birds exert all their powers, each striving his utmost to outsing the other, until the silence of the lonely night rings with bursts, and swells, and tender cadences of melodious song. Here and there, over the pasture, the intermittent green spark of the Firefly flits along, and at the edges of the bounding woods scores of twinkling lights are seen, appearing and disappearing in the most puzzling manner. Three or four Bats are silently winging along through the air, now passing over the face of the vertical moon like tiny black specks, now darting through the narrow arch beneath the steps, and now flitting so close over head that one is tempted to essay their capture with an insect-net. The light of the moon, however, though clearly revealing their course, is not powerful or precise enough for this, and the little nimble Leatherwings pursue their giddy play in security.

THE LIGUANEA MOUNTAINS.

Early in March, a day's pleasant sailing along the south shore of the island, in a little coasting steamer, carried me to Kingston. The aspect of the country generally, from the sea, is forbidding: very few traces of cultivation are seen; the harbours are few, and an almost interminable range of dark forest meets the eye, frequently degenerating to low, scrubby bushes, giving the impression of a very barren soil. This is especially the character of the scenery between the bold abrupt promontory called Pedro Bluff, whose broad front of chalk stands up almost perpendicularly from the sea, and the long peninsula of Portland, on which not a single plantation breaks the dismal uniformity of the stunted olive-brown bushes.

Once past this rugged point, the scene becomes more fair and interesting. We open a broad and deep bay, known as Old Harbour, dotted with beautiful islands; its shores rising up in an amphitheatre of verdant hills, bearing the marks of cultivation and residence. The wide mouth of the bay, about fourteen miles from point to point, is studded thick with little low kays, or rocky islets, breaking the waste of water with their refreshing greenness. This noble bay, when Columbus discovered it, was inhabited by thousands of Indians, the most intelligent and the most civilised of all the aborigines of the Antilles that he had seen. On the largest of these islets, embosomed in the sheltered lake-like harbour, dwelt the noble-minded Cacique who proposed to leave his country and to visit, in the protection of Columbus, the distant land of the wondrous strangers, of which he had heard such reports. No carved and painted canoes now steer out upon the glassy bay, from the bowery coves and verdant islets; but the mind delights to recal the picture so vividly, yet so simply, drawn by the illustrious World-finder, of the spontaneous homage paid by the confiding Lord of the Isles to the power and genius of the white men. The barbaric pomp of the stately procession; the richly ornamented canoes paddling in slow time and perfect order; the Cacique himself, in naked majesty, yet decorated with a coronet, a necklace, and a girdle of gold and gems; his sons and brothers supporting his dignity in loyal fealty; his lovely daughters, in native modesty, grace, and beauty,

——— "when unadorned, adorned the most;"

musicians in curious helmets of feathers, playing on tabors and trumpets of ebony; and the standard-bearer in the prow of the royal barge, clad in his mantle of variegated feathers, with a tuft of gay plumes on his head, and bearing in his hand a white banner that fluttered in the breeze; — all come up before the imagination, and combine with the lovely scene, the brilliant sun and sky, the sparkling sea, and the soft landscape, to make one almost wish that these fine and fertile lands could be put back again into their primitive simplicity and wildness, and the pristine inhabitants restored in their happy thoughtless independence.

While remaining at Spanish-town, enjoying the hospitality and society of my valued friend, Richard Hill, Esq., I made an excursion with him to the summit of the Liguanea mountains, an elevation of about 3000 feet. Little of interest occurred on the way up, until we arrived at the top, when a magnificent view of the interior to the Santa Cruz mountains, of the sinuous coast from Kingston to near Alligator Pond, and of the broad plain of Liguanea, spread out as in a map, with Spanish-town in its centre, — repaid the toil of ascent. This prospect has been sketched by Sir Henry De la Bêche, in the Geological Transactions, (vol. ii. 2nd ser.) from which an idea of its grandeur and vast expanse may be obtained. Turning northward, the sea was visible between the peaks on the opposite side of the island, near Annotto Bay; but on the east the view was shut in by a dark and frowning mountain, of greater elevation than that on which we stood, concealing the whole windward end of the island.

Six hours spent in ascending left but little of daylight remaining; and after the beasts were cared for, and our own weariness a little recruited by rest and refreshment, it was almost evening. I spent an hour, however, searching for shells and insects, though with little success; a specimen or two of a *Cyclostoma* hitherto new to me (*C. fascia*), and many of some of the kinds abundant around Bluefields, with the very common *Cylindrella cylindrus*, were all I could find; and insects there were none. Very few birds occurred, and none that were not familiar to me; the Hopping Dick (*Merula leucogenys*) was most numerous; but we heard once or twice in the evening, the long-drawn, clear, mellow note of the Solitaire (*Ptilogonys armillatus*), from the depth of the darkening woods.

The night was delightfully cold and quiet; the coldness made the unfamiliar appendage of a blanket quite agreeable; the stillness was peculiar, such a total absence of sounds as one never finds anywhere in the lowlands, nowhere, I think, except in very lone situations at a lofty elevation; *as if the silence could be felt.* In the morning we again looked with admiration on the extended prospect; there was something exceedingly interesting in the effect of the misty clouds that hung about the face of the towering mountain to the east.* Sometimes it would be quite hidden, overspread with a mass of grey cloud that joined its fellows in the sky; then it would appear dimly, as if covered with a veil of thin muslin; then openings here and there would break, revealing the dark green, almost black, surface of the mountain-side; these openings, rents in the cloudy veil, would flit along, borne by the breeze, now closing up, now enlarging, now coalescing with others; the towering mass at one time almost distinct and uncovered, then gradually dimming, and again brightening; — so that the changes seemed almost magical. The eastern sky was concealed by this vast mass, far up towards the zenith, so that we had no opportunity of seeing the sun rise, but we could witness the sudden irradiation of the plains and distant hills below.

* This must have been St. Catherine's Peak, or one of its mighty spurs.

Soon after sunrise we began to return, but not before I had taken many bulbs of a *Pancratium* with ovate leaves growing near the house where we had slept. It was not in flower; but the blossom was said to be fragrant. We returned by a different route, skirting the summits of the Liguanea mountains, and passing through smiling plantations, in order to descend into the romantic parish of St. Thomas in the Vale. After a while, we crossed and recrossed, many times, the winding Rio D'Oro, and at length entered the magnificent gorge called the Bog-walk (i. e. Bocâgua, *a sluice*), through which runs the Cobre, formed by the union of the Negro and the D'Oro. The road lay for four miles through this deep gorge, by the side of the river, and afforded at every turn fresh scenes of surpassing wildness, grandeur, and beauty. The rock often rose to a great height on each side, leaving only room for the rushing stream, which seemed to have cleft its course, and the narrow pathway at its side. Sometimes, across the river, the side of the ravine receded in the form of a very steep but sloping mountain, covered with a forest of large timber, and so clear of underwood, that the eye could peer far up into its gloomy recesses. Here and there the course of the river was dammed up by islets; some of them mere masses of dark rock, others adorned with the elegant waving plumes of the graceful Bamboo. But the most remarkable object was the immense rock called Gibraltar, which rises on the opposite bank of the river, from the water's edge, absolutely perpendicular, to the height of five or six hundred feet; a broad mass of limestone, twice as high as St. Paul's, in the crevices of which grow many small and slender Fan-palms (*Thrinax*), the plaited fronds of one waving against the roots of its next neighbour. It was now dry weather, but it is said that when the seasonal rains have been particularly heavy, a cataract pours down from the very summit of this lofty rock, breaking into a cloud of mist and spray before it reaches the distant river below. Surely this must be a spectacle of wonderful sublimity!

The rocky sides of the ravine were profusely fringed with the grass-like tufts of a Bromeliaceous plant (*Pitcairnia*), displaying spikes of long scarlet blossoms; and as we approached the termination of the gorge, a *Bletia* arrested my eye, agreeing in habit with that of Bluefields mountain, but with blossoms of a most rich and gorgeous purple hue. I secured as many bulbs of this fine terrestrial *Orchideous* plant as I could find, as well as roots of the *Pitcairnia.* Near Spanish-town, a small shrub growing by the road-side, displayed spikes of Verbena-like blossom of bright violet, which afforded me some ripe seeds. Thus terminated our very pleasant, and not unprofitable excursion.

This romantic ravine received additional interest a few days after our visit, by the occurrence there of a bird of singular brilliancy of plumage, the Scarlet Tanager (*Pyranga rubra*), which had never before been noticed as an inhabitant of Jamaica. The fact was communicated to me by Mr. Hill, after my return to Bluefields, in the following words.

"You will learn with much interest, that on the Thursday after we made our traverse from St. Thomas in the Vale, by the wild and embellished gorge of the river, called the *Bocâguas,* — in some garden grounds near the Flat Bridge where a fortress in ruins, with a block-house above, commands a sweep of the waters,— and where the waters themseves are rendered more picturesque by occasional placid spots, blended with the broken stream, in which detached tufts of Bamboos and Cocoa-nut trees stand reflected in the deep, dark and tranquil intervals,—a Scarlet Tanager in the most brilliant summer plumage attracted attention amid the verdure of the scenery. His sizeable figure, so bright and glaring, as he glanced from one group of trees to another, with his doubled image in the placid parts of the stream, made him 'the observed of all observers.' The knots of travellers along the gorge coming to the turnpike, spoke of the strange and beautiful bird they had remarked in the river-side garden above the bridge, and the turnpike-keeper's attention being, in consequence, drawn to a thing so curious, he went out and shot it. The specimen has been brought to me, after it has been preserved unskinned, and though the plumage has been a good deal deranged by the unskilful management of the person who has preserved it, it has sufficient of the true figure to enable me to make a very faithful drawing of it, and to add it to the list of our migratory visitors."

The female was with her mate, but her sober-coloured plumage possessing no attractiveness, she was allowed to escape unharmed, when the "leaden shower" fell upon her gaudy companion.

THE TRUNK TURTLE.

The following communication to the 'Morning Journal' of April 30th, 1846, relating to the capture of a Trunk Turtle (*Sphargis coriacęa*), though drawn up with somewhat of that vulgar wonder that is common to newspaper science, yet seems sufficiently accurate to merit preservation, while the measurements are valuable, so far as they go. The scene of the capture was at no great distance from Bluefields, being at the extreme west end of the island.

"The anxiety of the fishermen in this little village was aroused on the 30th of last month, by the track of a huge Sea-monster called a Trunk-turtle, which came on the sea-beach for the purpose of laying her eggs. A search was made, when a hole in the sand was discovered, about four feet in depth, and as wide as the mouth of a half-barrel, whence five or six dozen white eggs were taken out; the eggs were of different sizes, the largest the size of a duck's egg. On the morning of the 10th of this month, at half-past 5 o'clock, she was discovered by Mr. Crow, on the beach near the spot where she first came up; he gave the alarm, when all the neighbours assembled, and got her turned on her back. She took twelve men to haul her about two hundred yards. I went and measured her, and found her dimensions as follows, — from head to tail, 6ft. 6in.; from the outer part of her fore fin to the other end [to the tip of the other?] 9ft. 2in.; the circumference round her back and chest, 7ft. 9in.; circumference of her neck, 3ft. 3in.; the widest part of her fore fins, 18in.; her hind fins 2ft. 4in. in length. Her back is formed like a round top of a trunk, with small white bumps in straight lines resembling the nails on a trunk; her colour is variegated like the rainbow [probably the living skin displayed opaline reflections]; there is no shell on her back, but a thick skin like pump-leather:" &c., &c.

"Negril Bay, 13th April, 1846."

THE PEDRO SEAL.

In the Jamaica Almanack for 1843, Mr. Hill published a Memoir on a Seal inhabiting the Pedro Kays, a reef of rocks, lying off the south coast of Jamaicá. As it appears to be a species unknown to naturalists, and as the publication in which it was described had only a transient and local interest, I transcribe the Memoir at length, adding to it such particulars of the natural history of the animal, as have since been communicated to me by my friend.

"The differences which exist in the crania of the *Phocidæ*, and other discrepancies of structure which have been remarked as distinguishing the several genera into which the family is divided, would appear to make the Seal from the Pedro Shoal more allied to the *Ph. vitulina* of LINN. (*Calocephalus*, FR. CUV.) than to any of which we have detailed accounts, although very different from all.* The shoulders,

* From Mr. Hill's description it appears to have the incisors and nail-less hind feet of *Stenorhynchus*, with the molars of *Calocephalus*. The data are perhaps not sufficient to warrant the formation of a new

legs, and thighs, are concealed within the body, and the hand is extremely flattened and fin-like. The cranium is large, high, and convex: — there are ten molar teeth, and two canines in the upper jaw, and the same number in the lower; these, with *four incisors,* above and below, make in all thirty two teeth. The molars are *five-lobed,* and conical; and they terminate in a base of extremely rough enamel. The teeth are so disposed, that when the mouth is closed, there is no interspace above or below them, the points of the upper teeth filling the depressed intervals of the lower ones. Having no external auricles, and ears with foramina so small as to be hardly perceptible, the species belongs to the *Inauriculata* of Peron, or the earless division of Seals. The nostrils are narrow fissures, which appear like two slits in the nose, and are frequently and rapidly opened. The small orifices of the ears are in a similar manner rapidly opened and shut. The lips are full and fleshy, and covered with numerous strong bristles, very flexible, of a black hue with transverse bars of grey. The colour of the body is an intense and uniform black. The hair is short and stiff, and extremely and curiously close. This close bristly covering prevails every where except on the palms of the flippers, which are bare. The fore paw has much more the form of a foot than of a hand, the first finger, answering to the thumb, being the longest.

genus, but I may be permitted to propose the trivial name of *Wilkianus* for the species, in honour of George Wilkie, Esq., to whose courtesy I am indebted for the skin of an adult specimen probably of the same kind, shot by himself.

There are nails only on the fore paws, those of the hinder being rudimentary. The eyes are large, black, and full, and the irides crimson.

"When the specimen from which these notes were made first arrived, it was very lively, and so sensible to the slightest touch, that however lightly the hand might be placed on the fur, it felt the contact, and moved rapidly away, jerking the whole body forwards. When left unmolested it was playful. It ploughed the water with the nose, and snorted as it drew the head out. It grunted like a pig, and barked, growled, and snarled, like a dog. It was fond of turning upon the back and lying dozing. In this posture it slept and basked in the sun. It refused all food, and lived four months without eating. Symptoms of dulness only appeared in the last month, when it was found to be labouring under some disease of the head; and when it died it was discovered to have become totally blind, the dark pupil of the eye having disappeared, together with the crimson colour of the iris. It was surprisingly fat, notwithstanding its long fast. The fat was four inches thick, and yielded four gallons of oil. It was a male, but the organs of generation were not externally perceptible. This organisation is accordant with the peculiarities of the Seal tribe: in the female the teats are concealed in the skin, and the lacteal fulness swells with the rotundity of the body, so that the animal does not suffer pain or inconvenience when crawling on land; and the bifid termination of the tongue, another peculiarity, is an adaptation which enables the young of the *Phocidæ* to seize the nipple under comparatively difficult circumstances, attendant on lactation. The occipital aperture, which remains for a long time unossified in this tribe of animals, being still open, though reduced to a very small orifice, — this Seal may be considered to have been only just full grown. The unworn sharpness of the teeth indicated the same fact.

"The measurements of this specimen were as follows: —

	ft.	in.
Total length, along the back, from snout to tip of tail -	4	2
From snout to insertion of fore paw - - -	1	6
From insertion of fore paw to hind paw - -	2	10
Circumference of body near fore paws - - -	3	2
Circumference at hind paws - - - -	1	6
Breadth of back at fore paws - - - -	1	0
From one fore paw to the other, extended - -	2	6
Length of fore paw 10 in. Of hind paw 11 in.		
Breadth of head across ears, measured horizontally -	0	7
Length of head 9 in. Breadth of nose 4½ in.		
Length of tail 3 in.		

"The Kays frequented by these Seals are situated at about a degree south from this Island, and form portions of an extensive and dangerous line of rocks on a shoal about 100 miles long, the two extremities of soundings touching nearly the 77th and 79th meridians of W. longitude. These banks rise precipitously from the deep ocean, with reefs formed, like the usual rocks in these seas, of coral, with an accumulation of shells and calcareous sand. The depth of water varies from 7 to 17 fathoms. A scanty vegetation covers the principal group of islands, which are what are properly called the Pedro Kays.

The detached islets about 90 miles apart, known as the Portland and Rattlesnake Rocks, are nearly the eastern and western extremities of the bank. This shoal has always been visited as an excellent and inexhaustible fishing ground; and, probably from the variety and abundance of its aquatic animals and marine productions, it received from the Spaniards the name of Vivero, a word equally designating a warren or a fish-pond. The principal supply of turtle for the Kingston Market is derived from these shoals, and the rocks are numerously tenanted by sea-birds."

In the spring of 1846 George Wilkie, Esq., paid a visit to these Kays, and succeeded in obtaining a larger specimen of the Seal. Some notes with which he kindly furnished me, through the medium of Mr. Hill, of the peculiarites of the different islets, depict natural difficulties in the access to Seal Kay, sufficient to account for the meagreness of the information about Seals, possessed by the host of egg-gatherers, who annually resort to those rocks and shoals. Seal Kay lies about three miles to leeward of the principal group. It is about two acres in extent, and rises to twenty feet in height, but is entirely destitute of all terrestrial vegetation. Address, in landing, requires to be combined with strength, hardihood, and perseverance; and frequently before a footing can be obtained, the Seals, the objects of attraction, have escaped to the waters, and continue to avoid the shore as long as intruders remain upon the island. "When Mr. Wilkie's party first landed in their late visit, they surprised some five Seals on shore. They immediately succeeded in heading a 'Bull,' or Male Seal, both big and burly, and killed him. He proved to be an aged patriarch, with teeth nearly worn to the stumps, and a hide gashed and seamed with scars, got in many a fierce fight; — and about ten feet in length.

"In the scramble which the Seal makes to regain the water, nothing is to be remarked but the violence and impatience with which he jerks his body forward; but when he plunges from the shore into the sea, it is no small treat to see the suddenness with which the uncouth animal, so unwieldy and helpless on land, becomes gracefully alert in the ocean. The command with which he strikes through the water, the velocity with which he cleaves the flood, the ease with which he winds the mazes of the rocks, and dashes forward into the hidden recesses of the deep, are beautifully interesting in a creature looking so essentially a quadruped. When the boat is afloat again, the Seals come trooping out to reconnoitre. At a depth of about three feet they paddle about, gazing up through the clear liquid with an expression of countenance beaming with curiosity and intelligence. They dodge around the boat, occasionally ascending to the surface, to renew their inspirations of air, and to look upon their island home, to ascertain whether they may return thither and be at rest.

"A grown-up cub about four feet long had been taken by the people. One Seal was observed more persevering in her watchfulness and assiduity to regain the shore, than the rest. This was conjectured to be the dam of the slaughtered young one. The maternal instinct did not exhibit any stronger emotion than this anxious vigilance. The young one was sufficiently grown to be no longer dependent on the mother. Had it been still sucking, there was enough to show that the parental passion would have merged fearlessness into fury, and inquietude for the safety of its young, into unsparing vengeance for its fate.

"Without doing more than referring to Weddell's observation, that the jaw of the Seals he describes was so powerful in the agonies of death as to grind stones into powder, it seemed, from the condition of the teeth of some eight that were taken during the time Mr. Wilkie's party were on the Pedros, that their strength is exercised in more laborious work than crushing the bones of fishes. The opinion that the more experienced fishermen expressed was, that they fed as generally on molluscous animals as on fish, and that their teeth suffered much wear and tear in the work of breaking shells. Yet it is remarkable that the contents of the stomachs of those killed gave them no insight into the nature of their food:— they were invariably empty.

"I must not omit to mention that our friends had one opportunity of closely observing the progression of the Seal when ascending the beach. The advance was by a succession of zigzag movements. It was evident that the ground was first gripped by one fore flipper, then by the other, that the body advanced first to the right, then to the left, as one or the other flipper took its hold of the earth, and helped they body onward. The seemed to delight in basking in the sun, and to huddle together, and grunt out their pleasure in each other's company."

The skin of one of the specimens obtained in this expedition Mr. Wilkie kindly presented to me; a courtesy the value of which was enhanced by the fact of its being one of the chief of the *opima spolia*, a sort of trophy of his own exploits. It is now in the British Museum. As the skull was not preserved, the actual identity of the species with the smaller specimen described by Mr. Hill, cannot with certainty be established; and there seems a little discrepancy in the proportions, as will be seen by comparing the admeasurements of Mr. Hill's, already given, with the following, which were taken from Mr. Wilkie's specimen: —

	ft.	in.
Length from nose to tip of tail - - - -	6	6
Circumference at fore paws - - - -	3	4
Length of fore paw - - - - -	0	11¼
„ hind paw - - - - -	0	10¾
„ tail - - - - - -	0	2

The fur is of a nearly uniform dirty ash-grey, black at the base, and grey at the tips of the hairs; it is slightly mottled on the belly; it is very close and stiff, and not more than one-fourth of an inch long. The *vibrissæ* or whiskers are from an inch to an inch and three quarters long; white, with one on each side dark brown.

THE YELLOW BOA.

A serpent of the Boa kind (*Chilabothrus inornatus*) is commonly found around Bluefields, and I believe in most parts of the Island, and is distinguished by the appellation of Yellow Snake. It commonly attains a length of eight or ten feet, and a diameter of two inches and a half in the thickest part of the body. The ground colour is yellow, varying from bright golden to a clay-colour, marked with black in irregular spots and confluent bands. These are very few and remote at the fore-parts, but increase posteriorly, the yellow at length disappearing except as scattered spots on an uniform black ground. In the female there is comparatively little black colour; and the tail is very obtuse. The trivial name *inornatus*, which MM. Duméril and Bibron have selected to designate the species, must be considered as comparative; for this Boa, when seen alive, in its black and yellow livery, is, I think, far from unadorned, the contrast of colours being fine, and the purple iridescent glow that is reflected in the playing light from the dark parts of its polished armour is very rich and brilliant. (See Plate IV.)

From its size and aspect the Yellow Snake is looked on with considerable terror by the inhabitants, though no one pretends to have known of a fatal result from its bite. The worst evils that I hear of as consequent upon its attacks are flesh wounds attended with local inflammation, and rendered difficult to heal by the teeth sometimes breaking off in the flesh: terror, however, will occasionally induce fever in the patient. Instances are rare of even these results; the reptile usually retreating with precipitation if attacked. It is not infrequently found in houses and even in beds. I have been told of a gentleman, who on awaking in the morning felt a pressure on his head, which on turning his eyes he found to proceed from a huge Boa coiled up on his pillow: terror-struck, he neither dared to stir nor to cry; and thus he lay till his domestics, anxious at his non-appearance, looked through the window of his bedroom; and discovered the spell. They soon rushed in, and killing the dreaded intruder, released their master. A serpent of this species was discovered in my own bedroom one night at Content, as I was preparing to retire for rest. Though certainly not within the bed, it was but a few inches from my pillow; but the motive of its intrusion, which proved fatal to it, and afforded me the original of a drawing and description, was probably the pursuit of the rats that scampered along the rafters over the bed. A male which I dissected in February had a large mass of rat's hair in the stomach and rectum, consolidated by pressure like the pellets disgorged by owls.

Mr. Hill has recently communicated to me an anecdote illustrative of this Serpent's voracity. "Mr. Kelly of Tophill, Trelawny, showed me a fine skin of a Yellow Snake,—that most powerful of our ophidians. It measured, as I should guess by the eye, nine feet. The Snake was taken in a rat-trap. The pen-keeper, suspecting that the depredation committed in the fowl-house was the act of a Yellow Snake, set a rat-trap to catch him; and succeeded in fixing him by the neck. In his death-struggles he had nearly twisted his body from his head. When opened, he was found to have gulped down, whole and unbroken, seven hen's eggs. I forgot to inquire with what bait he had been enticed to the trap. If it was a recently-killed rat it would no doubt attract him;

but be it what it may, it was a manifestation of considerable voracity to take to the bait, after having swallowed seven eggs."

My negro lad Sam describes to me with much graphic force a scene which he once saw at Mount Airy. A Boa had just seized a hen by the head, and had enfolded her in its mortal coilings: then it proceeded gradually to suck in the whole until only the feet were exposed. Just then a fearless negro woman began to lay on lusty blows with a stout stick, which caused the reptile to disgorge its prey; but so slimy was its surface now, so lengthened and drawn out its form, while much diminished in diameter,—that it could scarcely be recognised. Not the least interesting part of the incident was, that during the whole transaction, the unfortunate hen's brood of young chickens, now orphans, surrounded the serpent, crying as they beheld the tragic fate of their mother. In this case the act of seizure was not witnessed; but Sam has seen a Boa ascend a mango-tree, on one of whose branches a fowl was perching, and when at some distance from the prey begin to dart out and vibrate its tongue, its eyes fixed on the fowl while it slowly and uniformly drew near; the poor hen all the time intently watching the foe, but without stirring or crying. Help came fortuitously, just as the Snake was about to strike, and the fowl was rescued. How strange it is that in widely remote parts of the world we should hear the same statements. Sam has never read what other observers have described about fascination; but he and others affirm from their own observation that some such power is exercised.

When a Boa is irritated by stones being thrown at it, it will sometimes rear up its body perpendicularly, until it appears to rest only on its tail; and then watching a stone, it will suddenly leap after it, darting to an incredible distance through the air, and drop upon it. Sam assures me he has seen a leap of this kind to a distance of full twenty yards, but it was down a declivity. The Black and the Grey Colubers will do the same, and will sometimes attempt even to leap on their human assailants in the same manner. I have been assured of these facts by both my servants, who aver that they have witnessed them, and on whose truthfulness I can entirely rely.

Dr. Palmer of Spanish-town has informed me that in 1829 a Boa of enormous dimensions was killed on land belonging to Sheldon, in St. David's, by Mr. M'Laughlan, the overseer. The people, to insure the death of so terrible an animal, had cut its body into pieces with their machettes or hangers; but the fragments were collected, and having been placed in contact, measured within a very few inches of twenty feet, and were as thick as a man's leg.

One about six feet in length which I kept alive for a short time, was very inert, lying coiled in the bottom of his box, and apparently as unwilling to be disturbed as those Pythons which we see wrapped up in blankets in zoological menageries. It was very offensive, the white creamy matter which it discharged from the cloaca, and which was probably the urine, being of a most overpowering fetor. For the same reason the skinning of one of these Snakes is a very unpleasant operation, as I have abundantly proved.

It is in addition a very tedious one; the skin adhering to the muscles with great tenacity, and needing to be detached with a knife or scalpel, the operator working round and round.

It is currently reported that when the Yellow Boa pairs, which is in spring, others of the same species approach, and twist themselves with and over the pair, until an immense knot or entwined mass is formed. Knots composed of many individuals are certainly often found, and killed without difficulty, as they are then very inert. Mr. Hill, of Spanish-town, once saw five Boas lying together dead on the road, which he was informed had been killed when entwined under such circumstances. This knotting is called by the negroes "cooting," perhaps from the Spanish *coito*. A black man, near Bluefields, going to his daily labour, found a large number thus contorted, and went on killing one after another, until the fetor proceeding from them made him quite faint, and compelled him to turn back homeward.

It is possible that the vast convoluted host of Serpents seen by Humboldt in the Savannahs of Izacubo in Guiana may have had a similar origin: the motive which he suggests seems scarcely consistent with the known habits of those reptiles.*

* Dr. Bancroft mentions this habit of congregating in twisted heaps, with some variation. "I shall also mention on the authority of some planters of credit, that a number of Yellow Snakes, as ten or twelve, are not unfrequently met with in the woody parts of the island with their tails twisted together, but the rest of their bodies free. This chiefly occurs about April and May, at their breeding season as is supposed. When thus surprised, they will raise their tails, and hiss, and it takes them some time before they can unwind

Sam observed, one day near the end of April, two young ones—not, however, *very* young—apparently caressing each other. The one was on a low wall, with its head hanging down; the other on the ground, with its head a little reared; so that their muzzles were in contact. The forked tongues of both were protruded, much vibrated, and seemingly engaged in licking each other's face. A large Boa was at a little distance, looking on.

April 30th.—A lad brought me, from Belmont, six eggs of the Boa. He had observed a Snake crawl out of a narrow hole in the side of a *yam-hill*, a low but wide heap of pulverised earth, in which the yam-tuber is planted. On examination, he found that the hole was the mouth of a narrow passage, just wide enough to admit the body of the reptile, and leading to a large chamber in the centre of the heap, where lay these six eggs. They were long-oval, $1\frac{1}{2}$ inch by $\frac{1}{2}$ inch; plump when first discovered, but now, through exposure to the air, shrunken in at the sides; adhering to each other by small portions of their surfaces, so as to form an agglutinated mass. Their surface was dull white, the integument thin, but tough, and resembling white kid leather. On snipping the skin of one, a clear glaire exuded, in which was a large whitish vitellus, stained with blood-vessels, and containing a young Snake about seven inches long, of which the tail was two and a half. The head was large; the two lobes of the brain promi-

themselves and separate; so that any active person armed might then easily decapitate or destroy them." (*On the Fishes, &c. of Jamaica. Zool. Journ.* v. 419.)

nent, and protected only by a soft skin; the eyes enormous, black, ill-defined; the scutation pretty well marked. It was of a pale flesh colour, but pellucid. One fœtus which I took out writhed.

At the mouth of the hole, the lad informed me, lay a heap of earth, excavated in forming the burrow. But how was it brought out? The boy suggested, "with its mouth." The chamber was well lined with trash, the soft strips of half-dried plantain leaves; these, I suppose, must have been carried in in its mouth.

On my communicating this circumstance to my friend Mr. Hill, he favoured me with the following note: "I should conclude that the Yellow Snake excavates the hole in which it deposits its eggs by loosening the earth with its head, and delivering, at the entrance, the crumbled dirt by the muscular movements of the trunk; the vertebræ and the ribs doing that for the transference of the detached earth from the snout to the tail, which they unitedly perform for the movement of the body forward in progression. The mechanical power is not that of the Archimedean screw, because the motion is not spiral; but it is a similar movement, alternating right and left, and left and right upon a plane; and it equally urges onward anything for delivery along the whole extent of the moving body within the perforated hole. The coil of the body at the extremity of this excavation would form the terminating chamber, where, in the midst of a bed of trash, it deposits its eggs. The spurs of the fig and the buttresses of the cotton-tree are favourite dormitories of the Yellow

Snakes. It is in such places alone that I have met with them in a state of repose."

A few days after this, Sam found a Boa lying in a nest of trash, made between the spurs of a fig-tree on Bluefields Mountain; the nest partly covered by some wood. The Serpent was coiled up, but there were no eggs.

The interesting circumstance of the *Python bivittatus* incubating its eggs, which took place in the menagerie of the Museum of Paris *, is thus shown to be characteristic of the family; the habit being common to the American and Indian species of the *Boadæ*. For the fact that the fœtus, in the case which I have recorded above, was fully formed, and capable of motion when extracted, sufficiently proves that some time had elapsed since the deposition of the eggs, while the exit of the Boa from the nest, which led to the discovery, shows that the parent was still fulfilling the duties of incubation.

Other persons have assured me that often, on killing a female Yellow Snake, they find *the young in her belly*. And this is curiously confirmed by a note from Mr. Hill, who thus writes me: "The Honourable Thomas James Bernard, Member of the Council, has related to me a very curious fact of the Yellow Snake. Lately, his labourers in the Pedro Mountain district, St. Ann's, killed a Yellow Snake containing some ten or twelve grown young ones, varying from eight to ten inches in length. The negroes expressed their surprise at this circumstance, because

* Ann. des Sci. Nat. (2nd ser.) xvi. 65.

they knew that this Boa produced its young from eggs.

"Is this to be received as a case of Snakes that retire upon alarm into the mouth and stomach of the parent? It is stated of the Rattlesnake in Hunter's 'Memoirs of a Captivity among the North American Indians,' that 'when alarmed, the young ones, which are generally eight or ten in number, retreat into the mouth of the parent, and reappear on its giving a contractile muscular token that the danger is past.' Credible eye-witnesses say the same of the European Viper. (See Charlesworth's Mag. Nat. Hist. vol. i. new series, 1837. p. 441.)"

In a recent communication to me, Mr. Hill describes the Yellow Snake as viviparous, on the authority of a young friend studious of natural history. A gravid female of this species, taken by him, brought forth, after some days, eleven young Yellow Snakes. If there was no error in the observation of this case, it must be considered as an aberration of habit. The generation of the *Boadæ* is well known to be oviparous, and the fact above mentioned as having fallen under my own notice, proves that, sometimes at least, our *Chilabothrus* produces eggs. Is it possible that a Serpent, normally oviparous, might retain the eggs within the oviduct until the birth of the young, when circumstances were not propitious for their deposition?

The scaly armour of this Serpent does not protect it from the attacks of parasitic insects. I found one infested by a flat Tick (*Ixodes*), in some numbers, the rostrum of the insect inserted beneath the scales of the body, by which they were nearly concealed. The form assumed by the pupil of the eye, a narrow perpendicular line, which gives so very sinister an expression to the physiognomy, indicates a nocturnal activity. The pupil is round in the *Colubridæ*.

Our Yellow Snake is replaced in Cuba by another form of the same family, very similar in appearance, *Epicrates angulifer*.

Probably other Ophidians remain to be described. Mr. Purdie, in his Botanical Tour (Lond. Journ. Bot. iv. 18.), thus speaks of a Serpent which he saw near Agley Gap, in the eastern part of the Island: "On a loose rock I observed a *large* and remarkable Snake, *striped like a zebra;* but on my attempting to capture the creature, it disappeared among the rocks.

THE PARDALINE SNAKE.

To the kindness of my friend Mr. Hill I am indebted for the knowledge of a pretty little Snake, placed by naturalists in the family of the Boas, yet possessing so many of the characters of the Colubers, that it must be considered as one of the connecting forms of the two groups. It was described under the name of *Leionotus maculatus,* by M. Bibron, in 1840, in the History of Cuba, by M. Ramon de la Sagra, a new genus being instituted for it. In the same year, M. Gundlacher, in the Archiv. naturgesch. von Wiegmann, described it as *Boa pardalis.* Mr. J. E. Gray has given it the generic appellation of *Ungalia.*

Two individuals were presented to me by my friend, of which one was adult, the other about half-grown. The former is grey-brown above, pale-yellow or white beneath; marbled here and there with reddish-brown; and marked with irregular round spots of black. The young one is much more beautiful, being of a pale clay-colour above, becoming cream-yellow on the belly. Four bands of reddish-brown run along each side of the back, and ten rows of round black spots, of which the two dorsal and the two ventral rows contain the largest. The tail is very short, and abruptly narrowed at half its length, where it becomes pale red, unspotted. On the hind head, there is a V-mark of brown, pointing forwards; and a brown band passes from the muzzle through each eye, and on to the sides of the neck.

If we except the *Chelonia* and the Crocodile, this is the only reptile recognised as common to the islands of Jamaica and Cuba. Mr. Hill informs me that it is rather common about Greenwich, in Kingston Harbour, and that it never exceeds sixteen or eighteen inches in length. The younger specimen mentioned above was taken at Mr. Wilkie's, near Spanish-town, in February, 1846. An excellent figure of this beautiful little Serpent is given in Sagra's Cuba; together with an elaborate description; but not a word of information about its habits or economy.

THE OWL-FACED BAT.

May 24th, 1846. — A pretty and interesting little Bat came into my hands, a species of the curious genus *Chilonycteris.** It flew in at an open window at Phœnix Park in the evening, but was not captured until after a very tedious pursuit, manifesting great agility on the wing. It looked much larger in flight

* This closely resembles *Ch. Macleayii* (GRAY), sent from Cuba by the zoologist whose name it bears. It differs, however, from that species in some of its admeasurements, and remarkably in colour. I therefore consider it distinct, and would thus describe it.

Chilonycteris grisea, MIHI. Expanse $11\frac{1}{2}$ inches. Length from nose to insertion of tail, 2 in.; tail 1 in., of which $\frac{2}{10}$ were free, above the membrane. Length of first finger, $2\frac{1}{10}$ in. Ear, measured up the longer side, $\frac{13}{20}$; up the shorter, $\frac{6}{10}$; tragus (front side), $\frac{3}{20}$ in. Muzzle to inner angle of eye, $\frac{7}{20}$; muzzle to ear, $\frac{5}{10}$; thigh, $\frac{8}{10}$; leg, $\frac{3}{4}$; foot $\frac{4}{10}$ nearly; calcareum, $\frac{3}{4}$ in. Dental formula, M. $\frac{5-5}{5-5}$; C. $\frac{1-1}{1-1}$; I. $\frac{4}{4}$; $= 32$. Molars very jagged, with sharp points; upper canines large; upper incisors, middle pair, large, wedge-shaped, like the human incisors, with one notch; lateral pair very small, pointed, leaning towards the middle pair: lower incisors small, with two notches, the points equal in height. Muzzle dilated and truncated, in the manner of a hog's, ending in four points, one above each nostril, and one on each side. Lower lip large, warty: below this there is, as it were, another lip, a thin projecting lamina. The skin at the sides of the head is a continuation of the ear, reaching almost to the muzzle, inclosing, with the ear, a deep hollow, at the bottom of which is placed the eye, the fur diverging in front and below, like the facial feathers in Owls; the eyes are hidden in a front view: probably this structure is to converge the sound, not the rays of light. The head is round, and prettily formed. The reproductive organs were inconspicuous, but it was a male.

Colour of the upper parts brownish-grey; of the under parts pale grey, silky; the whole fur very soft, short, and mole-like. The volar membranes black, much wrinkled. (Plate VI. fig. 1.)

than in the hand, probably from the great extension of the interfemoral membrane.

In captivity it uttered once or twice, very slightly, the peculiar short sound resembling the *clicking* of some delicate piece of machinery, which every one who is familiar with living Bats will remember as common to most of these animals.* It was very active, leaping up to flight from the table, and expanding the wings in a moment, though confined within a candle-shade. It bit fiercely at the hand that held it, but could not draw blood from the fingers. It usually carried the apical half of the interfemoral bent upward at the point where it ceases to embrace the tail, so that the tail seems to extend beyond the membrane. It is held thus by the calcarea, the tips of which, curving downward, carry down again the tip of the membrane, puckered into minute plicæ. This was usually (not invariably) the case, when held in the hand; how it might be in flight, I did not observe; but I suspect that the interfemoral would then be wholly expanded. It died sooner than I expected, and thus precluded my further observations of its living manners.

I never saw that very curious form, *Mormoops Blainvillii*, alive; but Mr. Hill was so kind as to present me with a specimen in fine preservation in spirits. It had been captured at Spanish Town

* I have heard the Marmozette Monkey (*Jacchus*) produce the very same sound; a curious little collateral evidence, if any were wanting, of the correctness of the principles, which impelled the sagacious Linnæus to place the Bats among his *Primates*.

in 1843. The development of membrane on the face of this species, especially in the involved and foliated expansions of the lower lip, exceeds any thing of the kind that I am acquainted with, and forms a very singular spectacle. The colour of the fur is not mentioned in the published descriptions; it is of a delicate light rufous hue; and its texture is particularly fine, soft, and silky.

In a letter dated February 27th, 1847, Mr. Hill mentions a Bat apparently of a well-marked, but new species; certainly very dissimilar to any that occurred to me. He thus writes: — "Among the specimens of Bats that you carried home, did you observe any with the membrane forming the wing in part transparent and in part opaque? I had a Bat given me a day or two ago, which I took to be a *Monophyllus;* — the ears being rounded and the nose-leaf sharp; but the membranes of whose wings were not brown, as is said of the *Monophyllus* assigned to Jamaica. (Pen. Cycl.; *art.* CHEIROPTERA.) Between the outer phalanges, answering to the first and second finger, the wings were of a transparent dirty white; and between the remaining membranes of the fingers and body of an opaque black. The Bat when flying looked like some birds with white-marked quills and secondaries. This specimen, I regret to say, escaped from me when I attempted to get him out of the handkerchief in which I had wrapped him; but I shall endeavour to recover a similar specimen from the place whence this one was procured."

My friend, however, had afterwards reason to think

that this diaphanous peculiarity was not a specific character, but a mere adventitious deviation from normal condition.

THE ALCO.

When I had the pleasure of visiting Mr. Hill at Spanish-town I made the acquaintance of a beautiful little inmate in his family, a white silky Lap-dog, with large melting black eyes. The interest which attached to this Dog was greatly enhanced, when I was informed that it was a specimen of the Mexican Mopsy, the white woolly variety of the Alco, or native Dog, found both in the islands and on the continent at the time of the discovery. To gratify my wishes for all the details that he could give me of the history of this interesting breed of dogs, my friend kindly drew up the following memoir, which he afterwards communicated to me.*

* In an accompanying note my friend thus alludes to the introduction of his illustrative reminiscences: I doubt not my readers will agree with me in thinking that they greatly embellish the subject by their graphic pictorial power, and need no apology.

"You will perceive that I have introduced an extract from my MS. notes of travels in Haïti. The Alco was so much a part of the domestic life of the aboriginal Indians, and this passage was so much an illustration of that life, that I thought it not inappropriate to the subject. In mentioning the fruits, I have set down the Indian and Spanish names generally. The Aguacate is the *Persea gratissima*; Avogada Pear is a corruption. The Guanâbana is the *Anona muricata*, our Sour Sop. The *Marañon* I have indicated by *Anacardium*, the Cashew. The Ciruclas are the *Spondias*, our Plums. The Maimon, as well as I can remember, is the largest of the *Mammeas*."

"The aboriginal inhabitants of these islands are represented as possessing at the time of the discovery a small domesticated Dog, which they called an *Alco*. This little inmate of their homes was remarkable for a gentle quietness of temper, and for an attachment that made it the constant companion of its master. When the indolent and pleasant life of the island was terminated by the servitude imposed by the Spaniards, and the dream in the shade, and the noontide slumber by the fountain, which had been exchanged for no ruder exercise than the choral song, and the evening dance, were succeeded by the labour by day, and the weariness by night; — and life became an endless repetition of the same toil and suffering, the hopeless despondency in which the Indian perished swept away also the attached inmate of his home.

"The domesticated Dog is every where the counterpart of his master. His instincts and appetites assume the passions and pursuits of the being he serves, and mould him into races as varied as the condition of mankind. From an animal of pursuit with the huntsman; of vigilant guardianship with the keeper of flocks and herds; and of sleepless watchfulness with the confiding household; — he becomes an indulged menial or a pampered favourite at the family board. Preserved and perpetuated in the several accidents of breed, to administer to the wants and pleasures of man, we may estimate the social condition of a people by the developed character of their Dogs.

"All writers agree in representing the Alco as a

small animal kept as a familiar pet by the Indian women.* It had yet so much aptitude for out-of-door purposes, as occasionally to return to a state of independence. The *Goschis* of Charlevoix, and the *Gasques* of Garcilasso and Peres, described as small dogs absolutely mute, with downy or silky hair of different and often of bright colours, possessed by the natives of St. Domingo, and the neighbouring islands, and used in the chase of their almost only quadruped the Agouti, before the arrival of the Spaniards, was a dog of the Alco race. The specimen which Mr. Bullock brought from Mexico and exhibited with his collection of Mexican curiosities at the Egyptian Hall, he described as an animal of the wild breed. Colonel Hamilton Smith represents it as having the appearance of a Newfoundland Puppy.† 'It was small, with rather a large head; elongated occiput; full muzzle; pendulous ears; having long soft hair on the body. In colour, it was entirely white, excepting a large black spot covering each ear, and part of the forehead and cheek, with a fulvous mark above each eye, and another black spot on the rump; the tail was rather long, well fringed, and white.' The island breed of this Dog is extinct. We

* Bryan Edwards, in his History of the West Indies (vol. i. p. 116.), when speaking of the Alco, quotes an author named Acosto, who says that "the dogs among the Indians of St. Domingo were a small mute creature, with a nose like that of a fox, which the natives called Alco. The Indians were so fond of these little animals that they carried them on their shoulders wherever they went, and nourished them in their bosoms."

† Naturalist's Library, *Mammalia*, vol. x. Dogs.

see occasionally specimens which come to us from the neighbouring continent, with the silky flow of hair common in the Spaniel, and the Maltese Dog, those caressed household favourites of Europe, but the best known variety of the Indian Alco is the woolly breed, so much sought after under the appellation of the *Mexican Mopsy*. It assumes the lanigerous character, we may suppose, in the colder atmosphere of the mountains of Mexico, — or it may be that it has been mixed with a breed remarkable for a woolly coating found beyond the Rocky Mountains. The Indians are said to spin and work the hair of a dog, along with other woollen materials, into garments.

"When these Islands first greeted the eyes of their European discoverers, the simple manner of life of the natives, free from toil and disquietude, strangely fascinated them. Existence in these Western Edens seemed like a delicious dream. Their holiday life was a reiteration of luxurious indolence. Under a serene sky and a voluptuous climate, they enjoyed the quietude of home, in which their little dog bore them company and shared their affection with their children. In Yasica, in Eastern Haïti, I found unchanged memorials of the ancient Indians. Peculiar slopes, cut like terraces along the hills of the vale, decidedly artificial, were gardens of the aboriginal inhabitants. The little stockade cottages upon these terraces, surrounded by groves of palms, and trees laden with fruit, and with fences enclosing a courtyard filled with flowering shrubs, had, to my eyes, a character so primitive, and so identical with the life of the Indians, that I could not fail to recur to their former owners. 'When, ten years after the discovery of the island, the intolerable tyranny of the Spaniards had driven the inhabitants from what Columbus had described as "the painted gardens of the plain," the homes of the Vega Real, and depopulated them, the natives fled in numbers to the glens and fastnesses of this adjoining district. The limestone rocks, which pierce the summits of the mountains, abound with caverns in which the bones of the unfortunate fugitives who preferred death to servitude, are still found accompanied by the remains of their domestic dog. It was in the hottest month of the year (July, 1831) that I visited this delightful valley. The rains which come with the vertical sun, and prevail all through the solstitial season, had that year been unusually late. They had as yet fallen only in gentle showers, and the constant sea wind that blows refreshingly between the mountains had scarcely heaped on the summits those white accumulations which were to fertilise the fields with seasonable moisture. But the air was agreeably cool, and all through the unprecedented drought of the year, the grass in these sheltered vales was fresh, and the ground overspread with flowers, and the crops of corn and pulse were affected by none of those heated blasts which had destroyed the hopes of the husbandman elsewhere. We found the nights beautiful: the stars shone with exceeding brilliance. Nothing could surpass the magnificence of the green parks of palm and heavy-laden fruit-trees of curious foliage in the villages. There were vestiges, — fragments of earthen vessels, stone hatchets, and chased ornaments, — that involuntarily led the remembrance to the household life of the meek and innocent race that once peopled them. The peculiar trees that shadowed their dwellings, — the calabash that supplied them with drinking bowls; the starry caymite; the russet nispero; the golden marañon, the anacardium; the guanabana; the anona; the rose-apple; the guava; the aguacate; the mammea; the orange; the cirucla; the maimon; the tamarind; the pine-apple, fruits that made part of their simple repast, were all growing, blossoming, and bearing, amid groves of palm. The nightingale * sings there, and the colibri visits the bowers at noonday, — but the people that sat beneath their shadows, where are they? Their cottages are in the village — the upright boards of the palma real, braced with the unhewn hardwood and tied with stems of bejuco, and covered with sheaths of jagua, are the same sort of huts they inhabited. The very hamac that swings there is theirs, but another race are dwellers within them. The glittering palaces of the Spaniard are crumbled to dust — earthquakes have buried them, and revolutions destroyed them. The golden dreams have faded. The anticipated future has deceived the avarice of kings and the venality of nobles. The empire conferred by the high-styled vicegerent of God is passed away. All has been cheated, — the cupidity of ambition, and the eagerness of power. Nothing remains but the sheltered hut and the shadowy garden, — the

* The Mocking-bird (*Mimus polyglottus*) is so called in the Antilles.

unambitious home of which they dispossessed the simple Indian. "There are no vicissitudes for the eternal beauties of nature," says Madame de Genlis: "while, amid blood-stained revolutions, palaces, marble columns, statues of bronze, and even cities themselves disappear, the simple flower of the field, regardless of the storm, grows into beauty and multiplies for ever."'*

"The wild race of Dogs of the Southern Continent which the Indians have reclaimed, and which, six and eight together in company, hunt agoutis, pacas, and wild gallinacea, with a solitary cry heard in the dense forest,—middling in size, light in colour, and close-haired, is the Aguara dog of Surinam. The species is not numerous. Among the aboriginal natives of the Northern Continent, there occurs a small dog of slender make, with broad pointed ears, covered with long white straight hair, and having the body clouded with blackish grey and brown intermingled spots, and the feet well clothed with fur. It is gentle and confiding in disposition, but mute; at least in its native land it is never known to bark. These and the lanigerous dog beyond the Rocky Mountains, are the only races of the native breeds of either continent, in which we trace any of the peculiarities assigned to the Alco, in the two races distinguished by that name.

"Of the three different species of Dog, included by Fernandez in his History of the Animals of New Spain, under the generic name of Alco, Buffon,

* My MS. Notes of Travels in Haïti in 1830 and 1831. (R. H.)

rejecting at once the Hairless Dog, identical with the Bald Turk of the old continent, admits readily the species called *Ytzcuinte Potzotli*, a dog, short-necked in an unusual degree, and humped in shape, with silky hair; and the *Techichi*, wild and melancholy in aspect; as true aboriginal dogs. The first of these two species he recognises as the same with the Lap-dog of *Peru*, — the little plump fondling of the Indian women; and the other as a larger and lighter-made species, but dull and spiritless, which the Indians make use of in the chase, — in which he sees the Forest Dog of *Guiana*. It seems exceedingly probable that there were two breeds of indigenous *Canidæ* in these Islands at the time of the discovery, and that their common name Alco was a generic appellation; — that one breed, the silky-haired Alco, came hither from Yucatan and Mexico, the seat of Indian civilisation; and that the other, the short-haired, was brought in from the Southern Main by the predatory Caribs. The variety now known as the *Mexican Mopsy* differs only in the woolly instead of the silky hair, from the detailed description of Buffon. Pendent ears, the sign of domestication; the forepart of the head white, save round the eyes and on the ears, which are tinged with rufous; the back, sometimes, inclined to yellow; the tail white and short, and flowing but half way down the thighs; the body marked occasionally with black spots; the legs white and toes long. These details are minutely descriptive of our little Mexican favourites, — the yellow-rufous prevailing on the face and ears, inside and out, but more especially on the inside; while the yellow on the back and the black spots on the body are often enough seen in the specimens about the streets.*

"Some four years and a half ago, a friend sent me a grown up pup, then about seven months old, a creole product of a stock originally procured from the Indian main. As this dog, which we still possess, exhibits marked traits of character which are said to prevail always in the breed, I shall set down some few of its peculiarities.

"The first act of *Prince* when brought to us, was to attach himself to a little niece of mine, twelve years old, — to whom, rather than to my sister or myself, he was a present, and he became so exclusively hers, as to disregard, and even receive with displeasure, the caresses of everybody beside. He did not long enjoy the beneficent eye of this mistress; — she was seized with a mortal fever which carried her off in a few days after his arrival in our house. Prince's place was, however, always by her pillow, — and he would rise from the soundest sleep at midnight to kiss her fevered cheek and be fondled by her, if he heard her voice. When she was in her coffin, Prince's place was under the head of it, where he sat silent and sullen, and seemed as much a mourner as any of the family.

"Prince's affection continues to maintain this marked

* "The passage in Edwards, quoted from Acosto (*see note on* p. 331.), represents the Alco as fox-nosed. Very many of the Mexican Mopsies are so fashioned; though Prince has not that feature sharp. A pointed, fox-nosed Mopsy is, however, more common than one with a short nose."

undivided character. His attachment has been transferred to the mother of his first favourite, since the child's death. No other object is permitted to participate in his regards. His whole heart is with his mistress. He sleeps at night by her bedside, — and he selects for his place of repose the spot where her shoes are put down. By day he lies at her feet. — No other living being is allowed to share in her caresses. If a child be taken into her lap, Prince leaps up immediately, and strives to thrust the object of his jealousy away. If he be checked or scolded for his presumption, his countenance assumes a character of unmistakeable displeasure, and he withdraws himself into some retired part of the room, and rejects every endeavour to reconcile him to his disappointment. His eyes being a deep unmingled black, his countenance expresses with distinctness all his varying emotions. We have felt some surprise that he should so passionately have taken to my deceased niece, child as she was, for his first love ; because we have observed that he has an unconquerable repugnance to the notice of all other children whatever : — he not alone refuses their caresses, but whenever they attempt to pat him, he snarls and endeavours to convince them of his dislike by trying to bite them. We have another dog in the house, but he refuses all companionship with him. He tolerates the cat, and will sleep on the same sofa with that fellow-servant, but we have never seen him familiar with any other canine, save a little dog of the same mopsy breed with himself which had been brought to us by some of the neighbours from the street on a supposition that it was our own little pet that had wandered away, and had been lost. This predilection was not induced by any sexual instinct, because this dog was also a male. This is a remarkable evidence of the disposition among animals, when they have become divided into races, to form companionship only with their own allied breeds, and to propagate with them. — 'The wild animal, preserving the same habits, nourished by the same food, sometimes in scarcity, and sometimes in abundance, and exposed to the vicissitudes of the seasons, assumes little variety which may not easily be traced to the operation of these causes; the same soil, climate, and subsistence continue to produce a similar race.'* Domestication, when it has established a peculiar character, maintains a preference for it, and by conformity to this preference removes a tendency to new varieties, until altered circumstances give rise to other qualities and modifications, when a new breed arises, and maintains its own uniformity by an adherence to the same preferences, and consimilarities.

"Prince's jealousy and selfishness leads him to eat up greedily anything he is disposed to refuse, if our other dog, *Spot*, be called to take it. This begrudging temper is very amusingly turned to account when we wish to force upon him medicine. If he is disposed to reject the food with which it is mixed, a call made for the other dog to come and take it, immediately induces him to swallow it hastily, to gratify his splenetic temper; but if anything be given to the other dog first, when he has been waiting or begging

* Dr. J. Bird Sumner on the Records of Creation. Appendix, No. ii.

for a share of it, he withdraws himself in a pet, and refuses to take it when afterwards offered to him. This jealousy, which allows of no rival in favouritism, induced a friend, when he first saw its humours, to observe how distinctly this pet of ours bore traces of the common failings of flesh and blood. The motives that influenced the dog-mind were as perceptible as if they had been declared in the words of some human sentiment.

"Prince's manner of expressing his marked regard is by rubbing his head into the bosom of the object of his affection, much in the way that a cat rubs itself upon a person when particularly disposed to fondness. He is very select in his food. He rejects all vegetables, and eats only cooked flesh; yet he is passionately fond of cakes, particularly those that are spiced, and is perfectly greedy of sugar. He rejects fresh fish, but is insatiably disposed for that which is salt. His sight is not quick, nor his vision distinct; — *this* is said to be the failing of the breed. He is usually disposed to silence and reserve; but he barks, and that vehemently, when he is roused so to express his emotions. But the temper, and the unsociable and passionate regard for a single person in a household, here described, to the exclusion of every one else, is the characteristic of every individual of this particular race.

"I should not forget to mention that there is a wild Dog in Eastern Haïti, very different from the Feral Hound of St. Domingo, delineated by Colonel Hamilton Smith. This wild dog is called a Xibaro (*Hībărŏ*), and at all points resembles the Aguara dog of Surinam. When I was journeying up to the mountains of Maimon beyond the Yuna river, our path in the pine forests was crossed by a solitary Dog of peculiar make. As it lurched past us, I heard for the first time the Xibaro spoken of. When we came to the spot at which the dog had skulked into sight, we found, a little way off our track, a *forest hog*, which had been run down, and just slaughtered. The entrails were torn open, and the Xibaro had been gloating on the reeking blood and steaming viscera. My inquiries led to some precise information about this race of wild dogs. Tradition makes them Indian. They maintain a uniform character in every district in which they are known. They are prick-eared, middle-sized, and light-coloured. A sketch supplied me by a Spanish friend, I find still preserved among my St. Domingo notes, and I close this account of the Alco with a copy of it, as the representation of a remanent aboriginal hunting dog."*

THE MANATEE.

June 13*th*. — A Manatee (*Manatus Americanus*) fortunately fell under my observation at Savanna-le-Mar; having been just captured by getting itself en-

* "I met somewhere in my reading a short time ago with the word 'Xibaro,' applied, as a South American term, to the Wild Dog of the Savannas. The animal spoken of by Col. Hamilton Smith at the conclusion of his notice of the Feral Hound of St. Domingo, as the Wild Dog of Mexico, and the 'Cimarron' of the Pampas, answers in all respects the description of the Xibaro; — small sized, with erect ears; bold and sagacious; not hostile to man, but destructive to the calves and foals of the wild herds; hunting singly and in troops; and burrowing in the open country."

tangled in a seine. It was still alive, and apparently uninjured, but lay with a dull inertness, manifesting little sign of vitality, except now and then a lazy flap of the broad tail, and the periodical opening of the nostrils. Its *pachydermatous* character struck me at once: the skin, rough and coarse, presented no resemblance to the smooth integument of a Dolphin of similar size, which is like the softest and smoothest kid leather, whereas this greatly resembled the hide of a huge pig. The form of the muzzle, also, though peculiar, has much analogy with that of a hog, and the remarkable character of its surface, semi-cartilaginous, plump, and lubricated, is the same in both animals. The development of muzzle, again, strongly marks the typical *Pachydermata*, as it does the Manatee. A few short hairs stand up perpendicularly from the head, especially on the front of the muzzle, on the lower part of which they become stout white tubes, about a line in diameter, and three lines in length. The nostrils are placed on the top of the muzzle, and consist of two tubular orifices about an inch in diameter, when open; but ordinarily closed by a sort of valve of semi-lunar form, so that their position is indicated only by a depression of that form, the horns of which point forwards. The action of breathing is periodical and sudden: I did not measure the intervals of respiration, but it might be several minutes. Suddenly the crescentic depressions become circular openings, and the warm breath is expired; in about half a minute they are closed as suddenly. The valve or stopper has a singular appearance: when its action is carefully watched, it seems to be *the front side of the tube itself* elevated by muscular action to close, and depressed to open, the nostril.*

The small eyes are deeply sunk, and hidden beneath projecting eyebrows: I could not see any eyelids or lashes. The region around and beneath these organs was maintained in a constant state of wetness, by the exudation of tears.

A slight depression down the mesial line of the back marks the course of the spine. The swimming-paws are little indeed like hands, and the nails could not have been detected, if they had not been looked for: they are broad, rough, and black, and are distinguished from the skin of the foot only by a depression around their bases: that is, as in the human nails, the base is lower than the surrounding flesh. I could find only three on each paw. I should much doubt the derivation of Manatee," or as the negroes call it, "Manantée, from *manatus*, handed. If these were indeed the "mermaids" seen by Columbus off the mouth of the Yaqui in Haïti, he might well say that they were by no means the beautiful beings that they had been represented, even though viewed under the influence of that *couleur de rose*, which his excited imagination was then casting upon every thing he saw.

The animal allowed itself to be dragged about,

* Looking at the Hippopotamus in the Zoological Gardens, I was lately struck with its mode of breathing, while in the water. The periodical and sudden opening of the valvular nostrils, and the explosive emission of the air, forcibly recalled to my memory the Manatee of Jamaica.

and even its muzzle to be handled, without the slightest attempt to bite, or any manifestation of impatience.*

After taking the foregoing notes, I made some careful sketches of it as it lay; no representation that I have seen conveying anything more than a rude approximation to its general appearance.† The great size and muscularity of the expanded tail, contrasted with the diminutive paws, at once strike the observer, and show how subordinate the latter are as instruments of progression.

* The following notes of the dimensions, and colours of this individual were taken from it while alive: —

	ft.	in.
Total length - - - - - -	7	8
From muzzle to base of tail - - - -	5	10
Length of tail - - - - - -	1	10
Transverse diameter of tail - - - -	1	11½
Girth of middle of the body - - - -	5	1
From muzzle to eye - - - - -	0	6
Girth of the head - - - - -	1	10
Breadth of body *measured in the curve*, from insertion of one paw to the other - - - -	1	11½
Length of paw - - - - -	1	2
Breadth of paw - - - - - -	0	5½

Colour, an uniform bluish black, rough-grained, with the cuticle peeling in several places, showing the colour brighter and cleaner beneath. Under parts slightly paler; front of muzzle dull grey. The eye was very small, not nearly so large as a man's; the iris very narrow, scarcely a line wide; the pupil comparatively large, circular, blue. The iris of a dull greyish white.

† The Society for Promoting Christian Knowledge have just published a figure in their Large Series of Natural History Engravings, the accuracy of which I can vouch for, as it was made from my own drawings above alluded to.

The next morning I rose long before day and proceeded to the place, intending to procure the skin for preservation. But when I arrived, the negro butcher had already killed the animal, and partly cut it up, so that my purpose was frustrated. I had the pleasure, however, of breakfasting on steaks of its flesh, which was of delicious flavour, without any oiliness: its taste was something between veal and pork, rather approaching the latter. The carcase was eagerly bought up in joints, as a delicacy for the table.

I am well pleased to be able to add some further notes of this little known but very interesting aquatic Pachyderm. My excellent friend, Mr. Hill, writing under date of the 8th December, 1848, from Spanish Town, gives me the following account.

"In the month of August last some fishermen from Old Harbour brought up hither a Manati, which they succeeded in keeping alive several days by taking it early every morning to the river. I saw it immediately after they brought it into town. Its periodically long intervals of expiration and inspiration, in which it opened out its otherwise closely-compressed nostrils, and snorted; and its convulsive flappings of the tail, broad, stiff, and horizontal, like the spasmodic jerkings of a lobster, were all very marked incidents in its economy. The snout, long and cylindrical, with a very decided turgescence, so as to make the diameter of the root less than that of the middle;—the exceedingly discal character of the extremity, with the remarkably shorn-like bristles about it, were very curious features, and never properly represented but in your exquisite little sketches of the Westmoreland specimen.

"This Manati was a female about ten feet long. It had become sufficiently familiar to grunt in answer to its name, Bessy. After exhausting curiosity here, the fishermen took it on to Kingston by railway; but its exposure to the midday sun in its journey so exhausted it, that if it did not die of fatigue, they found themselves necessitated, for fear it would,—at once to slaughter it, that they might keep it marketable at the shambles. It was readily bought up, and spoken of, as deliciously tasted meat."

In a previous letter my friend had communicated some notes of a Manatee that had fallen under his observation, during a sojourn in Eastern Haïti. "On the sands, which throw back the waters of the Yasica at its embouchure, and spread them out into a winding lakelet, a Manati that had been wounded up the river, had come and died. That was the first animal of the kind I had seen. In body it is shaped like the Seal;—but its face has a character decidedly cow-like. It is very obviously the ultimate link of the fluviatile *pachydermata* with the *cetacea*. Its fore legs are shapeless, being neither claws nor fins. Placed near the head so as very inefficiently to assist the motion of its great bulk on land, it was rather surprising to see that it had crept so completely out of the water as to lie dry upon the beach. Its eyes are small, and the orifices which form its ear-holes are narrow slits, scarcely perceptible. It has breasts like those of a woman, placed forward between the paws, and projecting with the swelling rotundity of a bosom. A mass of thick flesh forms its lips. It has molar teeth with square crowns and transverse projections, but no incisors, those of the jaw sufficing to crush the aquatic grasses, and the shore herbs on which it feeds.

"With locomotive powers little suited to enable it to seek the land, as the Seal does, the Manati is almost exclusively an inhabitant of the water. It seldom does more to relieve itself from this element, than raising its head above the stream. It coasts the green banks, and crops the bordering herbage as it swims along. The friend, at whose home on the Yasica I was staying, when I saw this specimen of the Manati, informed me that some time previous to my coming on this visit, he had surprised some six of these tenants of the river in an inlet which he used as a timber dock. He confined them there until they had cropped all the long aquatic weeds that lined the bottom of the inlet; and they grazed harmlessly and seemed to suffer no apprehension from their state of restraint. I tasted the flesh some years ago; — it was like beef: the fat was crisp and delicate. Its vegetable food might reasonably be expected to give it this similarity to the flesh of animals that graze the field. Purchas, in his 'Pilgrims,' gives an interesting account of a Manati, which one of the Caciques of Hispaniola had tamed 'in a lake,' as he expresses it, 'of standing water,' giving this narrative of the 'River Cow' on the authority of Peter Martyr."*

* This anecdote, though not resting on *scientific* authority, is worth transcribing. The picture was probably drawn from the life; and

A few months before the capture of the specimen which fell under my own notice, a pair were seen in the harbour of Black River. My informant, who saw them, assured me that they were playing at the surface of the water over the bar at the river's mouth, hardly a gun-shot from the bridge; and that they continued their gambols for a considerable time; yet no one cared to pursue them: a fair specimen of Jamaican apathy. These were about twelve feet long.

In Dr. Robinson's MS. volumes there are some occasional notices of this animal, and of the manner in which it was captured, about the middle of the last century. He states that a large Manatee would sell at Kingston for 30*l.* currency (= 18*l.* sterling),

even the docility, which would submit to the office of ferrying favourite attendants across the lake, I should by no means reject as impossible, nor as less likely than the same quality in the Elephant and the Horse. After a description of the animal sufficiently exact, the historian proceeds:—"There was a king of Hispaniola which put one of these animals (being presented him by his fisherman) into a lake of standing water, where it lived five and twenty years. When any of the servants came to the lake and called *Matto-matto*, she would come and receive meat at their hands; and if any would be ferried over the lake, she willingly yielded her back, and performed the office faithfully: yea, she hath carried ten men at once, singing and playing. A Spaniard had once wronged her, by casting a dart at her; and therefore after that, when she was called, she would plunge down again;—otherwise to the Indians she remained officious. She would be as full of play as a monkey, and would wrestle with them, especially she was addicted to one young man, which used to feed her. This proceeded, partly from her docile nature, partly being taken young; she was kept up awhile at home, in the king's house, with bread. The river swelling over his banks into the lake, it followed the stream, and was seen no more." (*Pilgrims*, B. viii. c. 14.)

and that at Port Royal it fetched 15*d.* currency (= 9*d.* sterling), per lb. In St. Elizabeth's at the same time it was worth little or nothing. He describes, on the information of a scientific friend, the legs (or swimming-paws) of this animal as resembling those of a negro that has the elephantiasis; and affirms that they are of no service to the creature in swimming, but only to "gravel" up the roots of some river-plants that it feeds on, and for the female to secure her calves when apprehensive of any danger. From the great silence and caution mentioned as necessary in order to approach it, it would appear that the Manatee is as timid as it is gentle. These qualities, added to its playfulness and docility, its love of society and care of its young, are amiable traits in its character.

CETACEA.

Mr. Hill has favoured me with some notes on the *Cetacea* which inhabit or visit the waters that lave the coasts of these islands. The species which I have already mentioned as occurring between Porto Rico and Haïti was probably distinct from the Dolphin alluded to in the following notes.

THE CACHELOT WHALE.

"The sea within this chain of islands is visited by the Cachelot Whale (*Physeter macrocephalus*). I cannot say whether they come hither yearly to soundings, as they are observed to do on the intertropical shores of the Pacific, — for there they *annually* congregate in the open bays and harbours; but here they are not unfrequently found upon the coast in pairs, and sometimes by three and four together. If my own experience would warrant me in fixing the period of their appearance with us, I should say that usually they come hither in spring.

"The muzzle of the Cachelot is so remarkably shaped, and its spouting is so distinguishable from that of the Arctic Whales, that there is no mistaking it for any other of the Cetacea, whether it be seen near or far. I was in a small coasting vessel in May, 1832, in the Bight of Leogan, going from Jeremie to Port au Prince, when at about some five miles from the Island of Gonave, with light winds just pleasantly stirring the sea into ripples, we observed about two miles from us a Cachelot Whale slowly sinking and rising, and spouting as he went along. All of a sudden to our great surprise we saw him '*breach*.' He leaped clear out of the water, showing his huge bulk in the air, and seemed half as long as the dark wooded cliffs of the island against which his form was relieved.* He rose upward, — and descended splashing the water to a great height. The white foam, where his broad length struck the surface, mantled like a cloud and gave out a sound like that from a distant gun. He then immediately fluked, as the whalers say, showing his tail uppermost, and sunk

* ——— "Leviathan,
Hugest of living creatures, on the deep
Stretch'd like a promontory." MILTON.

and was seen no more.* I learnt that Whales were not unusual visitors in these in-shore waters, and though rarely taken, a *then* recent instance of a mis-adventure which threw a Cachelot into one of the harbours on the north coast of Haïti was related to me.

"The bay of Fort Dauphin is a very remarkable one. The entry into it is by a narrow gullet, about three thousand yards in length, comparatively narrow, but having a depth of from fourteen to twenty-five fathoms. The water then expands into a land-locked space two leagues in length by a good

* In another communication my friend transcribed from his journal some accompaniments of this scene, which will be read with interest. "We had set sail on the 30th April at night, and the next morning found ourselves to the eastward of the Caimite island, with the low shores of Barraderes just before us. . . . . We lay becalmed nearly the whole day, May 1st, off the low peninsula, having the little town of Trou before us, beneath dark forest-clad steeps. Within the peninsula of Barraderes, the sea forms an estuary, seemingly very secure against the prevalent north winds, the worst weather on this coast. A little breeze towards the afternoon set us on to the Point of Tapion, which we reached at sun-down. There is little variety in the chain of hills forming the coast of L'Anse à Veau, but between Miragoane and Petit Goave, the scenery to one coasting near the shore, as we were, is sufficiently interesting. Here and there cliffs — an occasional border of sandy beach — a forest margin — green slopes — dark stream-enlivened dells — and, above all, a fine line of broken summits, presented an interesting coast view, rendered still more pleasing by the bluff of the Tapion in the distance. The wind came in flaws and spurts, so we passed on sometimes slowly, sometimes rapidly enough. It was altogether an agreeable evening's trip. A parcel of Dolphins and Bonito-fish gambolled around us; and to the north, against the blue hazy line of the Gonave, for we were in the passage between that island and the main shore, the whitened spray was dashed up in sudden splashes by the leaps of the Cachelot, escaping the attacks of either the Espadron (*Xiphias*) or the Sawfish (*Pristis*)."

half league in breadth; of variable depth; with five islets dotted about it. This fine harbour was in old times the stronghold of the Buccaneers, who fortified one of these islands, and used the neighbouring plains for a hunting ground. In the year 1828 or 1829, a Cachelot Whale entered it, pursued by an Espadron or Sword-fish. The reefs outside the narrow throat-like entrance I have described, have only a single channel for ships; the other openings in them admit of nothing much bigger than canoes to pass. Having got within these rocks, the Whale was driven to an extremity and had no alternative but to take the narrow channel into Fort Dauphin to escape his active pursuer. As the bay is so sheltered that the waters are as gentle as a mill-pond under the usual easterly breezes from the sea, it was soon perceived that the huge monster as he came rolling and blowing and spouting in, was wounded and bleeding. A host of canoes immediately pushed off to him, not to pursue him with harpoons, for they were not provided with the necessary weapons, nor with the requisite tackle; but to follow him and harass him until he should entangle himself in the shoal-grounds about the harbour. After he had been hunted about some time, he was driven aground in the south-east arm of the Bay, and when measured was found to exceed sixty feet in length. As the harbour of Fort Dauphin divaricates, and spreads into a form like the letter Y, the Whale first took one arm and then the other, swimming at a rapid rate, with his enormous head raised into the air, and taking breath. The water was lashed into foam in his course, and the shores reverberated the sound made by the dilated blows he struck upon the surface with his tail. He coasted the shore as if he looked for some outlet he had remembered, but which he could not find again; and some hours elapsed before he took ground in that circular convulsive sweep, which is described as being made by these animals when exhausted, and which whalers call the 'flurry.' This laid him fast stranded on the shallows, and rolled upon his side. This capture was related to me by General Kayer la Rivierre, the Commandant of a neighbouring arrondisement, who witnessed it. The Sword-fish was not taken, but the body of the stranded animal bore wounds, evidently inflicted by some such ocean enemy.

"Moreau de St. Meri, in his History and description of the old French Colony of St. Domingo, relates that in his time (1785), in the months of March, April, and May, as many as five and twenty vessels from the North American States could be seen on the coast off Sale Trou near Jacmel, fishing for Cachelot Whales, and, he adds, for Souffleurs (*Balænoptera*), and that this fishery was with equal spirit and success pursued within the gulf to the west of the colony;—that is, within the Bight, in which I saw the Cachelot breach. The whale fishers resorted to Turk's-island to boil their oil.

"I must not omit to mention that that rejectamentum of the Spermaceti Whale, '*odoriferous Ambergris*,' has been occasionally found on the coasts of these and the Bahama Islands, of very considerable size and weight.

THE DOLPHIN.

"Herds of a species of *Delphinus* have been often observed in Kingston harbour. The first and the last time they came under my attention was at Passage Fort; the last time was in the month of March, 1842. Regularly about an hour before sun-down for a succession of days, a company of some eight or ten would be seen scudding from sea-ward close in-shore, and taking up a station at the mouth of the Rio Cobre, within the river current, where it rippled out, half salt and half fresh. The ground, to some distance off, rose into dry sand-bars, and no portion of the water thereabouts was more than just deep enough to permit the Dolphins to toss and tumble in. From the perfect revelry with which they sported up and down, it was obvious that some particular food attracted them to these waters. They coursed about like terriers hunting rats. I inquired of Mr. Kirkpatrick, who was thoroughly acquainted with all the wild nature hereabouts, both on land and in the water, and who was an excellent observer of the habits and instincts of animals, what he concluded drew these Dolphins to this strange locality; he stated that the little fry at these freshets, were preyed upon by young Sharks from eight to eighteen inches in length, and that these young sharks had been found to be a favourite food with the Dolphin. The fishermen, he said, were so firmly fixed in this conviction, that, believing in their usefulness in thinning the number of sharks, they were extremely unwilling to molest them; the consequence was that they played about in these shoal-waters close in-shore in confidence and security, and had done so for the occasional years that he had observed them; hence it was, that although so common and so much within the power of any one curious enough to shoot them, they had never been disturbed, and no one knew anything more of them, than that they were Dolphins. The mingled river and sea-water that attracted them was a favourite fishing-ground of the Osprey. That bird would be seen at the same sunset hours hovering within the river mouth, while the Dolphins were busy with their chase on the outside. The young Sharks were prodigiously numerous on this coast. I once took the pains to calculate their numbers. At the rate at which the fishing-canoes brought them in with other fish taken in the seines daily, not less than ten thousand of them must be annually destroyed upon this beach alone. As far as a hasty glance at the Dolphins, when sporting, could be obtained with the spy-glass, they seemed to me to resemble the *Delphinus superciliosus* of Lesson in *shape and size.* The colour was decidedly of that hue called lead-colour: blackish-blue above and grey beneath.

"*September*, 1846. In my notes of a visit to Passage Fort, in 1842, I noticed the Lagoon between Fort Augusta and the Salinas. I spoke of it as the great stock-pond of the Kingston Fish-market for certain kinds of fish, and represented it as being a fine lakelet, interestingly varied with mangrove islets, and with clustered trees upon its borders. It is a spacious piece of water of two divisions, of very equal depth, and with a hard bottom of sandy loam. Mr. Robert Wilkie informs me that, in the month of April last year, he was spending some few weeks at Passage Fort for a change after a fever, and being induced to vary his boat-excursions by a row within these *ponds*, as the divisions are called, he was surprised by finding a couple of Dolphins sporting about them. The water being smooth and clear, never rising to anything more than ripples, and the ponds, though spacious, only filling the vision in any direction in which you look upon them, the Dolphins gambolling and rolling and tumbling from one end to the other, were always in sight, and their entire shape and magnitude visible in the water. Mr. Wilkie represents them to have been seven or eight feet long; lead-coloured, long-snouted, and gracefully proportioned. The fishermen said that they had been then occasionally visiting, in and out, for some time. The waters teemed with Mullets and Callepivas, Snooks and Snappers. It was just then a mullet-season. The fishermen represented the Dolphins as making their visits on a particular day in the week. I fancy they mistook their own 'particular-day-in-the-week' visit, on which occasion they found them already there, — for the Dolphins'. Mr. Wilkie repeated his excursion twice or thrice, and always saw the Dolphins. On the last occasion he made an effort to shoot them, but, though he had several shots, he did not succeed in their capture."

"*December*, 1848. When I was in the last week of my late sojourn at Ray's-town, that upper part of Kingston Harbour was regularly, for some successive days, traversed by a company of Dolphins, or, as the fishermen called them, Porpesses. They went rolling on, one after the other, very leisurely; keeping, when I observed them, which was at about sun-rise, in a line with a stretch of ripples that gave their thread-like motion to the surface from the upper to the lower end of the harbour, along the otherwise smooth-glazed waters. The fishermen described them as going to the upper end of the harbour to feed on the Mullet fry, just now very abundant there, and as not quitting them till they had stored their stomachs for the day. I thought their taking the wind was a very noticeable incident, and determined to set it down as something characteristic that had come under my own observation. I should have liked, however, to have ascertained, by repetition, whether it can be considered a trait of instinct or not. The fishermen of Ray's-town were hawling in a seine of prodigious dimensions at the time. This part of the harbour is shoal, and the nets are laid out far from the shore, being drawn in to land with a full mile of outward rope. The Porpesses crossed the lines of the seine, and, though they did not seem to regard the nets or the congregated swimmers within their sweep, their traverse had so frightened away the fish, that the hawl was altogether unproductive. The fishermen say they have occasionally caught them entangled, but are best contented when they never come near them, as they are altogether unmarketable prizes."

NOCTURNAL FOREST SOUNDS.

Various and strange are the sounds which strike the ear of one benighted in the forests of Jamaica. Some of these are the voices of night-birds, the rapid articulations of the Piramidig, the monotonous call or startling scream of the White Owl, the shrill wail of the Dusky Owl, the hoot of the Potoo, or the loud and reiterated cries of the Clucking-hen; and some are insect sounds. But, besides these, there are some which are certainly produced by Reptiles, though it is difficult to identify them. Nearly every night, at certain seasons, there ascends from the woods around Content a continual snoring of various tones, the voices of numberless Tree-frogs, or, as they are here called, Toads. They are said to reside in the large ventricose leaves of the greater Wild-pines, especially that fine one, *Tillandsia lingulata*, which, about the end of July, sends up a magnificent flower, somewhat like a huge carnation, with broad outer petals of a rich crimson hue and polished surface, and a cluster of smaller interior ones of pale yellow. In the coolness and moisture of these natural reservoirs, always half full of water collected from rains and dews, the Tree-frogs delight to lie, finding in them circumstances eminently congenial for the maintenance of cutaneous humidity, so essential in these reptiles to respiration. They are very rarely seen, and, but for their vocal powers by night, we should scarcely be aware of their existence; the number and universality of these sounds, however, in the mountain-woods, during the hours of darkness, prove that they are very abundant. Even when seen by day, their agility in leaping renders it a difficult matter to lay hands on them. The sounds in question bear a strong resemblance to the objurgations of an inveterate snorer, but are much louder; or sometimes remind one of the groaning and working of a ship's timbers in a heavy gale at sea.

These are probably the voices of some of the greater *Hyladæ*. But there are other and different noises still. While I am writing this note at Content, — it is a lovely night in June, — all around I am saluted with strange sounds. Now and then comes the singularly harsh and cracked voice of the Gecko, like the notes of a child's penny trumpet, or like a stick drawn across the teeth of a comb: — this I am familiar with. But I hear another voice, far more abundant, but quite unknown to me. It is now (about midnight) coming up from every part of the moonlit forest below me, with incessant pertinacity. It is a clear shrill note, so like the voice of a bird, and in particular so like that of the Solitaire, that it might easily be mistaken for it, but for the inappropriate hour, and the locality. Like that, it is beautifully trilled or shaken, and, like it, the individual voices are not in the same key. As I now listen to the mingling sounds, I distinguish two particularly prominent, which seem to answer each other in quick but regular alternation; and between their notes there is the difference of exactly a musical tone. I have little doubt that this is the sexual call of some Tree-frog. The groanings and snorings, which are sometimes so incessant, I do not now hear, except one such sound now and then in the course of an evening.

Some three or four species of *Anourous* Reptiles are all that Jamaica has produced to my researches, and all of these are of the family *Hyladæ*, whose dilated sucking-disks at the tips of their toes, enable them to cling about the foliage of trees and plants, and to leap from leaf to leaf with security and precision.

One of these was taken in a bedroom at Savanna-le-Mar, one night in October, having probably hopped in at the open window from the branches of a mango-tree only a few feet distant. I was surprised at its changes of colour, in this respect resembling the Chameleons and Anoles, or, still nearer, the Geckos. When I obtained it, the whole upper parts were of a rich deep umber-brown, with indistinct black bands. On looking at it at night, to my surprise I saw a great alteration of hue: it was paler on the head and back, though least altered there; on the rump, and on the fore and hind legs, it was become a sort of semi-pellucid drab, marked with minute close-set dark specks. When disturbed, it presently became slightly paler still, but in a few minutes it had recovered its original depth of tint. In the course of half an hour it displayed again the speckled drab hue, and now uniformly so, save a black irregular patch or two on the back, and a dark patch between the mouth and each eye. The belly, which was very regularly shagreened, was of a dull buff, not susceptible of change, The eyes retained their proverbial beauty, for the irides were of a golden brown tint, like sun-rays shining through tortoiseshell. This specimen was about as large as a middling English Frog, being two inches and a quarter in length.*

While in captivity, if unmolested, it spent a good deal of time motionless, squatting flat and close, with shut eyes, as if sleeping; but sometimes it was active. I kept it in a basin covered with a pane of glass, for facility of observation. It would keep its face opposite the window, altering its position pertinaciously if the basin was turned, though ever so gently. It took no notice of cockroaches, nor of a large flesh-fly, which buzzed about it, and even crawled over its nose. If taken in the hand, it struggled vigorously, so as to be with difficulty held: once or twice, while thus struggling, it uttered a feeble squeak; but if still retained, it would at length inflate the abdomen with air, apparently a sign of anger. It leaped, but not far.

* This seems to be an undescribed species, agreeing with *Hyla xerophylla* (Dum. and Bibr.) in many points, but differing from it in others. It may be thus described.

Hyla brunnea, mihi. The Brown Tree-frog. Length of head $\frac{3}{4}$ in.; breadth of head $\frac{8}{10}$; trunk, length, $1\frac{1}{2}$; breadth $\frac{9}{10}$; fore limb $1\frac{1}{2}$; hind limb $3\frac{7}{10}$; (thigh and shank, each $1\frac{1}{10}$; foot $1\frac{1}{2}$;) pallettes circular, those of middle finger and toe $\frac{1}{8}$ in. in diameter. Head broad, flat, semi-oval, rough with minute granulations; eyes large and very prominent; *canthus rostralis* forming rather more than a right angle; frenal regions not so high as the cheeks. Tongue very large, heart-shaped, notched behind; a longitudinal depression extends about half way from the notch to the front; another longitudinal depression is on each side near the margin; scarcely at all free behind. Vomerine teeth in two small semi-circular ranges, between the internal nostrils. Tympanum nearly circular, its diameter about half that of the eye. The palmation and other particulars as in *H. xerophylla.* Colours described in the text.

A much finer species, as to form, colour, and size, was brought me, in May, from the summit of Bluefields Mountain. It was of that curious genus named by MM. Duméril and Bibron *Trachycephalus*, having a distinct neck, a triangular trunk terminating in a point, and an enormous flat head studded with irregular sharp bony ridges. The three species designated by those learned herpetologists are assigned respectively to Brazil, Hayti, and Cuba; the present is manifestly distinct from either, marked by superior size, a peculiar style of coloration, greater development of the bony ridges of the head, and the prominent projection of the *sacrum*. It is a far finer, though more uncouth, species than any yet described. I therefore propose to call it, from its resemblance to the moss-grown bark of a tree, the Lichened Tree-toad (*Trachycephalus lichenatus*).* (See Plate VII.)

* Length from muzzle to cloaca 4·8 inches; breadth of head at rictus 1·7; breadth of inflated body 2·25; rictus from the muzzle along the curve, 1·5; horizontal diameter of eye ·4; arm from axilla to elbow 1; fore-arm from elbow to wrist 1; hand from wrist to tip of middle toe 1 7; breadth of middle pallette ·3; thigh along front margin 2·1; tibia 2·1; tarsus 8; largest toe 1·8.

Head, upper surface broad and flat, marked with bony ridges, the principal of which are the following: one from the front of each orbit passing to the nostril; one from the back of each orbit passing above the tympanum; and a sinuous one crossing the occiput transversely. Under surface of head very flat; neck constricted suddenly behind the rictus. Outline of face, from ear to ear, a semi-oval; posterior edge of occiput forming two salient curves, convex backwards, between which is a deep and sharp depressed notch; the extremity of the muzzle forming two vertical ridges with a deep pit between. *Canthus rostralis* forming a sharp and narrow ridge, running nearly straight from nostril to eye. Height of muzzle equal to half the interorbital space. Vomerine teeth set on the margins of two small projecting

Plate. VII.

P.H GOSSE DELT M. & N. HANHART IMPT

LITORIA LUTEOLA *N.S.* TRACHYCEPHALUS LICHENATUS $\frac{2}{3}$ *N.S.*

LONDON; LONGMAN & Co 1851.

During the short period that this fine reptile remained under my observation, it was impatient of confinement, leaping vigorously. Now and then it inflated its body to a considerable degree, which, as in the former species, was probably an expression of anger, or an attempt to intimidate. A moisture exuded from its skin, which took the form of a very fine froth, even while on the body, and when touched appeared of a gummy nature, adhering to the fingers, and stiffening them as it dried. It left shining marks on the table, too, like the trail of a snail.

I was unfortunately called away after having made the description below, and remained from home two days. I had placed the Tree-toad under a bell-glass, knowing that it would not suffer from fasting, but forgetting that it would need moisture. When I returned it was just dead. Decomposition, however, had not commenced, so that I was enabled to add to my verbal description a carefully coloured drawing, which I had commenced while it was yet alive.

ridges of semicircular form, very distinct from each other, and separated by about half the diameter of each. Web between all the fore toes short, but quite distinct.

Upper parts irregularly patched and mottled with dark brown and black upon a pale reddish ground; the dark hues chiefly aggregated in a large triangular mark between the shoulders. Muzzle and sides pale green, with spots of liver-brown running into each other. Under parts flesh-white; the chin speckled with liver-brown; the abdomen and thighs granulated like shagreen. The limbs answer to the body on their respective surfaces. Pallettes of toes pale dull green. Iris of eye golden, beautifully pencilled with reddish-brown; pupil sub-rhomboidal: the lower eyelid has the superior half as transparent as the purest glass; except that a narrow border at the edge is opaque and yellow, as is the lower half. Tympanum buff-coloured, pencilled with brown.

The peculiarity, mentioned in the note, of the glass-like transparency of the lower eyelid (which, in these animals, is the larger, and performs the office of closing the eye) is well worthy of notice. I have elsewhere* remarked a structure exactly similar in the Woodslave (*Mabouya agilis*), a pretty little Scincoïd lizard that plays about walls in Jamaica. In both cases, we may consider it a beautiful and effective provision, for the protection of the eye during the rapid movements of the animals, where sight would be indispensable. The Tree-toad dwells habitually among the sheathing leaves of the Wild-pines, always stiff and leathery, and often armed with sharp serrated spines at every edge. Among these it moves to and fro by violent headlong leaps, in which it needs to be guided by the sharpest sight. How interesting, then, is it to see that its gracious Creator has furnished it with a glassy window, which it may in a moment draw before its eye, for shelter from danger, without in the least hindering the clearness of its vision! This structure has not, I believe, been noticed by any naturalist; and, indeed, it is scarcely perceptible when the delicate membrane has become opaque by immersion in spirits. "All thy works shall praise thee, O Lord!"

That there is a second species of *Trachycephalus* found in the western districts of Jamaica, I infer from the following description of one in Dr. Anthony Robinson's MSS. "Rana palmis tetradactylis semipalmatis; plantis hexadactylis palmatis; pollice breviore. Weight, 2 oz. Troy. From the tip of the longest digit to that of the longest toe, 13

* See Proceedings of the Zool. Soc. for 1848, p. 60.

inches. All the toes joined by a membrane; the fore feet rather semipalmated; all the toes, except that answering to the great toe, which in this animal is very short, having broad, round, soft, and spongy extremities. Most of the joints on the inside of the feet had lesser sponges, which, being clammy, enable these creatures to ascend the loftiest trees with the greatest ease and facility. The toe corresponding to the thumb in man, answered much the same purpose in this creature, which by this member arrested one of my fingers. The head is of a cordate form; the eyes large and prominent; the irides speckled in a manner so elegant as to exceed description. The space beyond the eyes is hollowed, and rough like a file. Ears covered with a *purplish* membrane. The body, when inflated, taper and triangular. Body, above, *dull green*, as were the anterior parts of the thighs. Sides near the groin, *sky-blue spotted with purple*. Abdomen beneath the thighs covered with small tubercles of a dull flesh-colour. Feet, both fore and hind, *purple*, and clouded with the same as the hind part of the thighs. Mouth extremely wide; tongue fixed. Under the jaw whitish red, or incarnate, with *purple spots*."

From a carefully coloured drawing which accompanies the above description, it is evident that this is congenerous with my Lichened Tree-toad; but the colours are so different as to warrant the conclusion that it forms another species. The green hue of the upper parts, destitute, apparently, of the marbled intermingling of other colours, which is so characteristic of all the described *Trachycephali*, the sky-blue of the flanks, where mine is delicate green, and the prevalence of purple, are all strongly marked differences in coloration. I would therefore distinguish this as the Green-backed Tree-toad (*Trachycephalus anochloros*).

All that Robinson has recorded of its manners is contained in the following brief note: "It was given me by Dr. Anderson, of Hanover parish, who informed me that he held it in his hand almost half an hour, after which, having occasion to rub his eyes with the fingers of that hand, which was besmeared with a mucus left by the animal, he was seized with a violent pain and smarting that lasted nearly half an hour, but subsided on washing his hands [*query*, eyes?] in a little spring water." The learned Doctor compares this quality to the stinging of the Nettle-fish, by which he probably means the *Physalia;* but it seems to me much more analogous to the acrid secretion which exists in the cutaneous follicles of the true Toads; if so, it justifies the application of the term Tree-*toads* to these animals, by which they are provincially distinguished, rather than Tree-frogs.

A very little species of the same family, which the minuteness of the pallettes requires to be assigned to the genus *Litoria* of MM. Dum. and Bibr., is known to me by a single specimen which was taken in a tub of water, at Content, in February, Of its peculiar habits I can give no information.*

* *Litoria luteola*, MIHI. The Little Yellow Tree-frog. Length 1 inch; fore limb $\frac{6}{10}$; hind limb $1\frac{4}{10}$. Pallettes of all the feet very small. Hind toes very unequal, the fourth being extremely long; palmation scarcely perceptible. Head somewhat pointed: vomerine

Before I leave this subject I will extract from one of Mr. Hill's letters to me an interesting note on nocturnal forest sounds, heard in a very different part of the island. The remarks with which it is prefaced are in themselves interesting, as referring to a species of Owl hitherto unrecorded. "The Eared Owl," observes my friend, "which I had sent to me from Manchester, was taken on the wooded mountain-skirts forming the back-country of that and the adjoining parish of Trelawny. These back-mountains are opened in detached clearings, and planted with coffee, Indian corn, and esculent arums, and are usually exceedingly infested with rats; the cellular limestone which protrudes through their rich vegetable deposits being prodigious harbouring places for those rodents. I find among my loose memoranda, that a Mr. Walker, an overseer of a plantation situated among the furthest of the sugar settlements in St. James's, mentioned to me that he had observed narrowly the *Brown* Owl, and that he had acquainted himself with its great diligence as a destroyer of Rats. He had remarked a roost to which one of them resorted; and the extraordinary heaps of *casts*, deposited at the root of the tree on which this Owl usually devoured his prey, made piles of undigested bones, among which Rats' teeth were conspicuously

teeth arranged in two curved lines, whose convexity is forward, scarcely interrupted at their meeting angle. Tongue small; posterior half (or rather more) round and free; anterior portion oblong, and attached. Colour pale buff, studded with minute dark specks, irregularly scattered; accumulated in the form of bands across the legs and thighs. A band of deep brown passes from the muzzle, through the eye, and is lost about the middle of the side. (See Plate VII.)

prevalent. My memorandum sets these facts down as traits of the 'Brown Owl,' this name being used to distinguish it from every other Owl.

"A friend gave me a very interesting narrative of a benighted traverse that he made of the mountains between Manchester and Trelawny, in which the 'ohoo' moan of an Owl made a part of the night watches. He had gained the last and loftiest ridge of the intervening highlands, when a moonless but starry night, closing in upon him, compelled him to look for a convenient tree upon which to settle himself to repose. The first sound that saluted him was the dismal croak of the Tree-toad at long intervals; the croak near to him being answered by successive and repeated croaks more distant. Occasionally was heard the vehement hiss of some prowling snake; he concluded, our sizeable Boa, the Yellow Snake; then came every now and then the 'ohoo' moan of some Owl, whose voice was quite unknown to him. It was replied to by a similar moan afar off. I suspect that this was either the Eared or the Brown Owl."

As these syllables, however, convey the note of the Potoo (*Nyctibius Jamaicensis*), uttered in the darkness of the night as it sits on its lone watchpost, it is probable, as my friend afterwards suggested, that the voice heard was rather that of this great Nightjar than of an Owl.

GREGARIOUS TREES.

The overrunning of large tracts of land by some particular species of shrub or tree almost to the exclusion of every thing else is an interesting and curious phenomenon. In some cases it is doubtless the congeniality of the soil or circumstances that is the cause, the plant being in some way or other adapted to the place; but it would be difficult so to account for the fact in every case of its occurrence. In some districts at the eastern extremity of the island, and around Black River, the Opoponax, a species of Acacia (*A. tortuosa*, I believe), is said to have extended itself over such tracts as to be quite a pest. On the banks of the Cobre, just behind Spanish-town, species of *Inga* are particularly abundant. Between Kingston and the city just named the traveller sees nothing for long distances but *Lignum vitæ* trees, lovely indeed as they are with their compact dark-green forms, and splendid azure blossoms. To come nearer Bluefields, at Crabpond Point, there is a considerable tract along the sea-shore densely covered with a species of fan-leafed Palm (I believe a *Thrinax*) of small size, not more than twelve or fifteen feet high; and the open slopes of Culloden and the eastern part of Mount Edgecumbe are studded with multitudes of a similar Palm, but growing to loftier proportions. Near Belmont dense and perfectly unapproachable thickets run along the margin of the sea, formed of the Nickar (*Guilandina bonduc*), of which the long stems, the leaf-ribs, and the flattened valves of the seed-pods are alike covered with the most terrific hooked spines. I do not think so much of the beautiful and aromatic Pimento, that clothes these park-like estates, because its growth is cherished for the sake of its spicy fruit; though there is reason to believe it would naturally grow in the same extensive groves; nor of the mangroves and morass-withes that are dependent on the presence of water. But one cannot help noticing the Sea-side Grape (*Coccoloba uvifera*) with its round, leathery, crimson-veined leaves, and its bunches of red acid berries, that fringes the sea-beach in many places; not so much in groups or clumps, however, as forming a narrow belt or range of single trees, as close as they can conveniently grow together, along the line of high-water mark. Other species might be enumerated as having the same gregarious character, in the interior, as the Jointwood (a species of *Piper*) that thickly and exclusively covers large tracts on the summits of Bluefields Ridge. I shall however mention but one more, the Logwood (*Hæmatoxylon Campechianum*), originally introduced from the Spanish Main, but which now covers immense districts of the western portion of the island, and is fast extending its dominion on every hand, maintaining a too successful struggle with the feeble agriculture that characterizes the Antilles. The likeness of this tree to the Hawthorn of Europe is very striking, and has been noticed by many. Either growing singly, or in clumps, the resemblance is so exact that at a very little distance the stranger, if not corrected by his reason, would infallibly mistake it for that familiar tree. For hedges, it is equally adapted, and is much used; and in this form the similitude to the thorn is still perfect. I have admired scenes, such for example, as on the estate called Paradise, near Savanna-le-Mar, where, from the absence of characteristic tropical features, from the broad open fields, divided by logwood-hedges, and the green slopes studded with clumps of logwood trees, and crowned with neat-looking white buildings, I could have readily imagined that my eye was roving over English ground.

THE GREY SNAKE.

For what reason this little *Coluber* is distinguished above its fellows by a name expressive of peculiar venom, I know not, as it is perfectly inoffensive; but "Poison Snake" is its creole title. I find from Robinson's MSS. that it was so called in Clarendon parish, also, nearly a hundred years ago; though, he says, he has had it alive in his hands, and believes it to be a very harmless inoffensive creature. Though not rare, it can scarcely be called abundant, being much less frequently seen than either the Black Snake or the Boa. I have found it scattered in distant and varying localities, in the lowlands and on the mountain-tops. It does not affect walls so much as the other *Ophidia*, but is oftener seen in rocky places, upon the ground, or gliding through dead leaves and dry rubbish by the sides of roads. It sometimes takes up its abode in outhouses, where, lying on one of the beams within, or securing a narrow resting-place on the edge of a board beneath the shingles without, it waits patiently with its head hanging down, but its bright eyes wide awake, for passing prey. That this prey principally consists of the little Anoles and Geckos which always resort to such situations, I have no doubt; and indeed, I once saw this Snake taking such prey, though in a very different situation.

One day in June, being engaged on the Hampstead Road, above Content, in capturing the insects that were then so abundant about the blossoming trees,—I observed on a shrub two little Anoles playing; one had seized the other's tail in his mouth, and they were thus gamboling about the twigs in innocent sport, altogether unsuspicious of the lurking foe that was near. While I was looking at their gamesome tricks, all at once I became aware of a so-called Poison Snake, not a large one, silently and stilly watching them also. Suddenly he caught one, and throwing his head off the branch along which he had been lying, held the victim suspended in the air. It had been seized just behind the fore legs, but by an almost imperceptible motion of the jaws, the hold was gradually shifted forward until the Lizard's head was in the mouth of the Snake, and was rapidly sucked in by the alternate motion of the two sides of the jaws. There was no boggling about the fore legs; they were out of sight almost before I was aware; but one of the victim's hind feet had taken hold of a twig, and resisted the sucking in. The Snake had now drawn up his head again upon the branch; giving a sudden jerk, he made the Lizard relinquish its hold of the twig, and in the same instant the leg was engulfed. Of course, no impediment now was left; but when the last vestige of the tail was disappearing, I tapped the neck of the Snake smartly with a switch, and it fell to the ground disabled and dying. I transferred it to a bottle of spirit, and it is now in the British Museum;

the position of the Lizard sufficiently marked by the swollen and lumpy neck.

Robinson in his MSS. mentions his having found in the belly of one of these Snakes about fifteen inches long, "three inches of a Jack Lizard's* tail," and, by the dissolved substance, concluded that the Lizard had been swallowed whole, expressing wonder, when he reflected that the Lizard was twice the thickness of the Snake at least. This, however, is now well known to be nothing uncommon in the serpent tribes.†

* I believe he designates the Purple-tailed Anolis by this appellation.

† I would describe the Grey Snake as follows: — *Natrix capistrata*, MIHI. Scales sub-rhomboidal, imbricate, convex, and perfectly smooth, without any depressed points. Tail in young specimens, about one fifth of the total length; in adults about two fifths. Head long-oval, somewhat tapering in front. Body nearly cylindrical, not elevated in the mesial line; gently swelling to about the middle of the total length. Seventeen rows of scales, running very obliquely. Labial plates eight, of which the fourth and fifth form the lower wall of the orbit; they increase in size to the sixth, thence diminish. In the form of the crown-plates, in the form and situation of the teeth, and in the dimensions and direction of the gape, this species agrees with the Black Snake (*Natrix atra*). Abdominal shields from 170 to 176; caudal (according to my own observations) from 127 to 159 pairs; the higher numbers the more common.

This species is less than the Black Snake, and more elegant. The largest I have measured was twenty-six inches in length, of which the tail was ten. Colour of the upper parts uniform yellowish grey; the belly shields rather yellower; the whole marked with minute black specks. Each scale and shield has a pale border, scarcely observable, except with a lens. A black line runs from the muzzle through the eye as far as the gape, below which line the face is cream-white. Another black line passes between the two occipital plates. The skin between the scales seems mottled with black and white; so

THE CRESTED SNAKE.

Mr. Hill writes me thus: —"December, 1845. There are two living attractions in the Blue Mountains, a *Crested Snake*, and a sweetly mysterious singing bird called the Solitaire. . . . . The Snake is identical with one I was told of in Spanish Haïti, having *a red crest and wattles*, very much resembling the head of a cock. Strange fictions were invented of its crowing like the cock, and stealing into hen-roosts, and by this deception, when it was coiled up, and nothing but its crested head seen, surprising the poultry on the perch, and devouring them. The Spaniards narrated to me these particulars, with the words, '*Il canta como un Gallo,*'— and our people speak of it as crowing like a cock. Certainly it is the most wonderful serpent since the days that Eve was deceived in Paradise*, if it has a voice so much approaching to distinct if not articulate sounds.

"I believe there is a Crested Snake known here, I only reject the cock-crow story."

that the motion of the scales brings to view minute evanescent spots on the body, chiefly on the neck. The eyes are very beautiful, like some gems of a pale lustre, clouded. Robinson found each of the two rows of caudal scuta to contain 173, = 346 in all, more than double the number assigned by Linnæus to any of his species.

* ——— "On his rear,
Circular base of rising folds, that tower'd
Fold above fold, a surging maze! *his head*
Crested aloft, and carbuncle his eyes;
With burnish'd neck of verdant gold, erect
Amidst his circling spires, that on the grass
Floated redundant." *Par. Lost*, ix. 497.

In a subsequent communication my friend furnishes me with more tangible information respecting this mysterious reptile. — " Feb. 5th, 1846. I have conversed with Dr. Palmer, who, I remembered, had informed me, he had seen one of the Crested Snakes about which your curiosity has been particularly excited. He tells me that in the neighbourhood of a plantation called Drummond Castle, in St David's parish, without being more than rare, he has known specimens to have been found. He was present on a medical visit there some years ago, when the plantation people brought to the house one that they had just killed. Besides its remarkable crest, he says, he was particularly struck with its shape. It is the thickest Snake of its size he had seen. Although its length did not exceed four feet, it had the bulk of a Yellow Snake (*Chilabothrus inornatus*) of seven feet. It had a sort of galeated head, with a crest like that of the guinea-fowl. Its colour was that of dull ashy ochre, having large well-defined spots along the back. He states that the negroes, in speaking of its habits, represented it as making a noise, not unlike the crowing of a cock, and as being addicted to preying on poultry.

" Drummond Castle is about eight miles from Kingston, in the immediate vicinity of some interesting waterfall scenery; remarkable as the hiding-place of the freebooter, Three-fingered Jack, so familiarly known as a melo-dramatic hero. The scenery of *The Falls*, as this cascade district is called, is very romantic. The imagination of no painter of theatrical spectacles can surpass the wild wonders of the mountain hold of the *real* Three-fingered Jack. Part of the road by which you ascend the Falls is a subterranean passage; and caverns are entered by simple crevices which seem mere chinks in the irregular surface of the rock, all which natural peculiarities account for the mysterious disappearances, which the mountain Hero was enabled to enact from his pursuers. In this neighbourhood are the Mount Vernon Copper-mines. The whole district is interesting to the naturalist.

"I should not wonder if the mysterious Crested Snake prove to be an *Acontias;* a lizard without limbs, and with truncated tail, a species allied to *A. meleagris* of South Africa. The possession of what may be called a voice, and the unserpent-like form, strongly point this way."

Soon after this communication I had the pleasure myself of conversing with Dr. Palmer, who gave me a few particulars in addition to those contained in the above note. It was about the year 1829; he saw the animal lying by the road-side near Drummond Castle; decomposition had commenced, but the body was still entire and firm. Its thickness first struck him, and he then discovered its crest; he described it to me as a sort of a pyramidal helmet, of a pale red colour, somewhat entire at the edge, but with a kind of lobe or knob at the summit.

The suggestion of Mr. Hill, that the animal in question may be a Saurian, is felicitous, and removes much of the *primâ-facie* improbability of the received account. For though none of the limbless Lizards known to us possess appendages to the head, but are remarkable for the smoothness of their contour, many such are familiar to us in the Saurian Order. The inflatable hood on the occiput of *Basiliscus*, the serrated crest of *Iguana*, *Cyclura*, and *Anolis*, might present analogies to the coronal appendage of this mountain reptile of the Antilles; while the dewlap of *Iguana* and *Draco*, the protrusile bright-hued goitre of *Anolis*, *Dactyloa*, and their allies, and especially the cheek-frills of *Chlamydosaurus*, would afford some precedent for the lateral "wattles" with which it is said to be furnished.

So great was my curiosity to obtain all the information possible about this creature, that, some time before I left the island, I published in one of the periodicals an appeal to the well-wishers of science, intimating my desire to possess this interesting reptile in particular. I heard, however, nothing further on the subject at that time.

But since my return to England, my kind scientific correspondent, Mr. Hill, again alludes to this matter in the following words: — "Mr. Jasper Cargill informs me, that when visiting Skibo in St. George's, an estate of his father's, in descending the mountain-road, his attention was drawn to a snake of a dark hue, that erected itself from amid some fragments of limestone-rock that lay about. It was about *four feet long*, and unusually *thick-bodied*. His surprise was greatly increased on perceiving that it was *crested*, and that from the side of the cheeks depended some *red-coloured flaps*, like gills or wattles. After gazing at him intently some time, with its head well erect, it drew itself in, and disappeared among the fragmentary rocks. He inquired, and found that this, to him, very extraordinary Snake was well enough known thereabouts. He afterwards learned that it was occasionally encountered in the woods about Bath, in St. Thomas in the East. When your application to the public was published in the 'Agricultural Reporter,' he felt so anxious to meet your call for this particular reptile, that he offered a pound for a specimen; but although he had several promises from persons who professed to be acquainted with the Snake, he was not fortunate enough to obtain one. He has promised to be mindful still of the subject for you. 5th January, 1847."

While engaged in preparing these pages for the press, I looked over the magnificent volumes of Seba, "Thesaurus Rerum Naturalium," hoping that in the vast number of Serpents delineated by him, I might discover some parallel to the structure of this extraordinary animal. Nor was I wholly disappointed. In his second volume, plate 103, is the representation of a large Serpent, assigned to both Arabia and Brazil, about which, it is true, some rather apocryphal legends are narrated, but which seems to have been drawn from a real subject. It has the occiput enlarged into a bifid prominence, forming two rounded lobes. In the same volume, pl. 18, fig. 3, there is a Serpent from Amboyna, which is furnished with a lengthened process curving downward from the base of the inferior jaw on each side, the lower edge of which appendage is pectinated, or beset with a row of short bristles. But more to the purpose are the figures of two species of rather small size, in plate 40 of the same volume. Of the former, the text says: — " This was sent to us *from St. Domingo* with other animals. It is a beast truly extraordinary. It has a muzzle hard as horn, pointed like a bird's beak, with two fins under the lower jaw, that serve for swimming. The head is covered *with great scales*; all the upper part of the body is pale yellow, marked with oval reddish-yellow spots, like a tiger. *The fins are coral-red,* as those of a Perch. The ventral scales are yellowish grey, of irregular size, and set without order. This Eel is in almost every thing like a Serpent, and perhaps is a species of that tribe."

Of the second species, the author observes that it is " Like the former in all respects, except the colouring, which exhibits, instead of spots, a series of five sea-green bands, extending along the reddish scales of the back. The fins are coral-red as before. The eyes are minute, and placed near the anterior part of the muzzle. *The scales of the belly are large, and placed symmetrically, as in serpents.*"

It is evident from the expressions of wonder used by Seba, one of which I have quoted, that these were no ordinary forms. He is dubious, it is true, whether to assign them to the Eels or to the Serpents; and he calls them " marine," but whether on any other evidence than the fin-like appendages of the cheeks, does not appear. He calls these appendages " fins," and compares them (when speaking of their colour) to those of a Perch. The round pointed head suggests a *Muræna*, but *Muræna* possesses no pectoral fins, while in these fishes a long dorsal and anal meet around the extremity of the body; and the posterior parts are vertically flattened. The scales, too, in the Muraenoïd fishes are minute and inconspicuous. On the other hand, the scaling of these figures is decidedly serpentine; the belly of No. 2. is expressly said to have distinct large symmetrical scales, like those of Serpents, and the head in both is represented as shielded with the broad plates of a *Coluber*. The tail is drawn out to a long point, apparently in the form of a taper cone, without any compression or bordering fins. In No. 1. there is a little projecting point at the edge of the lower belly, which at first sight suggests the idea of the anal hook of a *Boa*, but which, from comparison with other figures, appears intended to represent the projection of the pre-anal scale; this, though a small matter, is yet an important mark of distinction between a Serpent and a Fish.

These confirmatory evidences, I of course communicated to my friend, whose observations in reply I subjoin.

"*Dec.* 24. 1850. . . . . You have dipped into the right Treasury in Seba's Thesaurus. His Serpent with the red gular appendages and mouth terminating like a bird's bill is certainly the curious Snake described to me as an inhabitant of St. Domingo, and mentioned in so many narratives of the people as sometimes met with here. Your minute notices of the delineation in Seba, the scaling of the back, the scuta of the belly, the plates of the head, the anal hook, and the absence of every thing that might represent flattening in the extremity, are all so many decisive evidences of its ophidian character. But be it what it may, it is indisputably the gilled and wattled serpent of St. Domingo, so amusingly described to me as a frequenter of hen-roosts, into which it thrust its head, and deceived the young chickens by doing its best to crow, since it looked so much like their own chanticleer. The Spanish friend who first mentioned to me this anomalous snake, and begged me to note it among my remarkable things of the country, told me that he had seen it when visiting in that far east of Hayti, known as the ancient Caciquedom of Higuey. The mountains there rise 'in terraces from ten to fifteen leagues in length and breadth, rough and rocky, interspersed with glens,'— remarkably fertile, and resembling our Red Hills, as much in their produce of Cassava-bread, as in their red soil. Las Casas, who supplies Washington Irving with his account of the physical aspect of this Indian territory*, in mentioning the mountain ascents, and their graduated rise, from 'terrace to terrace,'—'steep and precipitous,'— faced by 'rocks that resemble walls wrought with tools into rough diamond points,' — is describing mountain scenery very familiar to you, — the honeycomb limestone whose cellular surface is so remarkably rugged with cavities and angular spiculæ. I have been particular in making this reference to Higuey, because I did not visit that part of Saint Domingo, but was informed that the Serpent with mandibles like a bird, and with scarlet lobes or wattles, and voice that made what might possibly be a cluck like that of the clucking Lizard an imitation of the crowing of the cock, was commonly known there; so that if Seba should mention Higuey, or indicate the east end of Hayti as the locality from which he re-

* Life and Voyages of Columbus, book xvii. ch. iii.

ceived his specimens, you may the better apply your information to your conclusions, particularly when you shall take into account the Jamaica facts I am now about to relate to you.

"It was, I think, on Easter-eve, the 30th of March last, that some youngsters of the town came running to me to tell me of a curious Snake, unlike any snake they had ever before seen, which young Cargill had shot, when out for a day's sport among the woodlands of a neighbouring penn. They described it as in all respects a serpent, but with a very curious shaped head, and with wattles hanging on each side of its jaws. After taking it in hand and looking at it, they placed it in a hollow tree intending to return for it, when they should be coming home, but they had strolled from the place so far that it was inconvenient to retrace their steps, when wearied with rambling; but they had lost no time in relating the adventure to me, knowing it would interest me much, particularly as young Cargill's father had thought it a snake similar to the one he had seen at Skibo in St. George's, or to the crested serpent, for a specimen of which, when in St. Thomas's in the East, he had offered the sum of twenty shillings. The youth that shot the snake fell ill on the following morning with fever, and could not go back to the woodlands to seek it, but he sent his younger brother who had been with him; but although he thought he rediscovered the tree in which his brother had placed it, he could not find the Snake. He conjectured the rats had devoured it in the night. When this adventure was related to me, another youth, Ulick Ramsay, a godson of mine, who came with the young Cargills to tell me of their discovery, informed me, that not long previously he had seen in the hand of the barrack-master sergeant, at the barracks in Spanish Town, a curious snake which he too had shot among the rocks of a little line of eminences near the railway, about two miles out, called Craigallechie. It was a serpent with a curious-shaped head, and projections on each side which he likened to the fins of an eel, but said they were close up to the jaws. Here are unquestionably two of the same snakes with those of Seba's Thesaurus, taken near Spanish Town, and both about the honeycomb rocks that protrude through the plain of St. Catherine's in detached ridges, and cones, and hummocks, being points of the greater lines of limestone, which have been covered by the detritus of the plains, leaving masses of the under rocks here and there uncovered. These are the spots frequented too by the *Cyclura*; and are continuations of our Red Hills, — a country that so much resembles the terraced cliffs and red-soil glens of Higuey.

"Must we not take the horny coverings of the mouth of this Snake, which so much resemble the bill of a bird, as an affinity of the serpent with the tortoise, and the cutaneous appendages as indicating some relation to the Mata-mata, or *Chelys fimbriata* of Spix? It will be not much of a guess to suppose, that, like the Mata-mata, it conceals all but its head, and leaving that out, waits in ambush for young birds, and seizes them as they approach. I cannot consider the habits of this serpent fluviatile at all; for the gular appendage is no fin."

"*Jan. 6th*, 1851. — I showed young Cargill your

drawing [copied from Seba] of the eared serpent as he calls it. He says it is exactly that which he shot; the shape of the head angular in the same way, but not apparently so pointed in the beak. It was spotted all over, and had the lobes purple-tinted.

"I have just seen Ulick Ramsay, and shown him the drawing from Seba's Thesaurus. The head of the snake, which he saw in the hands of the Artillery Sergeant *from the Arsenal*, not the *Barracks*, was similar. The ears, as he calls the lobe-like organs, appeared higher up to the crown; the head was angular in contour, and the body without spots."

These comparisons, made merely from memory, and without the attention having been particularly directed to the points in question at the time of observation, are, of course, of little value. The accumulation of evidence, however, for the existence of this curious form, is, I think, irresistible.

THE SPOTTED-CHINNED SNAKE.

A pretty little Snake, hitherto unnoticed by zoologists, has occurred to my researches in the neighbourhood of Bluefields. It is much too rare to have allowed any opportunity for observation on its manners, but two specimens having fallen into my hands; and the common people are not acquainted with it, or perhaps confound it with the common Grey Snake. I describe it below.*

* *Natrix callilæma*, MIHI. (καλὸς, pretty, and λαιμὸς, the throat.) Head oval; snout obtuse, rounded; neck slightly constricted. body and tail slender; tail two fifths (in the young specimen one third) of the whole length: snout projecting; mouth curved, rising posteriorly: gape reaching as far as the rear of the occipitals. Labial

THE WILD HOG.

I have never met with feral Swine in the woods of Westmoreland: in the sombre, high-timbered forests around Shrewsbury, in St. Elizabeth's, they are occasionally shot; but it is in the remote lands near the centre of the island, and especially in the wild lofty districts of the windward end, full of mountain peaks and ridges, that they have chiefly multiplied; and it is to the experience and inquiries of my friend Mr. Hill, that I am indebted for all that I know of these animals.

"The character of the Indian wild Hog, which ours very much resembles, is 'a broad flat forehead; short pricked ears, rather round at their tips, and lying very close to the neck; the eyes very full, with

plates six, the third and fourth forming the lower wall of the orbit; the fifth large and long; the sixth small. Vertical plate large, five-sided, nearly as broad as long. Occipitals very large. Scales smooth, convex, hexagonal. Abdominal shields in adult 144; caudal 104 pairs; in the young, abdominal 135; caudal, 115 pairs. Length of the larger twelve inches.

Colour reddish-brown above, softening to white below. An oblong mark of deep brown passes along the summit of the head, somewhat dilated before and behind: from this a brown stripe extends all along the middle of the back, having a tendency on the nape to form confluent rhomboids, like our own Vipér. On each side of this a line of regular black dots passes down, and below these a narrow band of brown on each side. The dotted line, as well as the dorsal stripe, become indistinct towards the tail, but the lateral lines continue well marked. The latter pass through the eyes to the muzzle, and are succeeded on each cheek by two indistinct parallel lines. Chin and throat prettily spotted and marbled with dark brown on the white ground, the marks small and confluent. The body, and especially the belly shields, opaline. In age the ground-colour becomes much darker, and the characteristic markings less distinct.

much display of the white, when in action; the head short, thickly furnished with hair, inclining to curl; a very muscular neck; a high shoulder; the back very nearly straight; the loins broad; the bristles thick on the neck and shoulder; the tail rather short, and near the tip covered with lateral bristles, resembling the wings of an arrow.' *

"Our breed of Hogs was derived from the Canary Isles — a genuine African variety. We were stocked from the ships of the Spanish discoverers. The species has the prick-ears of the Indian, but not what may at all be spoken of as a short head, — the head being elongated, and the extremities having a correspondent peculiarity, — a proportional extension of leg and limb. Mr. Johnston, of Portland, tells me that all our hogs are prick-eared; and he has seen many with the *feathered* tail. They are generally black; but red swine, that is, foxy-coloured, have been caught in our forests. A lop-eared hog is very rarely seen, even in our streets.

"Buffon's description of tracking the wild hogs in the West Indian forest, and the caution against over-fast pursuit, is represented to me as exceedingly correct. Their turning about, however, to face the dogs, is not so striking a fact with us, as their directly running for large trees with deep salient roots, into which they thrust themselves backward, and stand steadfastly the assault made upon them. Many a time when the novitiate huntsman has supposed that in starting a hog in the open grounds, he would make a long run of it, he has found his pursuit terminated by his game ensconcing himself within the

* Col. Williamson's Oriental Field Sports.

buttresses of a Silk-cotton tree, and putting the dogs fairly at defiance. He has here set his back against the wall, and he fights in desperation and dies in his hold.

"I will relate a little adventure of my own with forest Hogs in Spanish Haïti. I was in the way, with the friend I have mentioned in some of my former notes, to visit a curious chambered rock in the savanna of Copey, near Puerta-plata. It was situated at the foot of a still more curious mountain, that gave out strange musical sounds; the warbling wind, struggling through the fissures and crannies, had conferred upon it a voice of enchantment. The Hatero's wealth in these parts consists in his woodland swine. A peon had been provided for us, to conduct us to these mountain wonders; and though we had to clear our way, as we advanced, through occasional underwood, the forest, which was of magnificent growth, was generally exceedingly open. The chambered rock was an immense mass, of 300 feet high, of detached stratified limestone, with its strata standing vertical. The obstructed waters of a small stream, over which it had been rolled, had cleared a way through its fissures, and formed curious galleries, whose roof, walls, and floors, were fretted with stalactitic incrustations. It was a grotto of crystal work, at once beautiful and strange.

"To ascend to the entrance it became necessary to pass along a narrow cliff-raised pathway, from which successive steps led to a labyrinth of galleries. The peon before, drawing his legs hastily up on a ledge above us, sounded out, 'Look out for the Boar!' In an instant up rose from a niche three forest Hogs, the central one of which was a well-tusked Boar; the other two being Sows. The peon and the friend with me were safe on some ascending ledges of limestone. I was on the edge of the cliffy pathway. In a moment the Boar, champing his mouth with rage, and rising to make one rush upon me, took some three steps in advance of his mates*; when, at the same instant, the two large blood-hounds we had with us, both together leaping in over some low bushes beside me, and the terrier in company smuicing it under the brushwood, with his short yap of a bark, stood between me and the assault of the excited Boar. It was all the work of a mere fraction of a minute. The Boar stopped short from his attack to defend his mates from the dogs, who now were covering me. They had run round the Boar and turned him, cutting him off from his two mates. It was immediately a chase, for the Sows scudded off, and the Boar followed; and quick as magic, I found myself in the midst of peril, and delivered from the onslaught of an irritated Boar in the forest. How intensely did my heart beat! for the danger, I was assured, was imminent. Many a time I have remembered the occurrence, and wondered at my deliverance.

"The original Hog of the Canary Isles, the parent of our wild animal, is said by Eyton to have the specific peculiarity of fewer dorsal vertebræ than the typical species.

"Though our woodland breed never attain a

* ——— ὁ δ' ἀντίος ἐκ ξυλόχοιο
Φρίξας εὖ λοφιὴν, πῦρ δ' ὀφθαλμοῖσι δεδορκὼς,
Στῆ ῥ' αὐτῶν σχεδόθεν. *Odyss.* xix. 520.

greater height than twenty inches at the shoulder, their thorough forest habits render the full-grown Boar a powerful and artful combatant. As long as he runs, and seeks his safety in flight, pursuit both to the hound and the hunter has little hazard attending it; but the moment he has reached some vantage spot, — an angular rock, or the buttress roots of some large tree, sufficient to cover him in the rear,— and he faces round, and stands, he is a dangerous enemy to encounter. His ability to cut with his tusks on the right hand and on the left, and his habit of striking a short upward blow in front by taking one step forward, render him in such a fastness unapproachable. As long as he maintains this position, he absolutely defies the assailant dogs to touch him. No animal can be conceived cooler, none more bold, nor more thoroughly intent on keeping himself unwearied in the labour of defence, by little exertion in the means of defending himself, than a well-toothed Boar, in such a hold. Let his back be in a corner, and there are but eighteen inches between him and death to any ordinary dog that will face him in the forest.

"The best developed domesticated Hog of the original African breed that I have seen, was a boar of the blue variety; and the best display of woodland instinct that I have witnessed was recently exhibited in some young pigs of this blue breed, brought from the commons and forest runs of a mountain farm, and domiciled in town. Three of these country pigs, a boar and two sows, had taken up with a black pig and some four young followers, evidently town born and bred. In tramping home, after feeding out for the night, some of the town dogs, of a good enough quality of the hound and terrier breed, set upon them. Instantly the country hogs turned round, and coolly taking up their position in the angle of a wall, put the black pig and four young ones within the corner in their rear, and threw themselves before them. They then commenced that peculiar short hasty grunt, with which the hog kind announce danger, or prepare for a resolute resistance. The dogs that came upon them being reinforced by a troop from the several yards round about, became a pack of twelve or fourteen in an instant. Among these were some five small curs. The three blue pigs were undaunted. They stood their ground with their faces to their enemies, and though the dogs beset them with a determination to fight in earnest, they successfully kept off their assailants. The curs barked and grabbed at them between the legs of the larger dogs; — the larger dogs rushed at them six in a line together. The young boar, with well developed tusks, stood in the centre, and stepping every now and then one pace forward, made his upward rips at the dogs, and effectively struck them without receiving a single touch himself. The assault continued some time, but the pigs were not to be moved from their position. The dogs received several severe cuts and grips. I never saw a better managed defence. No exertion or perseverance could force the pigs from their vantage ground. The conflict drew the neighbours out; and it was not until they had drawn away the dogs one by one to their homes, that the pigs could be enticed from their hold. On another menace from the dogs, they took up a second similar position, and firmly maintained a second onset of their assailants with the same successful resistance. No badger in his tub could have been more stout-hearted, resolute, and courageous, than these three mountain hogs in their corner.

"My father used to relate an encounter he once had with a full-grown wild boar. He had entered a forest where occasional rocks bounding the right hand and the left gave the character of a defile to a mountain pass. A hog traversing the glen, no sooner found himself, when on the road, encountering an enemy than he faced round, and assumed an attitude of opposition to all further attempts to approach him. There being no dread of anything in the rear, and all that he apprehended of danger being before him, he stood ready to strike with his tusks; and he continued this sort of threat, whenever any endeavour was made to advance upon him, for a full half hour. No menace could move him from this stand till he heard footsteps from behind, when his position being no longer secure, he crossed the road into the woods at full speed, and left an undisputed pathway to the travellers upward and downward."

In a recent communication, my friend thus returns to the subject.—"*8th February,* 1851. I have learned some new facts respecting hog-hunting. The present letter, you will perceive, covers some additional notes to those already sent you on that subject. They may be said to embrace the commercial details of our Forest Swine, with some notices of Maroon hunters. I find, in addition to these, that a very common mode of taking the wild animal is by *snares.* A sapling, in its place of growth among the dense and lofty trees of the woodlands, is bent down after having been stripped of its leaves and small branches. A noose, made of a withe, is securely attached to the end of the bent tree, and so adjusted with stakes slightly fixed, that, on a Hog thrusting its body or only a foot through the snare to seize the roots and fruits strewed about to attract it, the sapling rises up by the force of its own elasticity, and carries the Hog into the air, sometimes strangled by the neck, but as often caught round the body, or held by one of its legs. Your acquaintance with our forests will enable you to understand this species of trap, and to form a pretty correct conception of the kind of picture which might be made to represent it successfully, applied to Hog-snaring. . . .

"I mentioned to you that in Sloane's History of Jamaica, you would find some account of the Forest Swine. His graphic picture of the herds in the '*Crawles*'*, as the rude and extensive farms of the time were called, would form an interesting illustration of your notices of the Wild Hog. My birthplace, Montego Bay, formerly written *Manteca,* and Mantiga Bay, derived its name from the supplies of lard which were shipped from a district known in the old maps as Spanish Quarters. The historian Long, who mentions the wild hogs formerly abounding in this part, says, that the traffic of the Island at its conquest by the English in 1655, although small, consisted of supplies of fresh provisions to Spanish

* A corruption of the Spanish word *Corral.*

homeward-bound ships; that it was considered as the granary and victualling place of the traders; and that 80,000 hogs were every year killed for their *lard* alone, for which a constant market was found at Carthagena, on the South American main. When Sloane journeyed to the '*north side*' of the island, the cattle which the settlers had reclaimed in the southern plains, stored so abundantly with neat-kine the Savanna pens, that a single settlement possessed forty thousand head. The *north side* was the exclusive field for wild cattle and horses; and wild swine were plentiful in the same remote districts. Sloane, describing the two sorts of Hogs, one running wild in the woods, the other fed in crawles, says, 'the wild swine are brought out by hunters with gangs of dogs, and chiefly found in the most unfrequented, woody inland parts of the island. After pursuit, and they are wearied by the dogs, when they come to a bay, they are shot or pierced through with lances, cut open, the bones taken out, the flesh is gashed on the inside into the skin, filled with salt, and exposed to the sun, which is called jerking. It is so brought home to their masters by the hunters, and eats much as bacon if broiled on coals.' The hunters were both whites and blacks. The Indians, of whom there were then some in the colony, chiefly imported from the Indian coast, were 'exquisite at this game.' They pursued their business of hog-hunting far remote from the settlements, building huts 'in the places where swine came to feed on the fruits,' and where they remained marooning for several days, and preparing the flesh for market, smoked and packed in aromatic leaves. 'The swine fed at crawles were in very great plenty.' The crawles had clusters of sties built expressly for feeding and breeding them, and were committed to the care of the same class of men described as hunters, — white servants, Indian peons, and negro slaves. 'The swine came home every night in several hundreds from feeding on the wild fruits in the neighbouring woods, on the *third* sound of a conch-shell, when they were fed with some Indian corn thrown in amongst them; and let out the next morning, not to return till night, or that they heard the sound of the shell.' These remote plantations were very profitable. They afforded a constant and numerous supply for the market, which at this time must have still continued considerable. 'It was not a small diversion to me,' says Sloane, 'to see these swine in the woods, on the *first* sound of the shell, which is like a trumpet, to lift up their heads from the ground where they were feeding, and prick up their ears to hearken for the *second*, which, so soon as they heard, they would begin to make some movements homewards; and, on the *third* sound, they would run with all their speed to the place where the overseer used to throw their corn. They are called home every night, and also when such of them as are fit for market are wanted, and seem to be as much, if not more, under command and discipline than any troops I ever saw.'*

"Though cattle were so numerous in the savannas, and were running unreclaimed in the forests, and

* Introd.; xvi. and xvii.

the farms were well supplied with the several varieties of poultry — turkeys, hens, and mallards, and Muscovy ducks, with occasional geese — Swine's flesh was the 'most frequent dish at the table of the best inhabitants.' The 'barbecued pig,' which necessity had taught the huntsman to prepare in the forest, roasted in a rude oven of heated stones, and flavoured with native spices, — a more refined taste had transferred from the herdsman's hut to the plantation hall. Monk Lewis, whose taste was luxurious and fastidious, declares it to be one of the most delicious of viands. 'Several gentlemen of the country,' he relates in that very agreeable gossiping Diary, his 'Journal,' 'dined with me to-day' (January 26. 1816). 'We had at dinner a land tortoise and a barbecued pig, — two of the best and richest dishes that I had ever tasted, the latter in particular. It was dressed in the true Maroon fashion, being placed on a barbecue, or frame of wicker work, through whose interstices the steam can ascend, — filled with peppers and spices of the highest flavour, wrapped in plantain leaves, and then buried in a hole filled with hot stones, by whose vapour it is baked; no particle of the juice being thus suffered to evaporate. I have eaten several other good Jamaica dishes, but none so excellent as this.'

"When the Spaniards with their slaves retired to the north side of the Island on the conquest by the English in 1655, they left their negroes to hold the mountain fastnesses, and harass the conquerors and new settlers. These became the body of independent occupants of the interior, known as Maroons*, and afterwards recognised by specific treaties as a free people, governed by their own officers, and only in so much a part of the colony as that they received their appointments from the governors, and lived in villages under the superintendence of an European officer, commissioned and stipended by the government. In this state of freedom and independence, their characteristic habit, as mountain rangers, was made by express laws subservient to a sort of *police of the forest*. Their most stirring pastime was the hunting of the wild hog. This pursuit served the purpose of *chevies* for negro runaways, till trafficking in jerked pork and in rewards for apprehended runaways became a systematised business with them. In the days of slavery, the Maroon huntsman was a fine specimen of the athletic negro, on whom was stamped the impress of the Freeman. He was generally seen in the towns armed with a fowling-piece and cutlass, and belts that suspended on one side a large plaited bag, known as a *cuttacoo*, and on the other a calabash, guarded with a netted covering, in which he carried his supply of water. On his back, braced round his shoulders, and suspended by a bandage over the forehead, was generally seen the wicker cradle, that held inclosed a side of jerked hog, which he sold passing along, in measured slices, to ready customers, as an especial delicacy for the breakfast table. The accoutred Maroon, with this vendible commodity, was altogether a striking and characteristic figure in our streets. The abolition of

* From *Cimarrón* (Spanish), wild.

slavery having absorbed him in the general mass of liberated negroes, he has abandoned the huntsman's life for that of the husbandman. The pursuit of wild hogs has terminated with the rewards for runaway slaves; and in this age of railways and steam navigation, the flitches of American bacon in the provision shops have driven out of the market the jerked hog of the Maroon.

"When I applied to our agreeable noter of historical facts, Mr. Gregory Johnston, and asked him for characteristic accounts of hog-hunting and hog-hunters in the glens of Portland, where the Maroons still had settlements, I inquired of him whether the lairs of our forest breed exhibited what is a common habit of the wild hog in Europe, — the female covering her companion with litter, and after comfortably putting him to bed, slipping under the cover herself, and lying with him entirely concealed. I thought if this pretty trait of forest housewifery existed, the Maroons must have surprised them frequently in their connubial coziness. He promised me a graphic account of what he had learnt from a veteran hunter, but he did not live to write it out for me, though he had certainly jotted it down. His intelligent son, who was as enthusiastic a lover of woodland sports as his father, and to whom we are indebted for traits of the *Gowries* *, hardly outlived him. He perished accidentally: his gun went off and wounded him while shooting in the woods. I might otherwise have got his father's notes from him.

* See 'The Birds of Jamaica,' p. 58.

THE CROCODILE.

In some parts of Jamaica, Crocodiles (or, as they are usually called, *Alligators*) are sufficiently numerous; but in the neighbourhood of Bluefields they are so rare that, though I occasionally heard reports of one and another having been seen in the creeks and morasses around, I was never so fortunate as to fall in with a living specimen.

To atone for my lack of personal observation on these animals, I have pleasure in extracting from my correspondence with Mr. Hill, many valuable notes, containing much that is new and interesting, and tending to explain some things that have hitherto appeared discrepant and contradictory. There is no evidence, I believe, that any species of Alligator, properly so called, inhabits the Antilles: the Crocodile of Jamaica is *C. acutus*, the Slender-muzzled, possessed by this island in common with Martinique and Hayti, and represented in Cuba by the Lozenge-scaled Crocodile (*C. rhombifer*).

The first extract that I shall make refers to the mode of feeding; and though my friend used the term Alligator, he subsequently ascertained, on the authority of Descourtiltz, that the species is *C. acutus.*

"Mr. Waterton, I perceive, in his second series of Essays, contradicts Swainson's statement respecting the habit of the Cayman, 'to convey its food to some hole at the edge of the water, where it is suffered to putrefy before it is devoured.' He says, this statement must be rejected, because 'the mouth of this reptile is completely formed for snatch and swallow.' My experience and observation accord with Swainson's statement. I had so set down the habit of this reptile, on the authority of the Spaniards of Hispaniola, in the notes I made when in that part of Haïti; but when I returned to Cape Haïtien, I had an opportunity of proving it. The French consul, M. Barbot, had an Alligator, which he kept in a cistern, fed by a mountain spring, in a pretty garden of his residence. The Alligator would be found in the water all day, but at night he rambled over the garden, which was walled in; and was then a diligent catcher of frogs and toads. Usually in the morning, some half dozen *batrachians* would be found mangled and torn, and stuffed into all the crannies and corners of the cistern convenient for that purpose; their long disentangled entrails streaming out into the water. The Alligator would rise to the surface, take portions of the limbs and entrails *into* its mouth, and holding them with its teeth pressed close, with part of the entrails *on one and the other side of its mouth,* would gently squeeze the whole of the food, and swallow it most leisurely. Its action in eating most resembled a person chewing a 'quid' of tobacco. In this way it always fed, *at least by day.* This just agrees with the common account. The British consul, Mr. Heneiken, of Puerta Plata, an excellent observer in natural history, told me that in numerous Alligators he had killed and opened, he found nothing solid within them, except sticks and stones. The swallowing of these substances is an instinct they have in common with the *Phocidæ*. Mr. Heneiken conjectured that by this instinct they regulated the buoyancy and gravitation of their bodies with relation to the water."

"The verdant marshes of the Estèr in Western Haïti, where I first saw the Cayman, is the feeding ground of numerous cattle. Clumps of acacia and tufts of bamboos festooned with lianas, and embellished with blooming nymphæas floating on the waters, contribute to vary the aspect of these swamps. Egrets and gallinules inhabit them in numbers, and ducks frequent them in vast flocks. Large Caymans are to be seen there floating in the clear stream and prowling in the thickets; yet I saw naked herdsmen and fishermen navigating the waters in narrow canoes from six to nine feet long, and not more than eighteen inches or two feet broad. They had tied their camisettes of blue and pink and white check around their heads like turbans, having only the tanga or waist-cloth round the naked body, that they might wade the waters when necessary. Their singularly wild appearance, in these mere logs of boats, pushed along by poles, — the numerous cattle, and the multitudinous birds, with the frequent Alligators, in the midst of which man, bird, and beast were moving about, was altogether one of the strangest wild sights I had ever witnessed.

"In the calm long-enduring quality of its nature, the Alligator is really a timid animal. With the advantage of an impervious armour, impenetrable covering of scales, strength of limb, commanding shelter of the water, and jaws whose clasp is an inextricable hold, it fears to attack an animal it is unused to devour. The inhabitants, aware of its timidity, fearlessly cross streams even when numbers are swimming around; but this is not done without some *ruse*. The Alligator, they say, is greedily partial to dogs, and surprises them often when they come to drink at the river. The voice of the dog will always draw them away from an object when prowling. Those who would cross a a river without any risk from their attacks, send a scout down the stream to imitate the canine bark, yelp, or howl, when away swim the Alligators for their prey, leaving an unmolested ford for the traveller higher up. Instinct has taught the dog to secure himself by a similar expedient. When it has to traverse a stretch of water, it boldly goes some distance down the stream, and howls and barks. On perceiving the Alligators congregating in eager cupidity to the spot, it creeps gently along the banks higher up, and swims over the water without much fear of being pursued.

"This fact is universally asserted, and is unquestionably true, but I doubt the conclusion. The well known habit of the Alligator, never to eat its food until in a state of *putridity*, negatives the supposition that it has any particular choice in the flesh of its victim. The fishes on which it preys, it probably devours immediately after it has caught them; but all other things else, as soon as they are slain, are torn and mangled, limb by limb, and left to putrefy in the river, or in the sedge about its lurking place. It can have no particular preference, therefore, for the dog as food. I am disposed to ascribe this susceptibility to be roused at the canine yelp to the similarity of that sound to its *own peculiar cry*, under any species of excitement; — to the fact that it is the impassioned voice of its young — to the maternal solicitude of the female for its progeny when it hears that voice — and to the ravenous appetite of the male on the same occasion; for, like many of the rapacious animals, the male of this tribe preys upon its own offspring.*

"It is not very clear whether the male parent, after it has sought the attachment of the female, in which its passion is fierce and violent, assists her in the office of disposing the eggs in the earth. It is much more likely, from the necessity of her after watchfulness to guard against his reprisals, that he does not After burying the eggs in the soil, to be there matured by the sun, the female visits from time to time the place in which they are secreted, and, just as the period of hatching is completed, exhibits her eagerness for her offspring in the anxiety with which she comes and goes, walks around the nest of her hopes, scratches the fractured shell, and by *sounds which resemble the bark of a dog*, excites the half-extricated young to struggle forth into life. When she has beheld, with this sort of joy, fear, and anxiety, the last of her offspring quit its broken casement, she leads them forth into the plashy pools, away from the river, and among the

* Professor Buckland has discovered in the excrementitious fossils of the Plesiosaurus or Fish Lizard evidences of a similar rapacious appetite in those extinct animals. The bones of the young Plesiosaurus were found in the petrified dung of the older ones.

thick underwood, to avoid the predatory visits of the father. In this season of care and of watchfulness over them, she is ferocious, daring, and morose, guarding with inquietude her young wherever they wander. She turns when they turn, and by whining and grunting, shows a particular solicitude to keep them in such pools only as are much too shallow for the resort of the full-grown reptile. When I was in Yasica, a river district of that name, as many as forty had been discovered in one of these secret resorts; but in half an hour, when the boys who had found them out returned to visit their hiding-place, they saw only the traces of the coming and going of the watchful parent who had led them away to some further and safer retreat. In this period of their helplessness, the mother feeds them with her masticated food, disgorging it out to them as the dog does to its pups. In general it is rarely seen otherwise than crouching with its belly to the earth, and crawling with a curvilinear motion; but at this time it may be observed firmly standing on its feet. This is the attitude of anger and attack; and its spring is quick, a sort of agile leap, by no means short in distance. During all this time of protection and dependence, is heard the voice, by which the young makes its wants known, and the parent assures its offspring of its superintendence. It is the *yelping bark of the dog*, and *the whining of the puppy*.

"From all these facts I take it that when the sound of the dog's bark is heard, the Caymans press to the spot from which it issues, agitated by two several passions,—the *females to protect their young*, *and the males to devour them;* and to this, and not to their predilection for the flesh of dogs, are we to ascribe the eagerness with which they scud away, agitated by that voice which in the one case is the *thrilling cry of danger*, and in the other, the *exciting announcement of food.*

"This susceptibility to be excited by hearing a cry associated with peculiar instincts and appetites, has other parallels in other animals. A very striking analogy is to be found in Wilson's anecdote of the Cat-bird.

"The Alligator's motion when prowling is literally a crawl; and in their posture of attack they stand with their bodies off the ground, and make their onsets by successive leaps, the arched back mentioned by Humboldt being then a peculiar and distinguishing trait of their anger.

"An occurrence related to me, that happened to a Spanish priest on the banks of the Guayabino, will best illustrate at once the predaceous vehemence and lurking patience of the Alligator. The large savanna rivers in Spanish Haïti flow through wide but gently descending borders, carpeted with grass, and interspersed with thickets and clumps of flowering shrubs and forest trees. The grass has all the clean verdure of a lawn, and the clumps the variety and arrangement of ornamental shrubberies, and the earth is deep and loamy. These are favourite sporting grounds. Beside being verdant and beautiful, they are notoriously the game country. My friend and his companions, who counted some four in number, had divided themselves, trusting to the crack of their fowling-pieces to ascertain each other's whereabout. When they had finished their day's sport, the descending sun was already struggling through the lengthening shadows on the river. The friends assembled where they had parted in the morning, but the Spanish priest had not yet come in. No one had heard his gun from the time they had separated. They sought him through the darkening thickets, and along the stream, and found him at last, fast seated in a tree, into which he had been obliged to betake himself to escape an Alligator that had pursued him by a succession of leaps. It had run in pursuit of him, as he said, jumping rapidly after him, with its back crooked, like a frightened cat. He had sprung to the branches, and gained their security out of the reach of the reptile, who for a long time after he had got into the tree, crouched in a thicket close by, where it quietly watched and waited his descent from his retreat. I was not aware, until after I had heard this relation, that Humboldt had similarly described the attack of the crocodile when pursuing its victim on land.

"When Moreau de St. Mery, about 1790, collected materials for his work on St. Domingo, he noticed a Cayman that had been kept for ten years on a plantation at Gonaïves, not far from the Estèr, called Cocherel. When it was first taken, it was only eighteen inches long; but at the time he wrote it had grown to the dimensions of seven feet. This may serve to give one some idea of the progressive growth of this reptile. He mentions that it was kept in a sort of inclosure into which no other water than that which the rain supplied was received; and rain does not commonly fall out of the season at Gonaïves. It was fed on the dead animals of the plantation, and on sheep's entrails, but the people frequently neglected it; and it did not seem that in these intervals it got any food whatever, yet it steadily continued its growth. This notice of the penned-up reptile of Cocherel is most interesting for the fact of its living deprived of water. This is a circumstance, however, less remarkable of the Alligator than of the Crocodile."

The rapidity of growth in this reptile is mentioned also in a note which I find in Robinson's MSS., and which contains other interesting particulars. The species alluded to we have no means of knowing; but the writer evidently supposed it identical with the animal he was familiar with in Jamaica. "At a place in Oxford Road, London, I saw a vigorous young Alligator: I visited it several distinct times after, and each time observed it surprisingly grown. At first it might have been about 2½ ft.; the last time it was about 5 ft. long. It was kept in an open wash-tub just covered with water: the mistress of the house I saw feed it, which she did by opening its jaws with her hands, and chucking in sprats, which it swallowed instantly. It bore stroking or any kind of handling most gently and patiently: 'twas as tame as any of her domestic animals. The water, in the cold months, she informed me, she made milk-warm for him to lie in; yet, notwithstanding, she assured me, this familiar creature would several times in the day quit the tub, crawl to the hearth, and there bask himself till sufficiently warmed by the fire, and then return to his proper element in the tub."

To return to the communications of Mr. Hill: — After alluding to the specimens of *Loricata* preserved in the Jamaica Society's Museum, and to the various living individuals which had fallen under his own observation, — as all exhibiting the characteristics of true Crocodiles, not Alligators, — he thus proceeds:

"The Saurians of this genus in the Rio Apure noticed by Humboldt as exceedingly numerous, being seen by five and six at a time, he describes as 'real Crocodiles, not Alligators or Caymans, *with feet dentelated on the outer edge* like those of the Nile.' As he measured two dead individuals, one 17 ft. 9 in. long, and the other 23 ft., with their swimming powers so remarkably indicated, they must be formidable monsters. It is on the occasion of noticing the animals of the Rio Apure, that he describes what may be considered the characteristic movement of these reptiles. 'The motions of these animals,' he says, 'are abrupt and rapid when they attack an object, although they move very slowly when not excited. In running, they make a rustling noise, which seems to proceed from their scales, and appear higher on their legs than when at rest, at the same time bending the back. They generally advance in a straight line, but can easily turn when they please. They swim with great facility, even against the most rapid current.' On another occasion, when the same traveller mentions that even in the streets of Angostura, when the Orinoco inundates the quays, persons fall a prey to these reptiles, and relates the story of an Indian of Margarita, whom, when he had gone to anchor his canoe in a cove where there were not three feet of water, a very fierce Crocodile seized by the leg and carried off, and who, though he had the astonishing presence of mind to search for a pocket knife, and thrust his fingers into the animal's eyes, was yet so firmly held, that the reptile plunged to the bottom, and drowned him,—he describes it distinctly as a Crocodile. The fact mentioned of the animal plunging to the bottom of the river with his victim and swimming up again to devour him, is quite in accordance with the remarkable organisation of the Crocodile for diving and swimming; and I now suspect that the contradictions relative to the daring and ferocity, and the timidity and wariness of these monsters of the river and the lagoon, are the distinctive characters of the Crocodile and the Cayman. The protuberances which protect the eyes of the Cayman, with the feebler natatorial and diving powers of the feet, seem all to have reference to the prowling instincts of the crawler among morasses, rather than to the dashing fierceness of the bold swimmer in rivers and rapids. Humboldt says that the intrepid 'natives in contending with the *Crocodile* observe its manners as the torero studies those of the bull; and quietly calculate the motions of the enemy, its means of attack and the degree of its audacity;' while the animals of the Rio Neveri, and of the little river of Narigual, which he mentions as exceedingly infesting the fords, and speaks of both as Crocodiles and Caymans, he especially states were less ferocious than those of the Orinoco, for that "the people of New Barcelona convey wood to market by floating the logs on the river, while the proprietors swim here and there to set them loose when they are stopped by the banks, which he says they could not have done in most of the South American rivers infested by these animals." When he makes reference to the same animals in the marshes of Cuba, he distinctly speaks of them as two species of Crocodile, one of which he describes as "having an *elongated snout, and as being very ferocious.* Daring and power in the water seem therefore to be the distinction of the Crocodile, and timidity and stealthiness that of the Cayman; a difference which we might infer from the difference in the feet of the two reptiles."

"*February* 19*th*, 1849.— A Crocodile (the animal we usually speak of as an alligator) had been taken in the fish-nets at Hanson's pond by a fishing party, and brought into the King's House yard alive. It had just been killed by a pistol bullet discharged into its brain when I saw it. I attended whilst the negroes skinned it, and had an opportunity afforded me of observing the several peculiarities which are mentioned as characteristic in the structure of Crocodiles. I shall set them down as they successively came under my notice.

"On opening the jaws, the attention is taken by the sight of a conspicuous cartilaginous plate before the gullet, forming a ridge from one side of the fauces to the other, and expanding upward to meet a similar elastic fold depending from the back of the palate,

These are valves that shut in the throat. We are led to conclude on first seeing these valves that we are examining an animal that has no tongue, and that the underfold of what we are inspecting is the rudimentary trace of that member cut out. This, however, with the corresponding curtain above it in the roof of the mouth, forms an apparatus that closes the distending aperture of the throat, and permits the reptile to hold its prey and drown it, without being itself liable to be drowned.

"Between the branches of the lower jaw, a certain degree of muscularity is perceived in the yellow flooring of the mouth. This is the representative of the tongue. The thickened membrane shows its lingual analogue, though destitute of all approach to a red colour, by its rough glands and pores giving out saliva.

"The nostrils, placed at the extremity of the snout, terminate in a post-oral cavity, by passages that communicate with the throat behind the valvular apparatus we have been describing. This is a provision for respiration when the valves are closed, which at once renders intelligible and necessary a remarkable structure of the fauces by which the upper jaw seems to move upward, whilst the under one retains its horizontal position. The lower is prolonged behind the skull to a great depth. On raising the head at an angle, the upper jaw appears to move upward, and the under jaw to remain immovable. The upper jaw does indeed move upward, but not independently. On casting back the head, an acetabulum of the united skull and jaw acts on a condyle of the lower maxillary bone, and lifts the whole head like the coved lid of a caddy: by this mechanism the Crocodile, on elevating its nostrils just barely out of the water, is able to breathe. With the body and head sunk below the surface, it keeps the under jaw pressed upward, and holds fast its drowning victim, its own breathing all the while being carried on at ease. The mouth is open, but the throat is shut, the gular valve being closed against all access of either air or water.

"In some previous observations set down by me on the Cayman of St. Domingo, identical with our Crocodile, I had mentioned that, beside the habit constantly maintained by a young one kept in the garden of the French consul at Cape Haitien, of stuffing its mangled prey into the pond banks till it was putrid, it used to lie for hours together, with nostrils barely elevated above the water, keeping in its mouth junkets of frogs it had killed, without eating them. I now see that this habit was the young Cayman practising the art of drowning living prey. It did not eat what it had in its mouth while within the water, because its structure was as unsuited for feeding as it was for breathing open-mouthed in that element; and its feeble palatal organisation could scarcely do more towards gratifying its taste with the portion of frog it held, than keeping constantly present a sort of sensual consciousness of food. I have a lively and pleasurable recollection of the garden of Consul Barbot at Cape Haitien. A small plateau at the foot of the Haut du Cap Mountain, filled with clumps of shrubberies and scattered palms festooned and clustered with gaily-coloured bind-weeds, shadowed a little basin which collected the current of a spring that ran dashing and sparkling from the rocks. The fountain was always cool, for it was constantly refreshed from the mountain. The sun's rays played within the foliage; and the Cayman lay on the sunny waters, indulging his solitary passion in dreamy quietness. The nightly dews dripping from the herbage, and the oozy rivulet winding among the shrubs, tempted the frogs at nightfall within his prowl; for every morning saw the putrid food of the previous day devoured, and fresh carcases mangled and torn and stuffed into the crevices of the pond, and fresh pieces of meat in the Cayman's mouth, to afford him the imaginative enjoyment of holding a struggling victim between his teeth while he quietly rested afloat and killed it.

"When I was looking on at the skinning of the Crocodile, I related the foregoing among other traits of instinct to the Mosquitan Consul, General M'Chrysty, who was at this time a guest at the King's House. He mentioned to me that this explained what he had frequently seen on the Lake of Nicaragua. There he had observed the Caymans throw up into the air fresh-captured fish, which they afterwards caught in their mouth, and then threw up again. This they continued to do several times. This was another way of killing prey. It would be labour in vain to endeavour to drown a fish; the Cayman therefore killed it by keeping it in the air. This incident had been mentioned to me before, but the object of it was not clearly made out. I now see that it was another application of instinct to the one purpose of slaughtering prey. The victim that could not be destroyed by being kept under water was killed by being thrown out of it.

"In consequence of the manner in which the negro butchers detached the heart and other viscera, we lost the traces of the provision which exists in this reptile for circulating, by a trilocular heart, pure arterial blood in the anterior parts of the body, and mixed venous and arterial blood in the hinder members, by a junction of the pulmonary artery with the *aorta descendens*, low down in the back. The difference of the two qualities of the blood in the circulatory systems of the two extremities was, however, very perceptible, by the absence of all red blood in the hind limbs and tail, and by the presence of brilliant arterial blood in the fore parts and the head.

"The stomach of the Crocodile bears considerable resemblance to the gizzard of a bird. Several stones swallowed to assist digestion, render the similarity perhaps more obvious than even the form and structure. The Crocodile we were skinning and cutting up had evidently prowled along the coast before he had entered the pond in which he had been caught; for his stomach was filled with marine crabs half digested; and the stones found in it, if necessary on ordinary occasions to triturate its food, were essentially requisite now to crush the hard cases of crustacea.

"The dilatations and contractions of the thorax when sensitiveness was excited by the pain attendant on the first removal of the skin, showed the degree in which the ribs by their mobility assisted respiration.

"The absence of clavicular bones was evident; and the removal of the integument of the neck was attended by a perceptible odour of musk. The probable use of this odour is to entice prey, for it is well known that a paste perfumed with oil of Rhodium is a great attraction to fish.

"I should describe the tegumentary covering of the Crocodile to be, on the under parts, square tesseræ of horn set as close as they can be together; on the sides oval scales, with rather wide interspaces of tough skin; on the back oblong plates, some ridgy, others dentelated, and others bossy, with central tubercles.

"When stealthily swimming, the Crocodile nips its prey with its front lower teeth, two of which close up and sheath into sockets of the snout. The instinctive tenacity with which he holds whatever he has nipped and caught, was amusingly manifested in a Crocodile drawn out of a hole at Windsor Park pen near this town. He lay wounded; but by presenting to him a rope knotted at the end, with a small cross-bar driven through the knot-hole, he was enticed to nip it. The Crocodile held the rope closer as he felt the effort at resistance greater; and in this way he was drawn out readily, though measuring fourteen feet in length.

"The orifice of the Crocodile's ear is guarded by a plate firmly hinged, forming a moveable lid, rising and shutting at the pleasure of the reptile. This is a coincident provision with that for breathing by the nostrils, when all the head save the snout is under water. If the closing of the auricular valve be accordant with the act of shutting up the valvular apparatus of the throat, — as it must necessarily be, since both are provisions against the access of water under one and the same submersion, — then the lifting of the ear-lid must be simultaneous with the act of relaxing the gular cartilages. This seems to explain a well-known habit of the Crocodile in lying out of the water with his mouth open, — the opened throat is an accompaniment of the open ear-valve. The Crocodile is in the act of listening for an approaching prey. It is similar to that act in ourselves. The lips apart increase distinct hearing, — sound producing an aural influence as well internally as externally.

"The nostrils, like the ears, have moveable valves, shutting and opening according to the total or partial immersion of the snout. The power of the Crocodile of remaining under water without breathing is not considered to exceed ten minutes.

"All these provisions, with the linear contracting and dilating pupil to regulate vision according to the quantity of light, and the nictitating membrane of the eye with the two external eye-lids, demonstrate a nature most providently arranged for predacious life in the water.

"I close my present notes of the Crocodile by making reference again to its sympathy when apprehensive of danger to any of its species. This reptile is known to hasten eagerly to a spot where a dog may be yelping. I have mentioned elsewhere that the whine and bark of the dog closely resemble the cry of the Crocodile. As the maternal instincts are strong, we may expect that the arterial circulation in the fore parts is associated with endowments of a higher animal nature than that of a mere reptile. I mentioned, by way of analogy, the excitement of the Cat-bird, in Wilson's vivid description of alarm and sympathy in the forest when the cries of young birds are imitated. A very characteristic scene of emotion occurred the other day in the farm of a friend near this place. The calves had been penned up for the night, and the mother-kine were gathered about the adjoining common to be driven up for the morning milking. My young relative Peter, accompanied by another young friend, had gone to the pen-fold containing the calves, and had there indulged his musical predilections by carrying with him an accordion, from which at times he sounded those musical cadences, sometimes in an ascending, sometimes in a descending series of notes, at all times solemn and exciting. The cows, alarmed at the unusual sound from the pen-fold in which were fastened up their impatient and expectant young ones, ran eagerly from all parts of the field—so true was nature to her sympathies—to what they considered the cry of the calves in danger, as if one mind actuated them. The poor boys, alarmed at the menaces they saw on all sides, without divining the cause, ran away, sounding occasionally, as they ran, the still exciting accordion. The cows pressed in pursuit to ascertain the mystery of the clamour and cry of distress now turned from the cattle-pen to the field. The boys scudded on in great fright, and escaped the anxious curiosity of the kine with considerable difficulty. A friend related to me a still more remarkable instance of sympathetic emotion, in the simultaneous gathering together of the horses, asses, mules, and cows of an extensive mountain farm, attracted by a march played on a cornupiston. The retinue of animals quietly yet eagerly followed the new Orpheus, charmed by the martial music that so enchanted them."

I shall close these memoirs of the mailed Leviathan by two or three anecdotes. The first, illustrative of his mischievousness, is introduced by a description of the scene, portrayed by my coadjutor in his peculiarly vivid manner.

"A friend and I had arranged to rise at four o'clock from our place of stay, to visit the sources of a stream in the neighbourhood, and to trace its course to the sea, in order to ascertain the extent to which it might be rendered available for irrigating the adjacent plain. It was one of those intensely bright moonlights, which you know to be so exceedingly beautiful in these climates, under a calm and cloudless sky. Nothing stirred; not a voice sounded; not a watch-dog was moving in the slumbering villages; nor a cow lowing, nor a sheep bleating in the contiguous fields. The guinea-corn was in the full milky ear, but no flocks poached it; nothing was heard till we reached the green commons about the river, when here and there rose up, near and far, at distant intervals of time, the shrill wail of the plover, and the solitary call of the snipe from the dewy grass. The crozier stars, standing erect, were twinkling in the far south; and the silence of the world on which they shone, broken only by these brief intermittent voices, had a character of melancholy solemnity that I never remember to have remarked before.

"There was so much disparity between the temperature of the springs we visited, as they welled out and jetted up at the foot of great-grown and embowering Ceiba Cotton-trees, and that of the chill morning air (a difference of some dozen or fourteen degrees), that the surface of the stream reeked again. Folds of vapour rising through the sedges, and curling away in thin fleeces of clouds, prepared us to hear that the numerous Caymans which inhabit this river found its banks a very successful lurking-place for prey. We were shown a weir across the water-course, just within reach of the sea, which a large Cayman a day or two previously had been seen deliberately battering to pieces, by laying himself broadside against it, and lashing his tail at the stakes. The stockade fell slanting, making a sort of chevaux-de-frise. The Cayman was unable to get over in this position; so he bethought himself, in his new difficulty, of regaining the land, and passing the weir by the bank, re-entering the stream, and taking the river upward, after all his toilsome mischief."

The following is the repetition of a well-known and deservedly celebrated exploit: —

"Some time in the spring of 29 or 30 (most probably in March, 1830), a Cayman from the neighbouring Lagoons of Lyson's Estate in St. Thomas's in the East, that used occasionally to poach the ducks and ducklings, having free warren about the Water-mill, was taken in his prowl, and killed. All sorts of suspicion was entertained about the depredator among the ducks, till the Crocodile was surprised lounging in one of the ponds after a night's plunder. Downie, the engineer of the plantation, shot at him and wounded him; and though it did not seem that he was much hurt, he was hit with such sensitive effect that he immediately rose out of the pond to regain the morass. It was now that David Brown, an African wainman, came up; and before the reptile could make a dodge to get away, he threw himself astride over his back, snatched up his fore paws in a moment, and held them doubled up. The beast was immediately thrown upon his snout; and though able to move freely his hind feet, and slap his tail about, he could not budge half a yard, his power being altogether spent in a fruitless endeavour to grub himself onward. As he was necessarily confined to move in a circle, he was pretty nearly held to one spot. The African kept his seat. His place across the beast being at the shoulders, he was exposed only to severe jerks as a chance of being thrown off. In this way a huge reptile *eighteen feet long*, for so he measured when killed, was held *manu forti* by one man, till Downie reloaded his fowling-piece, and shot him quietly through the brain.

"You will perceive that this is precisely the feat performed by Mr. Waterton. He says his Cayman plunged furiously, and lashed the sand with his tail, but that, being near the head, he was out of the reach of the strokes of it, and that his plunging and striking only made his seat uncomfortable. This seemed really almost all the difficulty in David Brown's horsemanship; but as every plunge with him only drove the Crocodile's nose into the ground, whereas Mr. Waterton's Cayman was kept head-up by the

T 6

people tugging at the hook in his throat, that would make his chivalry a more desperate adventure than David Brown's, for *his* beast's efforts to get forward only more effectually set him fast where he was."

The last narrative is of a more tragic character, and bears out the statements of Mr. Waterton as to the ferocity of these powerful reptiles. The scene of the incident was Black River in St. Elizabeth's, where Crocodiles abound.

14*th July,* 1849.—"On the eastern bank of the river, just above the bridge, and right within a quay and jutting cranehouse attached to a long line of stores, a Crocodile, some twelve months ago, snatched off from the beach a young girl thirteen or fourteen years of age, who was washing a towel at the river, in company with an elder companion, at nightfall. She had been warned that it was dangerous to stand at all within the water after dark, for *Alligators,* as these Crocodiles of ours are erroneously called, would be then prowling, and fatal casualties had occurred. Just as the little braggart boasted that she heeded no such danger, a scream for help, and a cry, 'Lord, have mercy upon me! Alligator has caught me!' apprised her companion, intent on her own washing, that the girl was carried off. She was instantly snatched under water and drowned. The body was found some days after half-devoured, and two Crocodiles, one nine feet long and the other seventeen, were hunted down, and taken with portions of the flesh undigested within them. The bowels had been eaten away;—the lower limbs torn off; half of one thigh only remaining. The body had been carried away considerably up the stream; and the strange assertion, that the Manati, a cetaceous inhabitant of the Black River with the Crocodile, will remain watching a dead body, if brought within its haunts,—was witnessed in the case of this girl, by the body being found under the guardianship of a Manati, up at a place called Salt-spring, a tributary of the Black River, where Manatis abound."

LAND TORTOISES.

Some of the old writers mention, among the animals of Jamaica, a Tortoise, to which they assign the name of Hicatee. As they distinguish it from the Marsh Turtles, we may consider it to have been a true terrestrial species, one of the *Testudinidæ;* but whether actually indigenous or imported, is doubtful. Animals existing in a country in an independent feral state, have a right to a place in the local Fauna, even though the race has been originally introduced; but it was the custom of some of the earlier naturalists, as Browne for example, to enumerate and describe such animals as they saw in the country, though confessedly imported and preserved in confinement; on which principle the whole contents of Wombwell's menagerie ought to be described in a History of *British* Quadrupeds. The author just named mentions in his Natural History of Jamaica, "The Hicatee or Land Turtle," with the following remark: "This species is a native of the mainland, but frequently imported to Jamaica, where it is common." Long, in his enumeration of the animals of the island, also mentions it, without indicating whether he considers it indigenous or introduced, merely adding a note on its longevity. "Some of these have been known to live [in Jamaica ?] thirty years."

Mr. Hill has favoured me with a note or two on the subject of a Jamaican *Testudo*. "*April 2nd*, 1846. — I have learned that some time ago a *Land* Tortoise, — indubitably such, the limbs being short and stumpy, the carapace very convex, and covered with pyramidal angular plates, — was taken out of a pinguin hedge near this town. I must mention that Land Tortoises, brought from South America and elsewhere, frequently get away, and are in this manner found solitary. Some fifty of a prodigious size, brought by a Spaniard for sale, broke away near Kingston, and several, for months after, were found round about. One used to range the woodlands near the sea at Greenwich, where, in consequence of some injury it received after it had been a year out, it was found dead. I saw its remains under one of the trees there a little while ago."

"*May* 19*th*, 1846. — I have heard of two more instances of Land Tortoises being taken under such circumstances as would lead to the supposition that they were indigenous. In both instances they appeared to have quitted their usual haunts in search of water. They were taken in the midst of prevailing droughts; and one, which was captured after it had reached the stream where it had been drinking, bore the evidence of having travelled from far; the interstices of the scales on the legs being charged with a fine *red dust;* and red earth not being found any where in the plain in which it was taken. They were both of good size, being described as about thirteen or fourteen inches in length, measured over the carapace."

No clue to the specific identity of these specimens now remains, but the following note on a species certainly indigenous to the West Indies will be read with interest. I am indebted for it to my indefatigable συνεργός. "The other day, on my showing Mr. Millar my Marsh Turtle, he related to me some curious anecdotes of a Land Tortoise that he has now possessed for several years. His animal is a native of Antigua, where the species is indigenous. As the *Kinixys Homeana* is assigned to the neighbouring island of Guadaloupe, I should conjecture that his will prove to be that species also. A garden and grass lands, with thickets here and there covering a space of some three acres, are attached to the residence of Mr. Millar in Antigua; and over this inclosure his Tortoise is at liberty to pursue without restraint the impulses of his instinct. He shelters himself in the sultry hours of the day by sauntering among the shadows of the trees, or creeping within the covert of the house. He is extremely sensible of all the changes in the air that have a relation to rain. When a shower threatens, he shows more activity than ordinary; he then walks with stiffened limbs, — that is, he assumes then what may be characterised as his erect attitude, and moves nimbly, if we may so speak of the motions of a Tortoise. He disappears during the rainy seasons for a space of three months. It is not known where he conceals himself; but he reappears regularly, and, after these seasonal absences, resumes his customary walks, and comes and takes his food as formerly, and seeks those whom he affects, that he may be fed, and fondled, and petted as heretofore. He seems sensitive to the successive strokes of the hand upon the shell; for on every succeeding passage of the hand over his back, he extends more and more his head, until he stretches it at full, when he remains with it lengthened out, and eyes the person who caresses him with evident pleasure and satisfaction. The sensation in this instance is most probably electric, and must be referred to the delicacy of feeling which renders him so restless and stirring whenever rain threatens. His food is altogether vegetable, —pieces of potatoe (*batatas*), and yam, the peelings of fruit, and succulent leaves. He is as frequently the rough playmate of the children as the fondled pet of gentler hands; but he is alike reconciled to either notice, and never seems inclined for any other society, at least during his seasonal reappearance in the house, than the inmates of his accustomed home. He has been in the possession of Mr. Millar now twelve years."

If the species naturalized in Jamaica was introduced from South America, it is quite possible that it may be this same *Kinixys Homeana,* since this is common to the continent and the Caribbean chain. There are specimens in the British Museum, which were sent from Demerara.

MM. Duméril and Bibron refer the "Hicatee" of Browne to *Testudo tabulata,* which they attribute to tropical America and the Greater Antilles, citing specimens, in particular, from Cayenne and Guadaloupe. Spix and the Prince Maximilian found this species in Brazil, in the forests bordering the Amazon.

Testudo carbonaria, a species closely allied to *tabulata*, is assigned by the learned French herpetologists (I know not on what authority) to *Jamaica*, as well as to Brazil, Cayenne, and Chili.

In Dr. Robinson's MSS. I found the following brief notes of Tortoises (Hicatees) which he procured in Jamaica, but whether he considered the species indigenous or not, does not appear. — "One of these Land Turtles laid two eggs, an egg each day successively, on board the 'Friendship,' on my passage to England. The second-laid egg had an indenture, which makes me conjecture they are soft immediately after expulsion.* They are elliptical, and about one inch long."

"A Land Turtle had ten full-formed eggs, and ten small [in the ovary] when it died; and had laid two some days before, each of the eggs being indented."

THE AGUA TOAD.

About the latter part of the year 1846, Mr. Hill's attention was directed to some hitherto unrecognized tenants of the lowland ponds in the vicinity of Spanish Town. No native species of Frog or Toad had been discovered in Jamaica, with the exception of the Tree-frogs (*Hyladæ*), which have been already noticed, when a Batrachian was brought to

* I would beg the reader's reference to my own observations on the eggs of *Thecadactylus lævis*, in page 184. of this volume.

my friend possessing no slight claims to distinction. He thus describes it to me, under date of Jan. 5, 1847. "It is distinctively a Toad, but of prodigious size, being six and a half inches from the muzzle to the coccygeal extremity, and as broad as it is long. It was taken by some negro labourers while clearing land, in the neighbourhood of the Rio Cobre, at Passage Fort; and we are indebted to Mr. Robert Wilkie, our clerk of the peace, who had been seeking a sea-side change of air in that vicinity, for this interesting addition to the Jamaican Fauna. I can say little about any peculiarity it may have. Twice when taken out from the moist corner of the out-room in which I keep it, and brought into the house, it has discharged a large quantity of liquid from its vent, quite colourless; — and when I have had it by me in a reclined glass shade (one of the glass cylinders we use over candles), while making a drawing of it, I have smelt an odour, not very strong, but sufficiently offensive,— a compound of garlic and exploded gunpowder,—which I suspected came from it, but which I could not trace to it. The whole cuticle is tuberculous. The tubercles on the back, arms, and thighs are very large; but the surface of each is smooth, except those of the expanded eye-brows, which are rough and warty."

A few days afterwards another specimen came into Mr. Hill's possession, dead. — "Its hinder legs are longer, and its hinder feet are more deeply webbed than those of the former Toad; the connecting membrane in my living one being very little more than rudimentary. Though this last specimen is not so large as the first, it is still monstrously big, and in very many respects similar to the preceding. I now learn that they have been observed about several of the savanna ponds, and that they first attracted notice by the loud sort of modulated snoring noise which they made when the horses and cattle were led to the evening watering. None of the persons at present residing in the plains ever knew or ever heard of these reptiles; they were therefore unable to account for the noise when it first drew their attention. As they traced it to the pond, they were surprised to find it proceeding from the very water; and when they caught the creature that made it, and found what it was, they discovered that there were several others about the margin of the pond, which leaped in, and concealed themselves by diving as soon as they approached. We can only account for the present discovery of these full-grown reptiles in these places by the late inundations. Several gullies intersect the plains from the hills to the sea; and we may suppose that they had been swept down through these drains, and carried by the flooding waters into the savannas. If it be with these, as with frogs and toads in general, that the male alone is vocal, and that the voice is the call of courtship, we may expect, as long as rains keep our ponds unexhausted, that these reptiles will maintain their possession of the savannas. They bid fair to perpetuate their tenancy. 'Limosoque *novæ* saliunt in gurgite ranæ.'

"Just after I had set down the preceding remarks, a living specimen of the second Batrachian was brought me from a pond at the foot of the hills on the opposite banks of the river Cobre. The pond is of very considerable size, of full an acre and a half in extent, and not very distant from the river, but elevated as much as thirty or forty feet above it. The oldest inhabitant thereabout never knew of such an occurrence as frogs of this size or character in that district. The *Hylæ* or tree-frogs from time immemorial had mingled their loud croaks with the shrill tones of the *Gryllidæ* in the surrounding trees at nightfall; but hoarse bellowings from the ponds and pastures had never before saluted their ears. I know nothing of the specific peculiarities of the *Cystignathus ocellatus* of Wagler, or of the *Rana gigas* and *pachypus* of Spix, or of the *mugiens* of Catesby; but from the large size and sonorous voice of this unknown visitor, our people have called it the Bull-frog. Some of the American settlers in this vicinity have confirmed the propriety of this appellation, by recognizing a similarity between it and the great hoarse-voiced croaker of Carolina. I shall describe the appearance of our frog.

"I mentioned that the Passage Fort reptile had toad-characteristics, and that the skin was tuberculous. I would distinguish that of the present specimen as granular. Linear divisions, more permanently marked than wrinkles, pervade the whole cuticle as well below as above, and both upon the legs and toes. Each intersection swells into a distinct bump. Those on the upper surface are much larger than those on the lower, and are further distinguished by being minutely spotted with a little congeries of spicular dots, like the seed spots on a strawberry, only dark coloured instead of light. The tint of the upper parts of the body is a bright olive, yellowest in the hollow of the back, and greenest about the head. The belly is parchment white. The measure from muzzle to extremity of the vertebræ is five inches and a half. When this *Anoura* was brought me its throat was still inflated. It was puffed out like that of a pouter pigeon, and immediately on being set down on the floor it recommenced croaking. The sound was very greatly reduced from the reputed bull bellow it sends up from the marshes and meadows. It sounded to me not unlike the partridge call that I have heard country-boys imitate in Lincolnshire. I laid my hand upon it, when it filled its vocal sac, and croaked; and the convulsion, contraction, and jerk of the whole internal parts were as if every organ was pressed up to the back-bone. There were some then in the same meadow croaking to each other, and their voices were said to have been heard at half a league distant.

"I was told that some frogs I used to hear croaking at night in the boggy grounds about Port au Prince in Haïti, were not known there till a shower of rain brought them into the Island. The precise period of this occurrence was mentioned to me, and I have a note of it somewhere. If I could be satisfied that what the historian Moreau de St. Méry likened to 'tetards' (bull-heads) were tadpoles, I should say it was the same shower as the one he commemorates, of May, 1786. He mentions these tetard-like animals as exhibiting the same voracious savageness as that recorded by Professor Bill of the Tadpole. 'They killed and devoured each other.' He records no more than this prevailing ferocity. Had he carried on his observations to the surviving last one, he would have decided what their forms were, after he perceived they had tails terminating like slender oars, and great big eyes reflecting rays of silver green, and would have written grenouillettes for tetards. I have referred to this occurrence, from Moreau de St. Méry, because it possibly explains the source from whence we derived our strangers." *

" *Feb.* 6. 1847. —I have fathomed the mystery of the unknown toads and frogs, mentioned in my last two letters as recently and suddenly appearing among us. Mr. Anthony Davis, a proprietor in St. Andrew's, some two years ago (viz. in Nov. 1844) imported two dozen anurous Batrachians from Barbadoes, and established them in the ponds of Molina's estate. The great sandy gully that drains the uplands of Liguanea swept them, in the late rains, into the swamps between this place and Kingston, and from thence dispersed them through the savannas. These reptiles were imported for the purpose of

* " I quote the words of Moreau de St. Méry. — ' On éprouva depuis le mois de Novembre, 1785 jusqu'au 5 Mai, 1786, une sécheresse désolante.— Le même jour 5 Mai, 1786, il tomba, durant une forte brise d'Est, dans plusieurs endroits de la ville (Port au Prince) et de la plaine (cul de sac) une grande quantité d'œufs noirs, qu'on vit éclore le lendemain. M. Mozard, qui guarda une cinquantaine de ces petits animaux dans un flaçon à demi rempli d'eau, les vit changer de peau plusieurs fois. Ils ressemblaient à des têtards, et devinrent trois fois plus gros que ne le sont des cousins ou maringouins dans l'état de vers. Leur queue était terminée comme un aviron effilé vers l'extremité, et leurs yeux étaient grands, vifs, et réfléchissant plusieurs rayons d'un vert argentin. Ils étaient formés d'une manière visqueuse, et le 2 Juin ils s'entredévouèrent.'—*Descrip. de Saint Dominique*, ii. 413.

destroying rats in the lowlands. They had been considered in Barbadoes very important auxiliaries to the planter in this work, by their pursuit and destruction of the young rodents. I doubt the power of any cold-blooded reptile to do much in this way. Barbadoes, however, only availed herself of the experience of Martinique, from whence she derived her original stock of toads and frogs, Martinique herself having imported them from Cayenne. When you receive the specimens put up in spirits for you, you will be able to ascertain whether they are Batrachians of the southern continent. Some amusing stories are told of the terror they excited when their first outburst of bellowing was heard. Families sat up all night, believing their houses beset by persons who played upon their fears by these horrible noises only to pillage their premises successfully. They are now ineradicably established among us, and are to be added to the miscalculating delusions which gave us "*big rats*" to devour "*little rats*," and the *Formica omnivora*, the native ant of Cuba, to rid us of the accumulated pest of rats and vermin, and to become a more intolerable scourge than all the other plagues put together. The *Formica omnivora* is set down in our chronology of memorable things as an importation, by Thomas Raffles, of the year 1762. The big rats tradition gives to Sir Charles Price, as your inquiries informed you when here."

"*Feb. 27th,* 1847.—What with their noise and what with their depredations on the ducklings in the ponds, the imported frogs are found already to be a great nuisance. I suggest to our savanna farmers, who have convenient duck-ponds, like my young friend Mr. Thompson, of Tredegar Park, who complained to me of their depredations, that Marsh Tortoises should be kept where these plaguy Anouras resort. As they are carnivorous, and, in countries where frogs are common, prey specially upon frogs, as well as small fishes, and as their address in swimming and their prompt movement enable them to pursue their prey with avidity, the Emys or Marsh Tortoise seems the best counterpoise to their increase. Even if these should prove equally annoying to the ducklings, their voiceless existence renders them a more endurable nuisance. The frogs attack the feet of the ducklings, and lacerate them, and bite off their toes while swimming. Our farmers will be obliged to reduce to captivity some of our large Herons, to preserve their natatorial poultry from this pest."

The two specimens alluded to in these notes Mr. Hill kindly transmitted to me. They proved to be that enormous South American Toad described and figured under so many titles, the *Rana marina* of Seba and Linn., the *Bufo agua* of Latreille, Daudin, and Duméril. It varies exceedingly in form, proportions, and colours: Spix, in his great work on Brazil, has described and figured no fewer than eight forms, to each of which he assigns specific names, but all of which MM. Dum. and Bibr. consider as belonging to the present species. In its greatest development it is the most gigantic of all the Anourous *Batrachia*. The whole coasts of intertropical America produce it, and, according to the zoologists just cited, confirmed by Mr. Hill's researches, the Antilles; they mention Martinique in particular as the native country of some of their specimens. Jamaica will, for the future, be added to the geographical range; for, once established, they will, doubtless, fulfil my friend's forebodings, and maintain their ground.

DEER.

In the upland forests above the Caymanas, not far from Spanish-Town, there are occasionally seen small herds of Deer in a feral state, the descendents of some that had been imported and escaped. Tradition imputes their introduction partly to Mr. Dawkins, who possessed the Caymanas property about fifty years ago, and who at the same time naturalised the American Quail (*Ortyx Virginiana*) in the colony, and partly to Sir Charles Price, the owner of Worthy Park in St. Mary's, some thirty years earlier, who is said to have had several deer running at the Farm, now the property of Lord Carrington, within the Caymanas plain. These Deer are reported to have been obtained from the Spanish Main, and are considered to belong to the species known as the *Cervus Mexicanus*, which Humboldt describes as very abundant in the small uninhabited islet of Cubagua; a small Deer, of a brownish red hue, spotted with white, and of the latter colour beneath.

Mr. Hill, to whom I am indebted for all I know on the subject, tells me that about the year 1841, a buck was obtained in the forests referred to, which was slaughtered and sold in the Spanish-Town market. In one of his letters, dated 19th May, 1846, I find the following note. "Mr. Russell, barrister at Law, informs me that Mr. Townshend, a bordering proprietor on the uplands in which the Guazu-pita were turned out to establish themselves, has the antlers of the Deer killed some five years ago. Mr. Townshend himself ran down this buck with his dogs, having started it in a morning's stroll through his woodlands. The last batch of Deer were introduced thirty years ago by a Colonel Harrison at the Farm. Mr. Russell recollects as many as eight quarters of venison in the market at one time. There seems every reason to conclude that a colony still inhabits the impervious mountain-forests above the Caymanas Plain."

A few days after the above was written, Mr. Hill favoured me with the following note on the subject. "*5th June*, 1846. When your letter came to hand, announcing your arrangements for an immediate departure for Europe, knowing how important it was that you should early complete your notes of the Mammalian *Vertebrata*, I walked as far as Mr. Townshend's pen, and made the accompanying sketches of the Horn of the Guazu-pita Deer. I learnt from him a number of interesting particulars respecting the existence of these animals in the adjoining woodlands. I find that their number, if not considerable, are not a few;—that they are frequently seen in herds of several together, and that fifteen were surprised in the forest not more than two years ago, at which time a negro labourer of Waterloo pen, close by, a property of the Townshend family, brought down a young Fawn. The animals exactly agree with Humboldt's description of the *Cervus Mexicanus* brownish-red spotted with white, and were no doubt imported at the same time with a cargo of mules, sheep, and oxen from Cumana. Mr. Townshend says that the herds now in the forest are an importation of Sir Charles Price's eighty years ago; — Sir Charles at that time being the proprietor of the Farm pen close by. Mr. Townshend's father used to relate the occurrence of their getting from the Farm and taking to the Forest, as an incident that happened when he was a young man. He was familiarly acquainted with Sir Charles, who was a great lover of Natural History, and possessed many curious Mammalia and birds. An account of his pen in St. Mary's, called the Decoy, will be found in Long's History of Jamaica. It was there he kept his curious Ducks. Colonel Harrison, mentioned by me in one of my preceding letters as agent for the Farm, was an uncle of Mr. Townshend's, — the name of one of his brothers, the Father of the House of Assembly, an octogenarian, being George Harrison Townshend.

"The places in which our negro woodmen occasionally meet these Forest-deer is in some ravine track, or wooded pathway by which the herds descend to the springs to drink. They feed at night, resting still within the woodlands in the day-light hours. They have sometimes been surprised in moonlight nights crossing the highway to the Rio Cobre, by an offshoot of their mountain-hold; — a line of broken hills which gradually lower towards the stream, called the *White Marl*. It is only by accident that a herd can be detected: their acute instincts;—their power of smelling;—their quick eye and sensitive ear rendering it absolutely impossible, by any expedient, to take them by surprise, or, in the steep and cliff-bound hills in which they are located, to approach them by any precautions whatever. Under ordinary circumstances they are inaccessible to dogs. Though they have been tracked with the bloodhound, they have gained steep after steep with such facility that no untiring perseverance could circumvent them. Some short time ago, a negro searching after strayed cattle in the forest, in one of our recent droughts, crossed a herd marching in file along a path that they had been accustomed to take without molestation for a long period. Making sure of finding them in a locality so confidingly made their own, and so much frequented by them;—with a believing simplicity, very much in contrast with the doubting caution and careful strategy of deer-stalking, he gathered a host of adventurers with dogs to waylay them in their coming and going. The party lay out all night, expecting the herd to return from their accustomed feeding ground by the well-trodden trackway, but they had scented the footsteps of the human intruders and their hounds, and not only avoided that path for that morning, but quitting it for ever afterwards, shifted their haunts to other mountain fastnesses. The last locality in which they were encountered was in the upland forests of a friend of mine overlooking the Caymanas plain,— a place called Mount Gotham. As I am acquainted with this and the neighbouring holds, I should say that the wild-deer are destined never to be dislodged now they are numerously established in fastnesses so peculiarly difficult, and so congenial to their habits and instincts."

The antler in question at length came, through the kind offices of my friend, into my possession; but the result of a careful examination of published descriptions, and a rigid collation of the specimen with the extensive collection of Deer's horns in the British Museum, proved indubitably that it had belonged to no native American form of *Cervidæ*, but to our own familiar Fallow-deer (*Dama vulgaris*). Of course I immediately communicated to my esteemed coadjutor this unexpected conclusion, and I annex his note in reply.

"*Dec 24*, 1850.—I am quite satisfied you have got the right clue to Sir Charles Price's escaped Cerfs in the mountains of St. Catherine's. The information without doubt is quite correct that he got specimens of the Mexican Deer;—but, as he introduced at once *a park*, that is, about 8 or 10 pair, it is evident he drew his colony from Europe. The Mexican Deer he could scarcely obtain more numerously than buck and doe,—for the Indians have never so far advanced in the habits of a pastoral people as to herd their native animals. The Guazuti are very frequently brought to this island. A pair belonging to the 2nd W. Ind. Regiment have been for the last two years inhabitants of the barrack-yard in this town. They are frolicksome and mischievous;—their tint a dun-ash;—their ears large; their hair coarse and wavy; their body long and full-barrelled, so that the shape does not give one an impression so much of speed as of strength and endurance. When I examined these Guazuti, my mind quite misgave me about the history of the feral Deer of our near-by mountains. The spotless coat did not influence me in this uncertainty, for Humboldt had described the *C. Mexicanus*, without restriction to fawn-hood, as white-spotted. . . . I can account for Sir C. Price's preference of the Fallow-deer of Europe, supposing you are right in your conjecture. Their flesh is known to be good, whereas that of the American stock is coarse and not usually well-flavoured. Those who have tasted *our Deer* say they are delicious venison. If you can ascertain what description of Cerf Sir Bethel Codrington has in the island of Barbuda, you will probably learn that Sir Charles Price and he drew their colonies from the same herds. . . ."

In these notes my friend seems to speak of the Guazu-pita, the Guazuti, and the *Cervus Mexicanus*, as if these names were synonymous; they indicate, however, three species. The Guazu-pita is the *Subulo rufus* of Col. H. Smith, which inhabits the deep forests of South America from Brazil to Honduras. The Guazuti is the *Mazama campestris*, extending through the whole of the Southern Continent as far as Patagonia, and chiefly affecting open plains. Of the third, the *Mazama Mexicana* of Col. Smith, and, according to Pennant, the Teutla Maçame of Hernandez, very little is known, and we believe no specimens exist in European Museums. The term *Guazu*, in the native dialects of Brazil, is applied to any of the slender limbed *Ruminantia*, with an adjectival affix for discrimination of species.

That the Fallow-deer of Europe and some of the native species from America have been introduced into Jamaica, there is indubitable evidence; and it is highly probable that both have naturalised themselves, and are now existing in the forests in a wild state. We must probably look for the first importation of both races, to an earlier period than the time of Sir Charles Price. Long, in his History of Jamaica, published in 1774, includes in his enumeration of the animals, both the Red-deer and Fallow-deer: on the latter he makes no remark, but on the former he makes the following note. "These were *originally* imported from the continent, and *are now* grown scarce. They rarely grow fat here *in their wild state*, but their flesh has a good flavour." The terms of this note certainly imply that the race then in a wild state, had been introduced at a period of considerable remoteness from his own experience; and he was familiarly acquainted with the colony. Browne, writing some twenty years earlier, speaks in nearly the same language. He says, "The Red-deer are frequently carried to Jamaica from North America, and kept by many gentlemen in convenient inclosures, but they do not thrive well in the Island." The Fallow-deer he does not mention.

What particular American species has been introduced, or whether there are more than one, we cannot say. From Browne's note just quoted we should conjecture that he meant the common Deer of the United States (*Mazama Virginiana*), which is of a reddish hue in summer, and could be obtained with ease in any numbers. Col. H. Smith, the highest authority, speaks of having seen a female specimen of the Cariacou (*Mazama nemoralis*) at Spanish-Town, in Jamaica. He was in the habit of observing it daily for a long time in a domestic state; it would steal bread from the table of Sir George Nugent, the Governor. This individual was brought from Honduras; but the species is widely spread over the wooded regions of tropical America, and extends even into the United States.

We may venture to hope that this discussion of the subject, though imperfect, will awaken curiosity in the Island, and that investigations and examinations may be instituted by those who have the power, which will lead to the identification of the existing species of Deer in Jamaica.

WILD GOATS.

"I must remind you," observes Mr. Hill, "that on the Healthshire Hills there are Wild Goats, and that our Goat is a very beautiful variety of this animal. I have always understood that we obtained it, with our Hogs, from Africa, and that it was an early established stock from the Canary Isles. When, in my first ramble through the Salt Island marshes, I reached the crags where the sea separates the Healthshire Hills from the large island that shelters Old Harbour, called Goat Island, numerous Goats of the wild breed were seen sporting on the cliffs. It was very interesting to observe them ranging side by side, like a mustering regiment of horse at drill, on the edge of the uppermost rocks, and then, on a signal, — a significant bleat from one of the outstanding herd, — scampering off with a leap, along the ridge, inland."

THE RABBIT.

"Are you acquainted," my friend adds, "with the fact that Rabbits have long existed in the neighbourhood of Spanish-Town in a wild state? The furthest date back at which I have traced distinct notices of them is in Lady Nugent's Journal. As it is a work printed for private circulation only, and not likely to be accessible to you, I point out the date at which she mentions them. In an entry under the 29th May, 1803 (vol. i., p. 410), she relates the incidents of a fire which broke out in the pinguin fences of the Government pen, a country residence of the Governors, about three miles out in the Salt-pond plains. The fire occurring in the droughts that preceded the May rains of that year, she sets it down as an effect of the sun's rays. 'The grass and every thing were so dry that they seemed to burn like touchwood. The poor Rabbits ran out of the fence by dozens, and many of them were half-roasted.' We have the Rabbit mentioned in earlier writings, but so indistinctly that I could not venture to say whether the notice under that name does not rather relate to the Indian Cony, than the European Rabbit. It still inhabits the Government pen district, but, with all its known fecundity, is not common, yet not unfrequently met with. The pinguins are to these denizens of our fields what the furze covers are to the wild animal in England; but it is remarkable that while the prevailing colour of those of the European warren is what Buffon calls the grey, and which he defines to be a mixture of tawny, black, and ash-colour, these of our savannas are silver-sable, — the variety which he describes under the name of 'Le Riche,' — the slate-coloured rabbit, deeply tinted, with sprinklings of white on the neck, on the shoulders, and on the back, softening off to blue-white under the breast and belly.

"Our savannas in the rainy seasons are generally an inch or two under water, but the pinguins, being banked, form, through the most drenching weather, a secure asylum from inundation. Here they burrow and breed, and find more than usual supplies of food in the fields of Guinea-corn, the *Holcus sorghum*, which is a staple article of cultivation in salt-ponds, and in the customary roots and vegetables, particularly the *Convolvulus batatas*, or Sweet potato, which are grown in the negro grounds.

"'Our Rabbit,' says the Père du Tertre in his General History of the Antilles, 'was early introduced into the American Islands, and succeeded well.' 'Y ont très-bien réussi,' are words that imply they were established successfully, but whether in a state of domestication, or as inhabitants of the wilderness, I am not able to determine. I have not heard that they are common in any of the Antilles, great or small, but as Leo Africanus and Bosman represent them to be natural to Africa and Asia, and delighting in excessive heat, one wonders why so little progress has been made in stocking our fields with an animal supplying a delicate dish wherever it has been an object of home economy. The Guinea-fowl, though common as a wild bird in the southern plains of Jamaica, is only known on the north side in a state of domestication. I cannot find what is the prevailing colour of the Rabbit of the hot regions of Asia and Africa. If it be silver-sable, it yields to the influence which gives black as the predominant colour of our hogs, and brindle as that of our cows; while dark mingled tints more or less characterise the dappled and speckled covering of our goats and sheep.* In the moonlight nights our Rabbits are seen strolling abroad, and resorting to the open savannas; in the day they confine themselves to the shelter of the pinguin hedges, or to the thickets about their haunts."

RATS.

Rats and Mice are numerous in all parts of the island that I am acquainted with. The former present much diversity in size, colour, and proportions, and appear to constitute several species. They are bold and voracious; inhabit houses, hollow trees, holes beneath roots, and fences; and are common from the shore to the summits of the mountain ridges. I think I have identified the common Brown and Black Rats of Europe, and another species, among those which are abundant.†

* The wild Rabbit of Africa is of a russet hue like the European breed. The species is confined to the Mediterranean shores of that continent.

† My esteemed friend, G. R. Waterhouse, Esq., the highest authority on all that concerns the *Rodentia*, has kindly favoured me

But besides these, there is a species much talked of as infesting sugar estates, and doing much damage by devouring the growing cane; it is hence called the Cane-piece Rat. A single specimen has fallen under my observation, a male, which was brought me in February, from a plot of cane near Bluefields. It was manifestly a distinct species, as the description below will show*, easily distinguished by its

with his opinion on some twenty or more skins of the Jamaican *Muridæ*, brought home by me. The distribution of this ubiquitous group, whose members so closely resemble each other, and are yet so subject to individual variation, is peculiarly difficult; but he thinks that the specimens may be distributed into four species, all of Old World origin: *Mus rattus, decumanus, Indicus* (?) and *musculus*. The third of these is, perhaps, the most numerous; it is very closely allied to *M. rattus* in form and proportions, with the brown, brindled hue of *Indicus;* the belly is frequently, not invariably, of a delicate pale yellow tint. Its long tail and slender feet distinguish it from *decumanus*. The Mouse is of a warmer hue than common specimens of our *musculus*, being of an umber-brown above, and delicate pale bay beneath; but Mr. Waterhouse sees no reason to distinguish it specifically. It is a house-mouse in Jamaica, as in Europe.

I may mention a curious fact connected with the species supposed to be *Indicus*. A specimen, taken in a guinea-grass piece, was a male, but had been emasculated; the wound was not quite healed, when it came into my possession. The Squirrels have been affirmed to perform this operation upon each other in their combats, but the statement has been explained away. There was, however, no mistake in the present case.

All these four species are now distributed over the whole known world.

* *Mus saccharivorus*, MIHI. The Cane-piece Rat. Length from muzzle to base of tail, $9\frac{1}{4}$ inches; tail, 8 in.; from nose to inner angle of eye, 1 in.; from nose to orifice of ear, 2 in.: length of ear, $\frac{8}{10}$ in.; fore foot, $\frac{8}{10}$; hind foot, $1\frac{7}{10}$ in.; angle of muzzle, about 75°. Upper parts brindled. umber-brown; face, throat, breast, and belly, ash-colour, the tints separated sub-abruptly. Feet white; tail thick, little hairy, greyish above, whitish beneath. Incisors ferruginous.

size, by its rufous colour, and by the smallness of its ears, which gives it a peculiar physiognomy. Robinson was acquainted with the species, and has given an elaborate description of it in his MS. volumes (iv. 13.), as well as a coloured drawing. His specimen was considerably larger than mine, for he describes its length as 20 inches, of which the tail measured half.

I had heard also of a Rat said to have been introduced into the colony about a century ago, by Sir Charles Price, with the object of keeping down the smaller vermin; it is commonly spoken of as the Charley Price Rat; but whether it is distinct from, or identical with, the Cane-piece Rat, I could get no very certain information. Mr. Hill, as usual, endeavoured to aid my researches.

In a communication dated May 19th, 1846, he had remarked as follows: — "My inquiries about Rats have terminated in making out three species as distinctively known in the colony; — the Black Rat, the Brown Rat, and the Cane-piece Rat.

"Tradition relates that the Black Rat is the oldest of the family; that it was introduced by the Spaniards, and that it is the Rat of the South of Europe. It is now comparatively scarce, but is said to be sufficiently common in St. Mary's. The Brown

The skin of this specimen was unfortunately lost as soon as it was taken from the carcase, and I never obtained another; so that I have no opportunity, by examination of the skull, of determining whether the species belongs to that group peculiar to the New World, which Mr. Waterhouse has named *Hesperomys*. The probability is that it does.

Rat has a good deal of fulvous intermingled with grey in its coat, and though it is at this time by far the most numerous of these vermin, it is of comparatively recent importation. The negroes are said to distinguish it by the name of the *George* Rat. This name is a curious coincidence with the historic scandal that, under the name of the *Hanoverian* Rat, assigns a similar introduction to the ship that brought the Brunswick family to the British shores. It is larger than the Black Rat, and has a fur harsh and short. This is the common House Rat; out of doors it is the pest of the corn-field and the cane-piece; it inhabits the pinguin fences. A gentleman informs me that on a plantation of which he was overseer, the annual number taken, from accounts minutely kept in paying premiums for their destruction, was 12,000 of these Rats, through a succession of years.

"Of the Cane-piece Rat I can learn nothing, except that it is supposed to be the animal commonly spoken of as the Charles Price Rat. It is unusually large."

Further inquiries gave reason to believe that no connexion existed between Sir Charles Price's importation and the Rat of the cane-fields, or any Rat at all, beyond vulgar rumour. My friend remarks in a letter of April 22nd, 1847; — "You remind me that no satisfactory history has been given of the animal Sir C. Price introduced into the colony, out of which grew the story of the Charley Price Rat. I had a conversation some short time ago with his great-grandson Mr. George Price, of Worthy Park,

Luidas Vale, in St. John's, and he informed me that, according to the family tradition, the animal introduced by his ancestor was a substitute for the Ferret. It had been found that the European Ferrets were rendered useless by their inability to overcome the Chigoe infestment of the colony. Sir Charles Price bethought him that if he could find an animal in the country of the Chigoe, corresponding to the Weasel of Europe, he would accomplish the naturalisation of a Rat-destroyer with instincts capable of counteracting the plague of the parasitical insect. He accordingly procured something from South America, that in the eyes of the negroes had strong rat characteristics, but which was no Rat. It was of large size. Several were set at large about the house at the Decoy, in St. Mary's, and at Worthy Park, to establish themselves how they might. It would seem that nothing came of the scheme, for no animal allied to the Musteline group of quadrupeds has been found naturalised in the colony. The Poto, or Kinkajou, which Mr. Colinson communicated to the Count de Buffon in a letter of the 12th December, 1766, and described as having been taken in the mountains of Jamaica, and of which he gave Buffon a drawing, engraved in the 1st Edition of his Natural History (the 3rd vol. of his Supplement), is the nearest approach to an animal of this character found here in a state of nature.

"There are some three or four Viverrine animals of South America, which Sir Charles Price might have made experiments upon as substitutes for Ferrets, and which the negroes might have considered gigantic Rats. The animals which Buffon delineates, and calls, La petite Fouine de la Guyane*; le Grison de Surinam†; la Grande Marte de la Guyane‡; and le Touan§, might have been so mistaken: especially the first and last, which have tails for half their length naked and scaly, like those of rats. I would venture to add to these, also, that strange equivocal animal, known in Cayenne as the Crab-dog||, the tail of which is altogether naked and scaly. This animal M. de la Borde mentions as being in his time domesticated in the houses of the colonists, with the dog and the cat, and living on the same food with them, and doubtless performing common service with them in ridding the dwelling of vermin. This is the best history I can give of Sir Charles Price's connexion with the large species of Rat, which is said to be common here, and to bear his name. If we have any of these Rodents more than usually remarkable for size and ferocity, we [probably] owe them to the vessels of war, that, coming hither from distant cruises when the West Indies were the scenes of so many maritime conflicts, unloaded their stores at the naval magazines near Kingston, and round on the north side at Port Antonio, and so spread through the island the Rats of Asia and Africa, as well as those of Europe and America."

Browne and Long both speak of a Rat in the island bearing Sir Charles Price's name, but do not

* *Mustela Guayanensis*, LAC.; perhaps a young Coati.

† *Viverra vittata*, LINN.

‡ *Mustela barbata*, LINN.

§ *Didelphys brachyura*, PALL.

|| *Didelphys cancrivora*, LINN.

ascribe its importation to that gentleman. The former (1756) under the head of "Castor 1; cauda lineari tereti, the Water-rat, commonly called Price's Rat," says, "These creatures, though the natives *of some foreign land*, are now grown very common in Jamaica." He gives us no description, but distinguishes it from the "House or Cane Rat." Long observes, "Four different species of Rats infest this island. The largest is commonly called the Charles Price Rat, and obtained its name from having been first observed here about the time when the late Sir Charles Price, Bart. returned hither from Europe. It is said to have been imported by a Danish ship belonging to Sancta Croix, which was driven into Kingston Harbour by stress of weather." He considers it the same as the Water-rat of Europe, as it is manifest Browne had done before him, by his adoption of Linnæus' short technical description of that species. If the smallness of the ears and the red hue of the fur of *Arvicola amphibius* guided them to this identification, I should be inclined to conclude that my *Mus saccharivorus* just described was the species indicated by them. Long mentions the total length of one as eighteen inches, which would well enough agree with mine.

It is worth remarking, that there is a Rat of great size found in the Carribbean Islands, the *Mus pilorides* of Desmarest, which has the upper parts black, and the inferior parts white. It has not the dentition of the Old World Rats, but resembles, in the structure of its molars, the South American Rats grouped by Mr. Waterhouse under the generic name of *Hesperomys*. It is described as attaining the length of fifteen inches, exclusive of the tail, which is yet longer than the body.

In the above quotation from Long, he speaks of three other Rats besides the Price Rat. One of these he distinguishes as the Black House-Rat (doubtless *Mus rattus*); the other two are Field-Rats, which he thinks to be indigenous. "The larger is of a light ash, or greyish colour on the back and other parts, except the belly, which is entirely white. This subsists almost wholly on the sugar-cane, and is therefore generally called the Cane Rat." The other is described as much smaller, no larger than the English Mole. It is of a beautiful reddish colour, with a milk-white belly. Like the former, it takes up its habitation chiefly about the hollow roots of large trees, and the rocky acclivities of gullies and river banks. It is far from being numerous.

SUGAR-CANE INSECTS.

The ravages of insects on the various vegetable productions that are cultivated by agriculturists cannot but be a subject of deep and even anxious interest; an interest which heightens with the importance of the plant as an object of commercial traffic, or its value as furnishing food to man and beast. The Sugar-cane will probably always be the staple production of the West Indian isles, and whatever affects it may be said to touch the very life of those valuable colonies. Some notice of its insect enemies may therefore reasonably be looked for in this work; and though the part of the island in which I sojourned is not a sugar-growing district, the zeal and experience of my friend Mr. Hill will supply the deficiency of my knowledge.

The insects that have been noticed as injurious to the cane are principally the following:—the Borer-Weevil, the Borer-Moth, the Cane Aphis, the Cane Fly, and the Termite.

Of the first of these, the Borer-Weevil (*Calandra sacchari*), the best account is contained in a prize communication made to the Royal Agricultural Society of Jamaica, by Mr. Samuel Kell King, of Portland, and published in the Society's "Reporter" for March 6th, 1845. It is of considerable length, and being the result of great practical experience, is doubtless valuable. I shall make a few extracts from it, and condense the rest of the information. Mr. King suggests that the insect is not indigenous, but that it was imported into Jamaica from Tahiti; for it suddenly appeared in 1797, the year after the Tahitian varieties of the Cane were introduced into the island.

The insect Mr. King thus describes in its transformations. "An egg, the size of a small bead, in a considerable degree transparent, is deposited within the succulent vessels of the cane, where the adhering footstalk of the leaf retains the decayed foliage hanging to the germinating joint. . . . . . The egg deposited is hatched at the time when the growing bud, usually called the eye, exhibits the active influences of both heat and moisture. As soon as the maggot is formed, it commences its voracious injuries by worming its way from the verge of the footstalk where it had been hatched, into the very body of the succulent and vegetating shoot, where it grows with its growth, and strengthens with its strength. It then occupies the centre of the plant, making its way upward through the growing cane, but remaining within the sweet and perfected joints, and never ascending to the greener tops to devour the germ and destroy vegetation. It entirely exhausts the saccharine fluid in those joints in which it has lodged,—filling the excavation it makes with an excrementitious deposit, extremely injurious to the cane liquor from the mill; deteriorating it rapidly if it remain untempered while running into the pans. When the canes are cut, the grub-worm has already arrived at its second transformation. It has enveloped itself, within the gallery it has bored, in a shroud of decayed trash wrought with curious neatness; the shreds being plaited and wound together, and so closely fastened at the ends, that the air is excluded; and if exposed to the weather, no weather could injure it. I have watched the grub in the act of making this cerement. It first wraps itself all over with such of the rotting fibres of the cane as are near it. It tears the strips asunder with its forceps, and matting the pieces one within the other, it completely conceals itself within that kind of case usually called a cocoon, where it remains dormant for a little interval of time. It has now assumed its third or beetle state, and emerges from the excavated cane a weevil, bearing a rostrum or snout charged with fracticorn feelers, and wearing a splendid livery, striped yellow and brown, an insect about the size of the nail of one's finger. If the cocoon be opened before this last transformation, the pupa found within is of a dingy brown colour, and its bulky body is well supplied with the usual milky fluid, stored for that final change in which it comes forth from its temporary sleep, to become the parent of a succession of enemies to the planter."

The ravages of the weevil-grub do not, according to Mr. King, retard the growth of the cane, or in any way affect the healthiness of its outward appearance. The juice, however, is greatly deteriorated. "Though there is some quantity of really sweet juice still remaining between the rind of the cane infected by the borer, and the galleried centre through which it had worked its upward way, I have found it impossible to granulate this juice. It remains molassès, and to temper of any kind, whether lime or any more powerful alkali, can produce sugar from it. It might be possible to desiccate it by continued boiling, but the result would be charcoal, not sugar."

The writer recommends preventive and remedial measures. Among the former he deprecates the use of animal manure not sufficiently decomposed. A great number of the insects are carried in with the cane tops and leaves with which the cattle are littered. If this manure, unrotted, be taken to the cane field, the young sprouts become infected with the pest as soon as they begin to vegetate.

As soon as the sprouts begin to joint, the presence of the insect becomes evident to a careful eye by the sickly appearance of the whorl of terminal leaves. In such a state, "the sooner the whole of the field be trashed, and all such sprouts drawn out, the better."

When the canes have advanced in growth, and the joints are forming, repeated trashing, or removal of the lower leaves, must be resorted to: once every two months, or at intervals of twelve weeks at most.

"The sheathing footstalk is not only a shelter for this weevil, but it hinders the outer covering of the cane from hardening, and fixing that deposition of white powdery glass which resists the puncture of its proboscis. If with all this care, the planter finds himself overwhelmed by the numbers of his assailants, or by the success with which they have established themselves in his fields, nothing remains, but destruction by fire."

"But great as is the damage which this insect does in the field; it is a still greater calamity to have it in the mill-house. To mingle the juice of the injured cane with the uninjured, is to ruin a crop. The expedient of tempering the liquor, while running into the pans, may arrest the increase of the evil, but it does not get rid of it. If, however, the most pains-taking watchfulness has not secured the manager from an occasional bundle of infected canes getting into the mill, and if the pernicious consequences have begun to tell upon the proceeds of the boiling-house, I would recommend correcting the cold liquor with an alkaline mixture of potash and alum, in equal quantities. These, with the addition of as much boiling water as will hold them in solution, may be added to the temper lime commonly used in plain tempering. Such a measure of this mixture should be added to the liquor in the pan as would be deemed sufficient to correct the prevailing acidity. When the liquor has been boiled into what is called first syrup, it should be racked through the cock. By this timely remedy I have secured well grained sugar which has not deliquesced during the voyage to Europe, even when I had had the misfortune to have ground tainted canes."

The Rev. Lansdown Guilding, of St. Vincent's, who was honoured by the Society of Arts of London, with the Gold Ceres Medal, for his memoir on Insects which infest the sugar-cane, describes the *Calandra palmarum*, a gigantic weevil, known as the parent of the palm worms (the Gru-gru of the negroes), as a pest in sugar plantations, as well as its congener, the *Calandra sacchari*, the subject of Mr. King's description. The Palm Weevil, he says, is principally injurious to the cane-plants lately stuck in the ground, to which the female is allured by the juices which are exuded. These they sometimes attack so vigorously, that a fresh planting becomes necessary. They do not, however, seem to deposit their eggs in full grown canes, when palms are abundant in the neighbourhood.

The same zoologist also describes the Borer-Moth (*Diatræa sacchari*). "By far the most destructive and common enemy," he says, "is the smaller grub of the Borer-moth," belonging to the *pyralidæ* of Leach. "The sugar-cane is never exempt from this dreaded pest." "In the seasonable island of

St. Vincent, from improved cultivation, the animal is not very formidable; but in some other of our colonies, which, from the absence of mountains, or other causes, are subject to dry seasons, it has been known to blast the hopes of the year, to destroy whole acres of canes, and to ruin the unfortunate planter." The *Diatræa sacchari* is a small sized straw-coloured moth, with upper wings of a tint best described as an ochry-drab, varied with darkened lines and margined dots. The under wings are pale yellow. The caterpillar is of a yellowish colour, spotted with faint black dots, and in a slight degree hairy. This Borer-moth is indigenous to Jamaica, or at least is now ordinarily known to infest the Sugar-cane.

"The late Mr. Stephen Hannaford of St. Dorothy," observes Mr. Hill, "in a communication to me respecting this pest, so well known on this side of the island, wrote me that 'the system of trashing and of keeping the canes clean is the best and surest method, as well to prevent the depredations of the Borer, as to improve the juices. In seasonable districts, where this system can be fully carried out, the mischievous effects of the Borer seem to betray a want of proper attention on the part of the manager to his field. But in districts subjected to long spells of drought, the utmost caution is necessary. It is generally observed that the Borer commits the greatest injury to the cane after a rapid growth, which is followed by a spell of dry weather, whilst vegetation seems not only suspended, but the plant itself struggling for life. In this dilemma the planter chooses he least of two evils. To trash his canes under such circumstances would prove almost, if not entirely, destructive to his fields; he is, therefore, compelled to suffer the Borer to proceed unmolested, until rain has fallen, and the plant has again started into life. Then the removal of all the loose trash from the cane will check the progress of this insect, and by following up this operation as often as the canes require to be freed from superfluous trash the Borer ceases to effect further perceptible injury.'

"Mr. Guilding speaks of an undetermined *Aphis* as one of the insects that infest the sugar cane. Some sugar canes which a friend had growing in his garden in Kingston, were observed early in November 1844 to be infested with *Aphides*. They made their appearance in considerable swarms, and at once invaded both the leaves of the cane and those of the plantain (*Musa*). Beyond encumbering the foliage with substances that more or less obstructed their healthy growth, they did not seem to do any injury by lacerating the leaf, nor did they appear to waste the plant by diminishing its absorbent and excretory powers in any perceptible degree; they, however, continued increasing and multiplying without sexual intercourse, and were observed to be ceaselessly visited by the common small ants, which played around them, and touched them caressingly. These provident creatures were taking honey-dew from the abdominal ducts, those remarkable distinguishing appendages of *Aphides*. Nothing beside this good fellowship of the ants was observed in the colonies of our *Aphis*, till within some few weeks before they were brought under my notice, that is, in February 1845, when a very remarkable white insect with a bristled bordering of fleshy tubercles on the abdominal rings, a mark of the larva of the Aphis-lion (*Chrysops*), made its appearance among them. The Aphis-lions devoured the *Aphides*, and covered themselves with the fragments of what they slaughtered. Whether the white efflorescence, which makes its appearance on the vegetation infested by the Aphis, and the little white spots like patches of hoar frost, which mark the presence of the Aphis-lion, be what in other islands they call the *white blast*, which was said to have been distressingly prevalent in 1844, I am not able to say, but it is extremely probable that this plague had been general throughout the West Indies that season. On examining under the microscope the leaves infested with the *Aphides*, little pearl-like cases were seen, each containing an insect in its course of development. These cases were numerously spread about the foliage. Some of the cells were seen already perforated and empty. The *Aphides* were to be observed about in their active pupa state, some with their undeveloped wings glued to their sides, and with their bodies of a bulk somewhat exceeding that of the perfect insect; the perfect fly had wide reticulated wings marked with an obscure patch in one section of the outer part of the main nervure, while both possessed the two short conical abdominal tubes from which the ants that visited them gathered supplies of honey-dew.

"Some time ago the Agricultural Society of Grenada forwarded to the Council of the Society of Jamaica, elaborate delineations of a fly that had considerably damaged the growing canes in that island, which they call the cane-fly. It was a four winged fly that punctured the leaf, and deposited numerous black eggs; but when considerable apprehensions were entertained for the consequences to the yielding crop of canes, the planter was relieved by a little parasitic insect which made its appearance among them, and after fastening on the back of the perforated fly, devoured it, and rid them of this hitherto unobserved pest. This cane-fly appeared to be our *Delphax saccharivora*, a visitor as yet only known in St. Thomas in the East.

"I have yet to notice the Termite. However disastrous to the planter may be the visitation of the insects we have described, they seem to me all of limited injuriousness compared to the havoc committed by the *Ground Termite*. The common appellation of Borer has been given to two very dissimilar insects, the *Calandra sacchari* and the *Diatræa sacchari;* the one a beetle, the other a moth. The transformations which these two insects undergo in their respective seasons bring their depredations to an end before they have gone beyond some three or four joints of the mature sugar cane; but the ground Termites devour the entire stock, ascend from the root to the crown of the plant, change the whole vegetable mass from a saccharine to an acidulous pulp, and establish themselves in considerable patches in some of the most favourably growing fields. The destruction of the plant for all sugar-making purposes, is not however the sole evil attendant on their establishment in a cane-field. They carry into the body of the stock which they perforate a mass of pure earth, and add to the deteriorated juice of the cane a quantity of absolute filth, that must do incalculable mischief wherever only a single joint of the plant they have contaminated is unfortunately mixed with the sound and uninfected canes.

"We are all familiar with the Tree Termite, the *Termes lucifugus*, or white ant, that forms galleries in the trunks of trees: we are also well acquainted with the nest-building Termite, the *Termes arborum*, that constructs a large massy habitation in trees, and contrives a covered way from the ground to the nest, along the outside of branches, composed of small pieces of gnawed wood cemented together with a natural glue. Aversion to light is the common character of the whole tribe. All the species are not social, nor do they all exhibit the constructive instinct. The solitary ones simply destroy the bodies in which they take up their abode, and are to be traced by their perforations in books, and in closely packed and dry substances. Some of the perforating Termites, as they gnaw away wood, fill up the interstices with clay, tempered to an exceeding degree of hardness. The internal part of the timber perforated by this kind may be said to be transformed from wood to stone. One species uses red clay, another black earth, and a third woody materials. They adopt expedients to secure themselves against the attacks of the tree ants, the *Formicidæ*. As these would enter any breach of their habitation and seize them, and carry them away, they are alert in repairing the external breaches which accident makes in the walls of their galleries, by sticking into them patches of mud and compost.

" Smeathman, who has very minutely described and illustrated the tribes of Termites, and who considers their whole economy of life as a providential relation between the decomposition of vegetable substances and the means of accelerating vegetable decay, says they do not usually attack trees in a sound state: 'Vigorous healthy trees do not require to be destroyed, and accordingly these consumers have no taste for them.' When a large tree falls from age or accident, there are Termites that enter it on the side next to the ground and devour it at leisure. The inside is soon perforated and destroyed, and nothing but the outside case remains. It retains its form for a time after this process of excavation has been gone through, but it falls away and crumbles at a touch, and is so unsubstantial, that Smeathman very amusingly observes, 'you may as well step upon a cloud,' as set foot on one of these disem bowelled masses of the forest.

" The general remark that the Termite does not attack growing trees, is the fact to which I would direct attention. The depredation committed by the ground Termite on a field of sugar-canes contradicts this conclusion, and the general deduction that Smeathman would draw from the economy of life of the tribe. There is, however, a portion of the history given by him of the pursuits and habits of some of them, that bears upon their attacks upon sapid vegetables, and illustrates their depredations on our canes. There are royal chambers in the hillock-building Termites, surrounded by a countless number of others of different shapes and dimensions, all of them being either arched cells or galleries: these open into each other, or communicate by passages wide enough for the soldiers and attendants, of whom great numbers are necessary and always in waiting. These apartments lead away to the magazines and nurseries. The magazines are chambers of clay, and are always well filled with provisions, which to the naked eye consist of raspings of wood, and of *plants which they have destroyed*, but are found by the microscope to be principally gums and inspissated vegetable juices. These are thrown together in little masses, some of which are finer than others, and resemble the *sugar about preserved fruits;* others are like crystallised drops; one kind are quite transparent, another are like amber, a third brown and clear, and another quite opaque, just as we see ordinary gums are. The magazines are intermingled with the nurseries, which are buildings totally different from the rest of the apartments. They are composed entirely of wooden materials cemented together. Smeathman calls them nurseries because they are invariably occupied by the eggs and young ones, which appear at first in the shape of those called labourers, white as snow. We may perceive in this description of a colony of Termites, the purpose for which they establish themselves in our cane-fields. They are there abundantly supplied with those inspissated juices, which resemble the sugar of preserved fruits on which they feed. As they have an utter aversion to feeding in the light, they leave the outer portion of the stems of the canes they penetrate undevoured, retaining the intermediate parts as divisions and supports, so that, while the cane looks externally unblemished, it is internally destroyed. The earth is introduced to repair breaches and add resistance wherever decay has weakened the stock of the plant. Latreille says they are furnished with an acid for softening the wood, the odour of which is exceedingly pungent, and that they moisten their galleries with a gelatinous substance similar to glue. Whatever may be the substance they use, it is certain that the sugar disappears from the cane, and a powerful acidulated liquor only remains. The canes become as red as Braziletto-wood. As these insects are utterly incapable of subsisting exposed to light, the best remedy against being overrun by their colonies is the effective action of the plough. Tillage by horse or cattle implements, by which the soil is frequently disturbed and broken up, seems the only cure commensurate with the magnitude of the evil."

CURIOUS SPIDER.

In Sir Hans Sloane's Natural History of Jamaica, vol. ii. pl. 235. fig. 3., there is a representation of a Spider, which has the four pairs of legs set on the corslet by couplets, two pairs pointing forwards and two pairs backwards. In his delineation of the geometric web there are a few zigzag lines. These, as

Mr. Hill informs me, are the feeble notices of a very special instinct.

The creature itself, an insect of much beauty, fell under my own observation. It was at Culloden, an estate near Bluefields, the undulating and down-like pastures of which are thickly studded with palm-trees, chiefly of the Fan-thatch (*Borassus flabelliformis ?*), and the bristling Ebby-palm (*Acrocomia sclerocarpa*). Immense clumps of Prickly-pear have spread themselves over the rocky surface, and among the oval nodes of these were placed many irregular, perpendicular nets, geometric towards the middle. The very centre of each web was occupied by an elegant Spider, hanging head downward, the upper part of whose body was of a glistening satiny or silvery whiteness, the belly yellow spotted with black, and the legs marked with alternate rings of the same contrasting hues. It was the *Epeira argentata* of Fabricius, or a species closely allied to it. The interesting peculiarities detailed in the following communication from my friend, escaped my own observation, in my eagerness to secure the specimens, of which I took three. The season was May, but Mr. Hill's facts were observed in January.

"In a garden passage bordered with a hedge of the *Triphasia* (*Limonia trifoliata* of Linnæus), a spider's web had attracted my attention in consequence of its containing in the centre a thick tissue of zig-zag lacings, which gave it the appearance of being ornamented with a Saltier cross. On examining what there was peculiar in the Spider to lead to the peculiarity in the web, a little observation rendered the whole apparent. The arrangement of the legs of the species, an *Araneus cancriformis* as Sloane describes it, differs from that of the generality of spiders. They are set upon the corslet in double pairs, the doublets close together, two pairs before and two behind; the two forelegs of equal length, but the two hind ones slightly unequal. In the ordinary position of the Spider, when in a state of watchful repose in the centre of his web, the four double legs, representing the cross usually called St. Andrew's Cross, are extended out so as exactly to cover the zig-zag lacings, that form precisely a similar cross. The anomalous structure of the Spider's legs, in the mere necessities of wear and tear, rendered requisite this thickening of the web in the spots where the legs rested, and the whole economy indicated a relation of fitness, usefulness, and consistency, very special in character.

"All the web, except about some half inch of the centre, was composed of the usual radiated lines with concentric intersections. The centre was a sort of off-set disk of irregular meshes, to avoid the crowding together of the radiations before they converge to a point. Immediately beyond this disk, stretching through about an inch of the rays, the zig-zag lacings were extended, and so overlaid as to make up a mass of thickened tissue, about twenty or thirty times the thickness of single threads, or of the warp and woof of other portions of the web.

"By some accident the web I first noticed got destroyed; but, in the course of a single day, I found the Spider had rebuilt the web in the same hedge some ten yards further away; and though the new web was not so strong, nor so carefully constructed as the preceding one, it was yet formed with the same peculiarities of zig-zag lacings of a thickened tissue corresponding to the Saltier position of the Spider's legs. I have since, in the course of my natural history researches, met with some three of these Spiders (they are rather scarce), and the same peculiarity prevailed in all. Sloane evidently noticed it, but does not seem to have understood its purposes." *

STINGLESS BEES.

Everything relating to the marvellous instincts of Bees is so entertaining, that the following extract from the Journal of my esteemed friend, Mr. Hill, needs no apology.

"*November 8th*, 1847. — I was exceedingly interested this afternoon by the sight of two hives of indigenous Bees, shown to me by Mr. Garriques at Skelton Pen, on the banks of the Rio Cobre. The one hive, in the hollow of a Calabash-tree, had an entrance about half an inch wide at midway up the trunk, the cavity being supposed to descend some four feet down. The other was in a Cordia Cherry-tree, and was laid bare by a considerable portion of the tree being cut away. The cutting just disclosed the uppermost of the brood cells, but nothing of the sacklets that contained the honey. I take our bees to be similar to, if not the same with, the bee of Mexico, a *Melipona* or *Trigona*, called by the Spaniards *Angelitos*, from having no stings. They settled upon

* See his description at p. 196. vol. ii.

us, and we handled them; but they did no injury to us, though it was perceptible that they were excited, for they pursued the hand and clustered on it, when portions of the brood cells were taken up. The black Ants that infest forest trees had tracked the hive in the Calabash-tree, and had congregated around the entrance-hole, making an effort to gain access. A sentinel-bee, which was every now and then relieved from his guard, stood in a state of restless watchfulness, assisted at his post by two Bees behind. The Bees behind stood reversed, head-downward; and, clinging to the upper arch of the entrance, they gazed upward, and watched several Ants clustered above, in some two or three little groups within the crevices of the bark, prepared to rush in, if the sentinels remitted their vigilance for one moment. The active Ants paced upward and downward in lines, but found no opportunity of gaining a nearer access than a rapid reconnoitring of the doorway. The entrance, when occupied by the three sentinel Bees, admitted of no access by comers and goers of the hive, except by the centre Bee, that guarded the hole in front, momentarily stepping aside. This movement he performed with surprising quickness as often as a Bee came in or went out. The wax of these Bees is very unctuous and dark coloured, but susceptible of being whitened somewhat by bleaching. The honey is stored in clusters of cups, about the size of pigeon's eggs, at the bottom of the hive, and away from the brood cells. The brood cells are hexagonal, they are not deep, and the young ones, when ready to burst their cerement, just fill the whole cavity. The mother Bee is lighter in colour than the other Bees, and elongated at the abdomen to double their length."

Both of the stingless genera referred to, *Melipona* and *Trigona*, are found in the warm regions of the New World as well as in those of the Old. A species of the former, *M. fulvipes*, is found in Cuba; and the above may be identical with it. The species of *Trigona* build their nests at the extremities of branches of trees, in the shape of a large pear; the *Meliponæ* select hollow trees. The nest of a Mexican *Melipona*, exhibited at the Linnæan Society, Jan. 29., 1829, built in a hollow log, consisted of a number of irregularly-placed, black, oval cells, filled with thick honey of an amber colour, among which several of the Bees were lying dead.

THE UTIA, OR INDIAN CONY.

When Columbus discovered the greater Antilles in 1492, he found in them no quadruped of sufficient importance to attract notice save the Alco already noticed ("a species of dog which never barked") and "a kind of cony or rabbit, called Utia." The latter was the only four-footed animal that afforded meat to the simple natives, who were accustomed, according to the early Spanish historians, to hunt it during its nocturnal activity by the light of the *fire-fly*.

What this animal really was has been a matter of uncertainty until recently. It has been commonly confounded, by topographers and historians, with the Agouti, a species of the Cavy tribe, known to inhabit the Caribbean chain, and reputed to be indigenous to the larger islands. Thus Bryan Edwards, enumerating the quadrupeds of the West Indies, mentions the Agouti, the Peccary, the Armadillo, the Opossum, the Racoon, the Musk-rat, the Alco, and the Monkey.* Of these he concludes the first and the last to be the only indigenous animals now surviving in the larger islands, and thus alludes to the first. "The Agouti is sometimes called *couti* and *coati;* it was corrupted into *uti* and *utia*, by the Spaniards; and at present is known in some parts of the West Indies by the terms *pucarara* and *Indian Cony*. It is the *Mus agouti* of Linnæus, and the Cavy of Pennant and Buffon." (Hist. W. Ind. i. 90.)

A few years ago M. Fournier brought to Europe specimens of the animal which still bears in Cuba the name of Utia. It was found to be new to naturalists, and a genus was constituted for its reception under the appellation of *Capromys*, which Mr. Waterhouse arranges in the great Porcupine family. Two other species were afterwards added, all three being inhabitants of Cuba; and a fourth representing them in Haiti, but, on account of some difference in the dentition, erected into a genus by itself, and called *Plagiodontia ædium*. To these we have now the pleasure of adding a fifth, the Utia of Jamaica, which we shall see to belong to the Cuban genus *Capromys*.

The Indian Cony is quite unknown in those

* There is no reason to believe that any quadrumanous animal was indigenous to the greater Antilles.

western districts of the south side of Jamaica with which I am familiar. I often heard it spoken of as inhabiting the mountain regions of Manchester, a district characterised by red earth and honeycombed limestone rock. Mr. Andrew G. Johnston, in a communication to Mr. Hill, describes the animal as peculiar to a similar region in Portland:—

"I have not heard" he says, "of its inhabiting hollow trees. It is taken generally from under rocks; and that range of the John Crow Hills which furnishes the beasts in abundance, is broken limestone: high mountains on the north eastern side of the Rio Grande, from its source to Moore Town. The other range, which divides the valley of the Rio Grande from Blue Mountain Valley, has no limestone and no Conies. Both ranges are very lofty; from 3000 to 4000 feet in height. They form a sort of wedge at the sources of the Rio Grande and Island River (which last flows to Bath), from which wedge probably this district of wilderness derives its name Cuna-Cuna. This is conjecture only. The two chains unite at the point of the wedge and form a saddle; on the north rises Rio Grande; on the south, Island River. The broken limestone is in a great measure honeycomb rock, though some of it is compact limestone. De la Beche has misunderstood the whole chain. I suppose he never explored it. The other chain he has well described: transition and trap, with immense masses of conglomerate boulders;—but no Conies; nor did I ever hear of one in the lowlands at all, neither in woods nor hedges."

Mr. Hill, in his correspondence with Mr. Johnston

and with myself, was the first to see that the Cony of Jamaica had no close affinity with the Caviform Rodentia, but was generically identical with the *Capromys* of Cuba. The details of structure by which he convincingly proves these facts, want of space compels me to omit; but his polite transmission to me of specimens in skin and in spirit, has enabled me to confirm his conclusions, and to give to the Indian Cony the specific name and the distinctive characters contained in the note below.*

From the letters of this gentleman, however, I gladly make the following extracts, comprising many particulars of the economy and structure of this animal:—

* The Short-tailed Utia. (*Capromys brachyurus*, HILL.) Tail very short, about one-eighth of the total length. Fur dense and harsh, generally from ¾ to 1 inch in length, with a few longer hairs intermixed, but all of one kind: each hair is black, with a ring of bright bay or golden brown near the tip, imparting a brindled appearance to the fur, like that of a dark specimen of the Brown Rat. On the throat, breast, and belly the fur is yellowish, becoming white along the mesial line. The feet are clothed with blackish hair, short and stiff; the soles are black, roughened with rasp-like warts. The thumb of the fore-paws is a rudimentary tubercle, but armed with a distinct blunt nail. The great toe of the hind feet set far back, separable, and thumb-like Ears blackish grey, short, and fleshy. Muffle (or broad flat termination of the snout) blackish, clothed with a glistening pile of very short down, the extreme margins of the nostrils alone being naked. Moustaches long. Incisors white. Molars with two deep oblique folds externally and one internally. Tail stiff, taper, with rounded point; scaly, with thick short bristly hair, which is black on the upper surface, greyish-brown below: the base of the tail is nearly naked.

The following admeasurements are taken from two specimens; the one not quite grown, a stuffed skin in bad preservation; the other an

"*April* 19*th*, 1848.—Mr. Johnston's information respecting the district in Portland inhabited by the Indian Cony, showed that two conditions were necessary to constitute a befitting locality for it. Surface rocks, replete with cellular hiding-places, and abundance of succulent roots, fruits, and herbage. Our limestone hills bordering the plain of St. Catherine, are all well-marked with the first requisite, the cellular hiding-places; but it is only the Red-hills that supply the necessary exuberance and diversity of vegetable food:—there alone the Indian Cony is common. In the limestone districts, deficient in moisture and herbage, it is scarce, if not altogether unknown. Fruit trees, such as oranges, star-apples, nesberries, and Avogado-pears, together with plums, guavas, anonas, and plantains, abound on the Red-hills and tuberous and farinaceous vegetables

adult, disjointed, in spirits, in excellent condition, but wanting the central portion of the trunk.

	Young.	Adult.
	Inches.	Inches.
Total length, including tail -	16·0	18·0 (?)
Length of head - - -	3·0	3·5
From muzzle to orifice of ear -	2·6	2·8
„ to front angle of eye -	1·5 nearly	1·6
Length of ear (inner side) - -	- -	1·0
„ (outer side) - -	- -	·66
Width of ear - - -	- -	·7 nearly
Circumference of body - -	9·0	
Length of tail - - -	2·0	2·1
Diameter of tail at base - -	- -	·5
„ at tip - -	- -	·2
Fore sole to tip of longest claw -	1·5	1·6
Hind sole „ -	2·6	2·8

are cultivated there in profusion. Cellular labyrinths riddle the rocks through and through, and their hiding-places are intricate and difficult. As they are secure in such localities against the pursuit of dogs, the Conies brought for sale are always obtained by trapping. They go out to feed, and return to their lodgments, by a constant pathway, observed to be well-trodden, and strewed with pellets of dung, and leading into adjoining provision grounds. As they feed at night, they are not commonly met with abroad; but if they are surprised, and the track-way to their holes be at some distance off, they may be dodged, and run down, and taken. I have seen live Conies brought into town, but they were always young animals, and were said to have been run down and caught before they could retreat to the rocks; they wanted the experience and cunning of the old ones, though they are naturally of quick vision and of acute scent, and their taste is delicate and fastidious.

"I mentioned Fournier's description of the Utia of Cuba. He seems to have been the first naturalist who described the true characteristics of this animal, and assigned it its proper place in classification; hence the Cuba Cony has received his own name, *Capromys Furnieri.* The distinguishing characteristic of our species is its shorter tail. That appendage is about half the length of that of the Capromys of Cuba. Whatever may be the differences of power and of habit which accompany this distinction, there seems to be enough of a common character to make the following extract from Fournier's account of his animal applicable to the Jamaica species. 'Their walk is almost absolutely plantigrade, and their movements are slow, their hinder parts appearing when in motion to be as embarrassed as the bear's.' 'They take occasional leaps, suddenly turning round from head to tail, like the field mouse; and they gallop when at play, and make a noise with the soles of their feet.' 'They climb with ease, assisting themselves with their tails as a support, and they use the same in descending.' 'They often raise themselves to a listening posture, sitting erect, with the hands hanging down, like rabbits and hares; and in eating they employ sometimes both, at other times one, of their hands only.' When the substance is small enough to be held between their fingers and the tubercle at the base of the thumb, the single hand is then put in requisition. If the Jamaica species should not prove to be the second Cuba one to which Pöppig has given the name of *Capromys prehensilis*, and about which there exists some doubt among naturalists, I should suggest *C. brachyurus* as a distinctive name for our Cony.

"I cannot much depend on the Negro statements respecting its prehensile powers; they represent them as feeble, or not frequently put in requisition. They seem to me only to have seen the Cony under the hurry of alarm; for they tell me, they will attempt to scramble up trees, but generally fall back, unable to make good their hold. I have observed the *Capromys* of Cuba in ascent. Some three or four of them confined in a capacious pen in which trees were growing, were very familiar; and being exposed to no apprehension when looked at, or even when handled, they pursued their wonted frolicking, and moved about as when free and at large. They planted their feet and hands on the limbs like bears, and climbed leisurely, but with great facility, along the branches, and descended rather rapidly. We may suppose they would hardly take to trees at all, however they might fail in making good their retreat up them, unless they were familiarly acquainted with them. They only seem to fail to do hurriedly that which they can at any time effect only deliberately.

"Although the extremities of the Indian Cony exhibit such fitness for holding and grasping, they appear never to be applied to the purposes of digging and burrowing. They may extract the roots, which they make part of their food, by scratching the ground and drawing out such portions as their strength may manage; but they never remove the earth to any depth to reach the esculent underground provisions of the Negro garden; yet when they are enticed to take the rude traps and springes which the Negroes set for them, the temptation is always some piece of the large farinaceous roots, which by their own labour they can never procure for themselves. Their delicacy of scent enables them to perceive when any substance has been previously touched and handled. It would be quite hopeless to entice the Cony to the trap by setting it at once. It is necessary to reconcile him to feed on the roots which are to attract him, by permitting him for a succession of nights to eat without danger, by strewing the food about the unset trap. When it has been ascertained by what has been devoured in the previous nights, that the Cony's confidence is well secured, the springe is set, and the morning scarcely ever fails to exhibit him caught and strangled.

"*Jan.* 20*th*, 1851.—I procured the other day an Indian Cony, and have the head, feet, and tail in spirits, to be transmitted to you by the first vessel for England. I am led to think you will find, on examining these several parts, that our animal is the *Plagiodontia ædium* of your extracts. In pressing the limbs to stow them, I observed the tight manner in which the fore-paws closed up, showing great power of prehension, and the extent to which the hind toes under a similar degree of force expanded and stretched out. The hand, though thumbless, has evidently all the grasp peculiar to the Squirrel, and the hind feet more of a capacity for pressing rugged surfaces than belongs to the feet of the Rabbit. Every observer concurs in representing the Indian Cony as frequently sitting up while eating, but few assign to it that ability in using the hand that is given to the Agouti and the Capromys, and all doubt whether it has much power to hold the food and carry it to the mouth single-fisted. I could have wished I had seen a living specimen, not alone for making observations on this uncertain point; but *specially* for another matter in which this *particular* animal is interesting *above all others whatever*. The teats are not, like those of rabbits, or other allied races, placed on lines from the belly to the breast. They are situated midway between the *belly* and the *back*, that is, just at the extremity of the short-ribs along the sides, where they shoot out backward as protuberant nipples to the right and left of the body, two on each side. This is altogether without analogy in the kindred Agouti. In that animal Buffon says, "Il m'a paru que cet animal avoit douze mamelons, sept sur la poitrine, trois à droite et quatre à gauche, et cinq sur le ventre, trois à droite et deux à gauche; mais je n'ai pas pû m'assurer que tous les tubercules que j'ai pris pour des mamelons, fussent en effet de vrais mamelons, ou qu'il n'y en eût encore d'autres que je n'aie pû distinguer de tubercules qui se trouvoient à la racine de chaque poil. Il résulte de la position de douze mamelons dont je viens de faire mention, qu'il devroit y en avoir deux de plus, l'un sur le côté droit de la poitrine, et l'autre sur le côté gauche du ventre." — (1st edition, 1760, vol. viii. page 386.) This is certainly a curious arrangement of the teats in the Agouti, not, however, as to place, but as to the numerical distribution of the nipples along the belly and the breast. Yet it makes no approach to the anomalous position of those of our Cony. I wonder whether there be anything of this in the Utia of St. Domingo (*Plagiodontia*) or in that of Cuba (*Capromys*). Rudimentary tubercles in the male, the representatives of teats, are found in the same situation.

"To conceive the animal in the act of suckling, we must picture it seated in the posture of a Hare in a form, pressing the body to the earth and filling the udder with milk, while the young ones draw the nipple on both sides at once. We must conclude from this organisation that the cavities in rocks and trees in which it lives and breeds, are not narrow holes like the burrows of rabbits, but capacious cells, the young requiring room to the right and left of the parent to stretch out in the act of lactation. The exposed nipples would be liable to be torn by prickly shrubs, so abundant in our wastes, if the Cony fed in open commons like the wild rabbit. Its movements are confined to the provision ground, where the herbage is soft and succulent; and as these are ordinarily well-weeded, with clear space for running to and fro, however massed and matted the vegetation may be, it courses through it without much danger of lacerating the teats . . . . You will observe that the nails are quite pointed, and that the edges of the under groove are sharp, showing that the feet of the Cony move over the ground without much wearing and rubbing. . . . . .

"As the capability of the Rodentia to raise objects between the fore-paws is determined by the form of the clavicles, perfect clavicles not alone enabling them to exert them with this effect, but giving them the ability to climb trees, the uncertainty respecting this power in the Indian Cony will be determined hereafter by an examination of that part of the skeleton. Hares and rabbits, whose clavicles are incomplete, have no such power; they sit on their haunches, but they do no more than play with their fore-feet; rats and mice, with perfect clavicles, are powerful climbers, and are not without ability to use the hands, but their clench is not considerable. Squirrels, on the other hand, have the climbing and hand-folding capacity in perfection. With these researches completed, and with what we have of other peculiarities, we shall no longer be at fault in classification."

This strange position of the teats is not quite singular, though it has never before been remarked in the *Capromys*. Mr. Waterhouse in his beautiful work on the *Mammalia*, now in progress, thus alludes to the subject: — "The female Coypu (a large South American Rodent of the same family) . . . . . . swims with her young on her back until they are sufficiently large to follow the parent. This habit helps to explain the singular position of the nipples in the female Coypu. Of these four were found by M. Lereboullet on each side of the body, and situated rather above the mesial line of the flanks, the foremost being placed behind the shoulder, and the hindermost in front of the thigh." Mr. Waterhouse considers that this position of the nipples, rather on the sides than the under part of the body, will be found a common circumstance in the Hystricine division of Rodents, since he has found them so situated in the genera *Lagostomus*, *Octodon*, *Habrocoma*, and *Nelomys*. — (*Nat. Hist. Mammalia*, ii. 299.)

The following notes by Mr. Johnston will be read with interest. After a minute detail of some particulars of structure distinguishing the animal from the Cavies, he thus continues: —

"The animal treads on the whole of the lower joint of the hind leg, from the hough, as does the rabbit; it is in fact plantigrade, but its usual motion is very unlike that of either rabbit or rat: the former always, and the latter generally, moving the pairs of limbs together, galloping, in fact. The Indian Cony, on the contrary, walks like a horse, or, applying its plantigrade feet as the bear, moves onward like a dog trotting. In its gait it waddles rather grotesquely. I am describing a living specimen before me, which is, as I understand, about eight weeks old, and would weigh ten ounces. It is of a dark-brown colour, very familiar, and almost affectionate; gratified by kind notice, and utterly void of fear. It sleeps with one or other of the children in the house, and eats anything which they feed on, vegetables and flesh, and it is very fond of salt-fish. If handled by a stranger it is more shy, squeaks, and even threatens to bite and scratch if displeased, but has scarcely as yet proceeded to these extremities. It was taken when very young from under a limestone rock on the northern ridge of the Carrion Crow Mountains, and was raised with little trouble on milk and vegetables. It now waddles about the table when permitted, picking up crumbs, or any fragments of food, and when it can find no more it jumps down without fear, although this effort has a very awkward appearance. The soles of the feet are black, soft, and raspy. It had a respectable moustache, but being a little too inquisitive about the flame of a candle, it singed these almost entirely off. Though it waddles in its common motion, it jumps like a rabbit when hurried or impatient.

"The Cony is easily procured, being abundant up here [Portland]; it is inoffensive until attacked; then it becomes as savage as a badger. It 'chaws' the dogs. The flesh of the young is very good indeed."

Mr. Hill has lately favoured me with the following notes, in addition to the above, which conclude our present information concerning the economy of the species.

"At Two-mile Wood, a run of stunted forest on the road to St. Dorothy's, is a perfect warren of Indian Conies. Patches of *Bromelia pinguin* occur among the woodland. This is an aboriginal cover in the lowlands, and is almost the only place in which they occur in the plains. The overseer, Mr. Russel, tells me that when he surprises them out in the adjoining sward, and his dogs cut them off from the cover, they run up the Cashaw trees, and grapple the rough bark easily. This is another point of resemblance to the Cuba *Capromys*. I have applied for living specimens to send to England."

"I learn that our Cony is fond of browsing the Logwood (*Hæmatoxylon*) and the Bastard Cedar (*Guazuma*); or rather that it barks these trees just above ground. It is never seen out in wet weather; and it has never yet been traced by its foot-prints."

The species appears to be the "Small Indian Cony, *Mus 3; major, fusco-cinereus, caudâ truncatâ*" of Browne; of which he gives no distinct information, except that it is a native of Jamaica. Sloane takes not the least notice of it: Long merely includes it in his enumeration of the wild animals.

ORCHIDEÆ.

Though Jamaica cannot boast of Orchideous plants so curious or so gorgeous as some that are found in

Continental America or in the great Indian isles, yet it has many species of much beauty. *Epidendrum ciliare* and *Brassia caudata* are curious forms, the former for the delicately-fringed white lip, and the latter for the length and slenderness of its spotted yellow petals. The shell-like form of *Epid. cochleatum* and *E. fragrans* is interesting, as is the delightful perfume widely diffused by the latter. The lilac spikes of *Ionopsis*, and the crimson ones of *Broughtonia sanguinea*, yield to few in beauty; and of terrestrial species, a rich purple *Bletia* that I found in the *Bocâguas* near Spanish Town, and a *Phajus* from the summit of Bluefields Peak, are particularly noble.

I have not found that the *Orchideæ*, any more than other tribes of plants, are confined to one particular season of flowering. It is commonly considered that the dry season is a period of rest for these parasites, and that the return of the periodical rains stimulates them to push forth their leaves and flowers. In Jamaica, however, there are many exceptions to this, so many that it can scarcely be deemed the rule. Thus in February and March, the time of drought, *Broughtonia sanguinea* flowers profusely and magnificently all along the shore of Westmoreland. *Epidendrum umbellatum* I also found in blossom, not far from the shore; and a little *Polystachia* (?) appeared to have just done blossoming. *Angræcum funale* continues to throw out its elegant flowers successively all through this period of drought.

In the mountain woods, at the same time, I found the dense spikes of *Epid. nutans* depending in abundance from the forks of the trees, and the fine *Phajus* that I have already spoken of rearing its magnificent head in the gloom of the bush. The rose-coloured *Bletia* of Bluefields Mountain, which at the commencement of the drought showed only the withered leaves crowning the round compacted bulbs, was in full flower at the latter part of this period; and it was in March of the following year, a season even more arid than the former, that I met with the profuse blossom of the lovely purple *Bletia*, growing on a precipitous rock on the banks of the Rio Cobre. Towards the end of the dry period I saw *Epid. fragrans* and *Ep. cochleatum* in blossom on open trees in the beautiful park-like pen called the Kepp, in the Luana Mountains; and, a little later, *Brasavola nodosa* flowered at Bluefields, where it is abundant. It is worthy of remark that I had found this species in blossom at Alligator Pond in company with *Broughtonia sanguinea*, during the early part of December, when it rained nearly every day; yet around Bluefields, while the latter, as already mentioned, was flowering profusely, the former was flowerless until the beginning of March.

At the same time many kinds were out of bloom throughout this season, some of which flowered soon after the commencement of the vernal rains. *Maxillaria Barringtoniæ*, whose great wrinkled bulbs had been conspicuous on the mountain trunks, threw out its fleshy flowers from among the roots about the end of March, and continued flowering through April. A little later, the long spikes of *Oncidium Carthaginense* were waving in the breeze all through the

Y 2

high woods that bordered the shore. Bulbs of *Epid. fragrans* and of *Brassia caudata* that had been brought from the mountain in the dry weather, and planted out in the open air, blossomed, the former at the beginning, the latter towards the end, of May About the middle of the same month, in the tall, dark, and humid woods of Shrewsbury, about halfway up the mountains of St. Elizabeth's, I saw several racemes of a beautiful *Ionopsis** in rich bloom. The irregular tuber-like bulbs of that terrestrial Orchid, with a Bletia-like habit, which grows abundantly in the dense bush on the summits of the Bluefields Peaks, had thrown out their tall but not very inviting panicles of flower through the month of June. The rains were at that time descending copiously, and continued to fall until the middle of August; about which time I met with *Epid. fuscatum* in blossom in the tall woods of Basin Spring, a little lower elevation than Bluefields Mountain. About this time also the singularly fringed blossom appeared on *Epid. ciliare*, which had hitherto displayed only its long spindle-shaped bulbs, each crowned with its pair of leathery leaves. Soon afterwards the autumnal drought commenced, but I have no further record of the flowering of *Orchideæ*.

If this irregularity of flowering, or rather apparent independence of seasonal rain, had been confined to the recesses of the mountain woods, it would not have been surprising, since there actual dryness seems unknown. On the summits of Bluefields Peaks, and

* Swartz, if I mistake not, describes *Ionopsis* as affecting the driest open pastures.

especially in the shallow intervening valleys, whether from the closeness of the woods preventing the escape of vapours, or the luxuriance of the rank vegetation, the air feels at all times surcharged with moisture, even during the dry seasons. In the night this moisture is concentrated very copiously, so that in the concavities of large leaves, sometimes as much as a wine-glass full of clear water may be seen collected, especially on such leaves as those of the *Heliconia*, or wild plantain, and of the great esculent Arums, called Cocoes, of the Negroes' grounds. I have many a time been refreshed by drinking from these leaves. The ferns and other herbaceous plants that fringe the narrow paths, and the *Lycopodiums* and mosses that form thick cushions around the roots of the trees, are always found heavy with dew, and as it were saturated, when one visits this locality in the early morning.

With respect to the *choice* of trees exercised by *Orchideæ*, it certainly seems to exist, but is not, I think, very exclusive. In the lowlands, the lovely *Broughtonia sanguinea* is found in very different situations; on the Palmetto thatch (a small species of *Thrinax*, I believe), that forms an extensive grove, at Crabpond Morass, it grows abundantly, at the height of four feet from the ground. At Alligator Pond I took it from a Hog-plum (*Spondias graveolens*), and around Bluefields it is common on the Calabash, both trees of moderate dimensions. But on the road to Savannah le Mar, bunches of its brilliant blossom are seen depending from the lofty forks and limbs of the towering Cotton-trees (*Eriodendron*), at an elevation of eighty or a hundred feet.

Brasavola nodosa has perhaps an equal range as to elevation, and is found on the same trees as those just mentioned, frequently in company with *Broughtonia.* It also affects the Yokewood (a species of *Bignonia*), the Birch (*Bursera*), and many other trees; being one of the commonest of the lowland Orchids. The leafless *Angræcum funale*, as I have before intimated, clings by its tortuous roots to the trunk of the Calabash, at the height of a yard or two from the earth; the great mass of its roots depending, in a tangled plexus, in the air.

Oncidium Carthaginense also prefers the Calabash; but it is found likewise on the Fiddlewood (*Cytharaxylon*) and other trees, always on the branches, or in the forks, at an elevation of from fifteen to thirty feet. The flat bulbs of a *Maxillaria* (as I believe) cling in abundance to the trees growing in the morass that borders the shore at Cave; and I have found what I suppose the same genus, with a little *Epidendrum*, on a Star-apple tree (*Chrysophyllum*) half way up the mountain. On Bluefields Peaks, *Max. Barringtoniæ* is numerous, affecting the trunks of various trees, *close down to the ground.* I have taken the bulbs also from a *fallen trunk* near Kilmarnock, in the mountains of St. Elizabeth's. *Epidendrum nutans* clings to the trunks of large trees in the mountains, twenty and thirty feet from the ground; and the habits of *E. umbellatum* and *E. fuscatum* are similar, the latter at a moderate elevation, the former near the sea-shore. *Epid. ciliare* prefers the Avocado-pear (*Persea*); and is confined, I believe, to the mountains; I have found it most common on such trees in open pastures, growing on the trunk and larger branches in great luxuriance, at the height of fifteen to twenty feet; one bunch that I procured comprising forty-two bulbs, cohering together. *E. fragrans* and *E. cochleatum* also grow on fruit-trees in the mountain pens, eight or ten feet from the ground. *Ionopsis* I found only in the situation above-mentioned, on the trunks of forest trees, a few feet above the ground.

In the tall woods on Bluefields Mountains, almost every tree, from the thickness of one's arm upwards, is found to bear its bunch of *Orchideæ*, frequently four or five species growing on the same tree. The trunk is the most common situation on the tree, but in very large trees the forks and great horizontal limbs are likewise studded with these and other parasites, *Tillandsiæ*, sessile and caulescent *Ferns*, *Jungermanniæ*, &c. I was surprised and delighted at the number of minute species, some with tiny bulbs, others with small oval, alternate, almost pinnate leaves, and others long and grass-like, which, in company with the larger and more common kinds, crowded the trunk of an enormous Fig-tree that had been recently felled on the top of the Bluefields ridge. The massive, pillar-like stem, sixty or seventy feet long without a branch, was studded from end to end, and on all sides of its surface, with these delicate little parasites, which also spread themselves upon the great arms. On the lower mountains the huge Cotton-tree (*Eriodendron*) forms a perfect nursery of *Orchideæ* as well as *Bromeliaceæ*. Of terrestrial species, both the kinds of *Bletia* were growing on the bare rocks, exposed to the unmitigated glare of the sun. The tuberous kind with a similar habit, and the magnificent *Phajus Tankervilliæ*, on the other hand, occurred only in the deepest shadow of the dense humid mountain woods.

Specimens taken from one tree I found would grow very well on another. I was accustomed to transplant many of the masses brought from the mountains and other distant localities, affixing them either to logs of wood, or to fruit-trees in the pasture, or else merely laying them on the flat top of a stone buttress, with a brick or some other weight on them to prevent their being blown away. The smaller specimens I pushed into crevices between the stones of the same buttress. Those which I planted on fruit-trees were fastened by string passed several times round the trunk and the Orchid. In all these cases the specimens grew healthily. The roots, which had been torn from their attachments in procuring the specimens, never adhered to the new surface, but fresh roots were soon pushed out from the base of the bulbs, which clung to the support by their flattened under sides with the same tenacity as if they had been on the original trees; and after a few months needed not the assistance of strings or weights to enable them to maintain their position. Sometimes the mere laying of a bunch of bulbs in the fork of an orange-tree was found sufficient. I do not think the parasite is dependent on its tree for anything but support; the roots do not penetrate the bark in the least degree, but derive all their nutriment from the moisture with which the air is charged, or which, in the form of rain or dew, trickles down among their matted masses from above.

At the same time there is an unaccountable preference of certain localities to others. Thus of the vicinity of Bluefields in Westmoreland, and of Content in St. Elizabeth's, including both lowland and mountain, the former district is rich in species and individuals; while the latter, embosomed in tall humid woods (the pristine forest extending from the level of the sea to the summit of the first range of the Luana Mountains), possesses, as far as I have seen, *scarcely a single specimen of any species.* Again, the low level land around Savanna le Mar seems equally destitute of these parasites; but this is less to be wondered at, since so large a portion of the district is overrun with logwood (as is indeed a large part of the once-cleared land in St. Elizabeth's), a tree on which I do not remember to have found an *Orchideous* plant, (though *Tillandsiæ* are common enough on it), with the exception of *Oncidium Carthaginense,* which occasionally grows on the hedges that are made of this thorn-like tree.

DEPARTURE AND RETURN.

At length the day arrived for my departure from Bluefields. On the 19th of June, 1846, I embarked on board a little *drogger,* or coasting sloop, for Kingston, and cast a farewell glance, not without regretful yearnings, on the sunny slopes, and wooded glades, and mountain-peaks, where I had spent so many pleasant months. Seven dreary days were swallowed up in slowly beating along a hundred miles of coast, against the fierce daily sea-breeze. Nothing of any note occurred to break the monotony of the voyage, except that on the morning of its last day — as if I were not to be allowed to leave the island without a taste of its bitters as well as its sweets — I was stung by a Scorpion. While lying in the berth of the little close cabin, I was awakened by a severe twinge on one side of my neck: on putting my hand to the place, I took hold of some object which had pierced the flesh, and which, requiring some force to make it let go, I fancied to be a beetle that had nipped me with its mandibles. There was a dim lamp in the cabin, and on holding up my prisoner against the light, I found that it was a large Scorpion, which I had fortunately seized by the tail, so that, though it sprawled, it could not do any further injury. The pain was severe, but the old skipper kindly applied some camphorated rum, which he well rubbed on the part. The flesh began to swell and form a lump; but very soon both this and the pain subsided, so that in two or three hours only a slight soreness was perceptible about the region, and even this was gone before night. Altogether the affair was not equal to the sting of a wasp. One of the most curious of the results was a numbness of some of the nerves of the tongue, perceptible in the *papillæ* of the surface, which felt as if dead: this was soon after the sting.

On my arrival at Kingston on the 26th, I found that the length of the coasting voyage had lost me a passage to England, the steamer having just sailed.

This however gave me a few days' leisure here, which I should have been loath to be deprived of. At intervals, when the breeze had been too strong to be stemmed, I had landed at one or two points of the coast, and obtained a few specimens, chiefly in entomology. A little to the eastward of Black River there is a dreary, rocky, and inhospitable shore, marked on the charts with the sufficiently appropriate name of Starvegut Bay. While windbound here, I took a walk on shore, climbing over the immense masses of fragmentary rock, against which the surf was beating and boiling with furious violence, and shooting up ever and anon white jets of vapour-like spray through the sea-worn holes. In the woods, which consisted largely of the Cashaw (*Prosopis juliflora*), intermingled with some species of *Inga* and the great *Cactus Peruvianus*, — a vegetation totally different from that in the neighbourhood of Bluefields, — I observed a Vanessa-like butterfly, of brilliant blue iridescence, and some white spots near the tip of the fore-wings, which was, I doubt not, *Cybdelis Hyperipte*. I had never met with it before, and as I had no net with me, I did not capture any specimens now. It was however in some abundance; flitted along close to the ground, in the shadow of the woods, allowing an approach within a distance which would have rendered its capture with a ring-net an easy matter. Its manners bore some resemblance to those of the *Satyridæ*. I also saw here *Anolis maculatus*, that zebra-marked Lizard, which is so common around Kingston.

An arid plain, just behind Pedro Bluff, afforded

me some insects differing almost totally from those of the leeward part of the island. The elevation is scarcely above the level of the sea; the soil is sand; the trees, chiefly Lignum vitæ, scarcely attain a greater height than twelve feet, and the heat of the sun is on these accounts peculiarly intense. The collection of gum-guaiacum, the produce of the trees just mentioned, helps the poor inhabitants of this spot to obtain a livelihood. I found here a second specimen of *Myrmeleon Leachii*, and on the twigs of the stunted prickly trees, *Diaprepes Spengleri*, and a much finer beetle, *Psiloptera* (*sp. nov.* near *torquata*), both in considerable abundance.

At Alligator Pond, my negro lad tried the net with little success, the only novelty here being a species of *Libythea*, apparently new, but closely allied to the North American *L. motya*.

While I remained at Kingston, the lad attended to insects in the environs, particularly at Greenwich, near Passage Fort. The beautiful *Lachnopus aurifer* was the most common beetle, occurring in great numbers upon the bushes. The most abundant butterfly was *Callidryas Neleïs*; as it had been near Bluefields in the preceding April and May. A *Chrysopa*, much resembling our *C. perla*, and a great bee with violet wings, resembling *Xylocopa violacea*, also attracted notice.

My own time here I chiefly spent in consulting the valuable MS. volumes and drawings of the late Dr. Robinson, on the botany and zoology of the island, preserved in the Library of the Jamaica Society, and in comparing notes with my valued friend Mr. Hill, of Spanish Town, to whose kind courtesy I have been indebted for so much information, and who, at parting, added to his other kindnesses the gift of many zoological rarities, which would otherwise have remained either totally or nearly unknown to me. On the 9th of July at noon, I took my place on board the mail-steamer Avon, and bidding adieu to kind friends, finally left this lovely island.* A last glance at it, about 9 at night, revealed it dimly fading into the darkness, with the friendly light on Point Morant shining brightly.

The mountains of Hayti were in sight at daylight the next morning, and during the whole day we ran along the great promontory of Tiburon, the ancient province of Xaragua, once the happy domain of the beautiful but unfortunate Princess Anacaona.

On rising at early day on the 11th, I found the steamer at anchor in a little bay, environed by abrupt mountains, at the foot of which is the town of Jacmel. The boat was just putting off with the mails. There

* To some of the "courteous readers" who have accompanied me through the details of the preceding pages, it may not be altogether without interest to know the amount of my collections in zoology and botany, made during eighteen months' residence in Jamaica. They were as follows: Mammalia, 41 specimens; Birds, 1510; Reptiles, 102; Fishes, 94; Nests and Eggs, 34; Shells (marine), 1276; (terrestrial and fluviatile), about 1850; Crustacea, 100; Insects (including Arachnida and Myriapoda), about 7800; Echinodermata, 57; Zoophytes, &c. 42; Sponges, 550. Dried Plants, about 5000; Living Plants (Orchideæ), about 800; (Bulbs and Suckers), 932; (Cacti), 32; (Ferns), 222; (other Living Plants, young Trees, &c.), 117; large Capsules and Seed-vessels, 383; Seed of flowering Plants, 170 packets; Palm-seeds, 14 boxes; Gums, 24 specimens Woods, 50 blocks.

had been rain in the night, and the shaggy hill-tops were partially robed in fragments of cloud, undefined and changing, which contrasted finely with the dark surface of the forest. Inland the mountains in the morning sun looked inviting; and I noticed that they displayed the same singular resemblance to crumpled paper, as those in the eastern part of Jamaica. Our stay here was short; the harsh, deafening sound of escaping steam was succeeded by the crashing of the paddles on the water, and off we rushed on our homeward course. By nightfall we were just within the singular insulated rock called Alta Vela, or the lofty sail, from its very deceptive resemblance to a ship in the distance.

At sunrise on the 13th, we were under the little island Mona, between Hayti and Porto Rico, and through the day we steamed along the northern shore of the latter island. The land, thickly strewn with cultivated estates, spotted with clumps of trees, has a very beautiful appearance, contrasting in this respect with both Jamaica and Hayti, whose forest-covered coasts display little traces of cultivation, and look rude and repulsive. Soon after mid-day, the Moro, or fortification which protects the port of San Juan, was in sight, like a white wall projecting into the sea, and at 4 P.M. the steamer moored under it.

A few of the passengers went on shore for an hour or two. Everything showed we were in a *foreign* country. The town gloomily walled, and strongly fortified, the turret-like houses, and little balconies to each window, the well-paved streets, the ladies in black mantillas, opening and shutting their fans as they walked, solemn priests in black robes and shovel-hats, the children, the men, the *posadas* (taverns), everything wore a character so novel, so unlike any thing in our colonies, that I was greatly entertained.

In an hour or two we were again afloat, and steaming away for St. Thomas, where the sun on his rising the next morning found us snugly anchored. It had again been raining hard, and the mist hung over the town and slopes behind; yet the beauty of the town, rising from the sea on the sides of three conical hills, in the form of three pyramids of buildings, could not be concealed. I walked on a hill at the entrance of the harbour, covered with stunted bushes, and spent an hour or two entomologising. The insects were almost entirely different from those with which I had been familiar in Jamaica.

On the 16th we left St. Thomas, arrived on the 20th at the little isles of Bermuda, with their English-looking scenery and the clear transparent sea around their rocks, as still as the water of a Polynesian lagoon, as little answering as might be to the anticipations one might have formed of the "still vexed Bermoothes." We found the steamer Clyde lying here, migrated from the Avon to her, and reached Southampton on the 5th August, 1846.

HERE I take a respectful leave of my readers and of tropical natural history together. If I have succeeded in imparting to the former a small portion of the delight, admiration, and enthusiasm, which invest in my own feelings the things I have essayed to present to them, I shall not have lost my labour. Eminently pleasing that labour has been: the compiling of these pages from my journals and papers, and from the correspondence of my beloved and honoured friend, has recalled in vivid power the lovely Eden-like scenes through which I wandered, and has made me live over again those months of unwearying delight that I spent in beauteous Jamaica. I can echo with fullest truth the experience of Bishop Heber:—" In every ride I have taken, and in every wilderness in which my tent has been pitched, I have found enough to keep my mind from sinking into the languor and the apathy which have been regarded as natural to a tropical climate." Nay, I may truly say, I found *no tendency* to apathy or ennui; every excursion presented something to admire; every day had its novelty; the morning was always pregnant with eager expectation; the evening invariably brought subjects of interest fresh and new; and the days were only too short for enjoyment. They were not days of stirring adventure, of dangerous conflicts with man or beast, of hair-breadth escapes in flood and field; their delights were calm and peaceful, I trust not unholy, nor unbecoming the character of a Christian, who has his heart in heaven, and who traces, even in earth's loveliest scenes, the mark of the Spoiler. The sentiments expressed on this subject by my friend and fellow-labourer are those which I would ever associate with the study of science. " If the sight of nature," observes Mr. Hill, " were merely the looking at a painted pageantry, or at a spectacle filling the carnal mind with wonder and delight, the spirit would be overpowered and worked into weariness, but it is admiration at the wisdom, and reverence for the beneficence of Almighty power. He who 'dwelleth in the light which no man can approach unto; whom no man hath seen, nor can see,' is yet visible in his perfections through the works of his hand, and his designs are made manifest in the purpose of his creatures. Wherever our lot is cast, into whatever scenes our wayward impulses lead us, the mind-illumined eye gazes on divine things, and the spirit-stirred heart feels its pulses bounding with emotions from the touch of an ever-present Deity. The habit that sees in every object the wisdom and the goodness as well as the power of God, I may speak of, as Coleridge speaks of the poetical spirit, 'It has been to me an exceeding great reward; it has soothed my afflictions; it has multiplied and refined my enjoyments; it has endeared my solitude; and it has given me the habit of wishing to discover the good and the beautiful in all that meets and surrounds me.'"

"GREAT ARE THY WORKS, JEHOVAH! INFINITE
THY POWER! WHAT THOUGHT CAN MEASURE THEE, OR TONGUE
RELATE THEE?"

FINIS.

APPENDIX.

I.

THE Nurse (*Scyllium cirratum*), p. 241.—"On some of the particular shoals and banks within the Bay of Old Harbour, these fishes congregate in great numbers at special seasons, probably to deposit their egg capsules in the sunny waters. At these seasons nothing more is necessary than to get among them upon the shoal with sticks and strike them. [Hence the expression used by Sam, "*strike* Nurse," which I had supposed to imply the use of the harpoon. P. H. G.] They may be taken, and are taken in this way in large numbers. This habit may account for the shipwrecked Richard Falconer, on the Alacranes, being able to take this fish, when he had no means of securing any thing else for food."—(*Letter from Mr. Hill, 28th March*, 1851.)

II.

The Silk-cotton Tree (*Eriodendron anfractuosum*), p. 279.—"There is a profusion of magnificent Ceiba-trees around us, and all are more or less in their winter nakedness. The aspect they have assumed after shedding their autumnal foliage, is that of young leaves in some trees, and that of flowers and seed-pods only in others, the buds containing the leaves not having yet opened. Now as the Ceiba does not bear flowers and seed but in alternate years, though there is an annual hybernation, biennially it produces only foliage. In alternate years it blossoms, but never throws out a single leaf till those blossoms have expanded into seed-pods as big as walnuts. In one case it produces leaves in January, in the other it does not assume that livery till April. If I can conveniently walk out with my sketch-book, for I cannot ride on horseback, and I am scarce able to bear the jolt of a chaise, I will make a sketch of some tree that has got out of the equilibrium of leaves, — and flowers and fruit,—in alternate years, and exhibits the biennial succession in one half only of its branches, so that the east side will be all foliage up to April, and the west bare stems and twigs with only terminal seed-pods to the same month, and *vice versâ* next year. The *Bursera gummifera*, though deciduous, does not, so far as I have observed, bear in alternate years. In St. David's, where I was when I last addressed you, that tropic Birch is the commonest of the trees, among the scrubby forests that line the mountains towards the sea, and they were all leafless alike. I narrate these facts exclusively from my own observation. Mac Fadyen does not notice them." — (*Letter from Mr. Hill, dated Lawrencefield, St. Catherine's, Jamaica, 27th March,* 1851.)

III.

Vegetation on the Pedro Kays, p. 310. (See also "The Birds of Jamaica," p. 435.)—" I must not omit to let you know that the low shrub, the only vegetation on the Pedro Rocks, if we except the single Cocoa-nut tree, — the memorial of a dead seaman, — is the *Suriana maritima*, a plant which Humboldt says the South American Spaniards call *Romero de la Mar.* It puts forth its leaves in tasseled tufts, and produces small yellow flowers. I sent and procured the plant. R. H." This plant, which is common to the sea-coast throughout the tropics, presents some anomalies in its structure, which render it very difficult to classify.

IV.

Viviparous generation of the Yellow Boa (*Chilabothrus inornatus*), p. 323.—The case of the Boa which produced in captivity twenty-three living young ones (not *eleven*, as stated by mistake in the text) appeared to me so anomalous, that I sought a further verification of the circumstance. Mr. Hill accordingly obtained from his informant the following clear and interesting details of the matter, which render the fact indubitable, however strange.

"Cumberland Pen, [Jamaica,]
8th May, 1851.

"My dear Mr. Hill,

"According to your desire I sit down to give you the particulars of the Yellow Snake, that I told you had produced young in my possession.

"It was on the 3rd of July, 1849, that I caught the Snake, on a sunny bank at Halfway-tree Pen (its companion, the larger of the two, escaped, and I lost it in the grass). I put it into a box with a wire front, and with it imprisoned a mouse and two ground-lizards: the Snake made no attempt upon the lives of its fellow prisoners, who were fortunate enough to regain their liberty in a week or two.

"I could never induce the Snake to eat, though I offered it everything I could think of; and it was more savage than most others, and bit me several times, each bite drawing blood like a severe scratch from a cat.

"It measured 6 feet 1 inch (its tail short and blunt), and 10½ inches round the body. It was very inactive, lying all day in a corner of its cage, or coiling in graceful folds about the perches.

"On the morning of the 19th of October, I was surprised to find my captive had produced twenty-three young ones; they were all perfectly formed and of much the same size. I measured six of those that died first, and found them 16 inches long and 1½ inch in circumference. The last of the young ones died on the 24th, and the mother on the 28th, of the same month. Since that time I have only caught two or three more, and they have always got away by some accident or other; but I hope soon to find more, as I am anxious to try them again, for I had always supposed they laid eggs like other snakes, though this one certainly brought forth her young alive.

"I am, my dear Sir,

"Yours very truly,

"F. R. Griffith."

INDEX.

THE END.

LONDON:
SPOTTISWOODES and SHAW,
New-street-Square.

CPSIA information can be obtained at www.ICGtesting.com
Printed in the USA
LVOW08s1922071213

364249LV00002B/192/P

9 781108 063739